T0138348

Sex Itself

SEX ITSELF

The Search for Male and Female in the Human Genome

Sarah S. Richardson

The University of Chicago Press
Chicago and London

Sarah S. Richardson is assistant professor of the history of science and of studies of women, gender, and sexuality at Harvard University.

The University of Chicago Press, Chicago 60637
The University of Chicago Press, Ltd., London

Printed in the United States of America

22 21 20 19 18 17 16 15 14   2 3 4 5

ISBN-13: 978-0-226-08468-8 (cloth)
ISBN-13: 978-0-226-08471-8 (e-book)
DOI: 10.7208/chicago/9780226084718.001.0001

Library of Congress Cataloging-in-Publication Data

Richardson, Sarah S., 1980–
Sex itself : the search for male and female in the human genome / Sarah S. Richardson.
pages cm
Includes bibliographical references and index.
ISBN 978-0-226-08468-8 (cloth : alk. paper)—ISBN 978-0-226-08471-8 (e-book)
1. Sex chromosomes. 2. Human genome. 3. Sex differences. I. Title.
QH600.5.R53 2013
611′.0181663—dc23
2013014688

♾ This paper meets the requirements of ANSI/NISO Z39.48-1992 (Permanence of Paper).

For Rich

CONTENTS

CHAPTER 1

Sex Itself

Human genomes are 99.9 percent identical—with one prominent exception: instead of a matching pair of X chromosomes, men carry a single X, coupled with a small chromosome called the Y. Today, with the genome sequences of the human X and Y chromosomes at hand, geneticists are searching anew for the elements of maleness and femaleness—for what the famous American fly geneticist Thomas Hunt Morgan in 1916 called "sex itself."[1] With knowledge of the genes animating sex differences, researchers hope to aid medical understanding of sex-specific diseases and uncover the foundations of male-female differences in cognition, intelligence, and behavior. Some suggest that we will at last discover what it means to be a male or a female. Some hope that genomics will help us measure, quantitatively and precisely, differences between men and women. According to commentators in the March 2005 issue of *Nature*, which announced the publication of the complete sequence of the human X chromosome, these differences will likely prove to be greater than ever thought.[2]

Tracking the emergence of a new and distinctive way of thinking about sex represented by the unalterable, simple, and visually compelling binary of the X and Y chromosomes, this book examines the interaction between cultural gender norms and genetic theories of sex from the beginning of the twentieth century to the present postgenomic age. The past century has seen enormous shifts in the biological model of sex. At the turn of the twentieth century, the metabolic theory ruled the day. Biologists believed that metabolic rate mediated masculinity and femininity and determined whether a newly fertilized egg became male or female.[3] The hormonal model emerged in the 1920s and dominated the midcentury. The sex hormones, pharmaceutically powerful, quantitative, and malleable agents of sexual behavior and secondary sex characters, have been the overwhelming focus of histories of twentieth-century scientific ideas about sex.[4] The genetic account of sex has received far less attention.

The genetic binary of the X and Y chromosomes was discovered at the turn of the twentieth century. It entered human biomedicine with revelations of cases of human males with extra X chromosomes (XXY) and females with only one X (XO) in the 1950s. Now, with the completion of the Human Genome Project, genes and chromosomes are moving to the center of the biology of sex. As this book shows, the X and Y chromosomes, little symbols of unbreachable sex dimorphism, came to anchor a conception of sex as a biologically fixed and unalterable binary, the very conception premonitioned by Morgan's "sex itself."

From the earliest theories of chromosomal sex determination, to the midcentury hypothesis of the aggressive XYY supermale, the longstanding belief that the X is the "female chromosome," and the recent claim that males and females have "different genomes," cultural gender conceptions have influenced the direction of sex chromosome genetics. Gender has helped to shape the questions that are asked, the theories and models proposed, the research practices employed, and the descriptive language used in the field of sex chromosome research. Analyzing the history of human sex chromosomes as gendered objects of scientific knowledge, this book adds gender to our intellectual histories of human genetics, and genetics to our histories of scientific theories of sex and gender.

Today, scientific and popular literature on the sex chromosomes is rich with examples of the gendering of the X and Y. Humorous maps of the X and Y chromosome—pinned up on laboratory walls and always good for a laugh in an otherwise dry scientific talk—assign stereotypical female and male traits to the X and Y, from the "Jane Austen appreciation locus" to "channel flipping" (see fig. 1.1). The X is dubbed the "female chromosome," takes the feminine pronoun "she," and has been described as the "big sister" to "her derelict brother that is the Y" and as the "sexy" chromosome.[5] The X is frequently associated with the mysteriousness and variability of the feminine, as in a 2005 *Science* article headlined "She Moves in Mysterious Ways" and beginning, "The human X chromosome is a study in contradictions."[6] The X is also described in traditionally gendered terms as the more "sociable," "controlling," "conservative," "monotonous," and

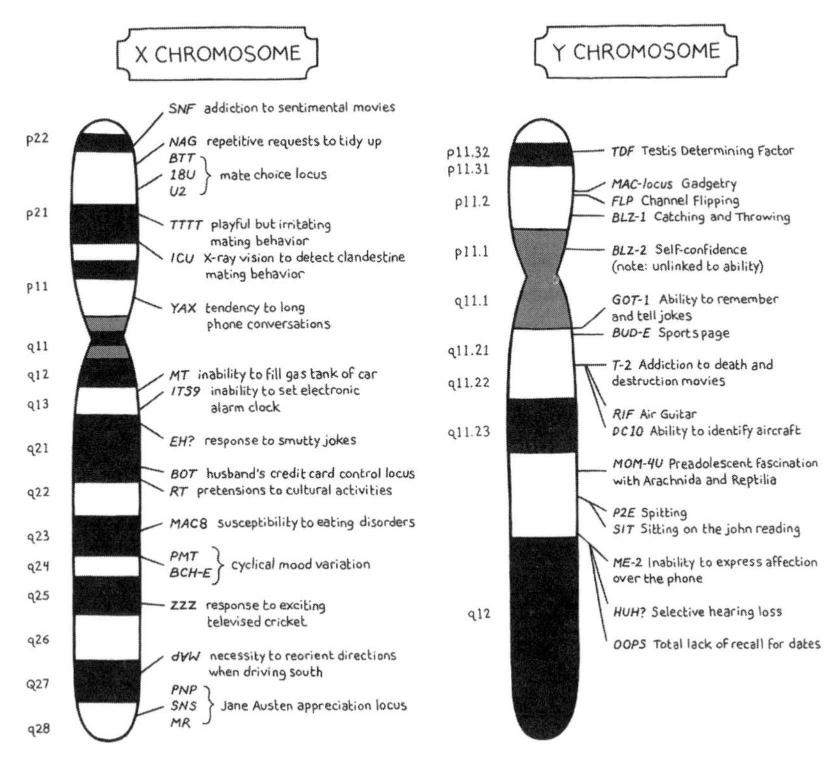

Figure 1.1. Humorous maps of traits on the X and Y chromosomes. Recreated by Kendal Tull-Esterbrook, with permission from the authors. The X chromosome is after an original designed by Dr. Jennifer A. Graves; the Y chromosome is after an original by Dr. Jane Gitschier.

"motherly" of the two sex chromosomes.[7] Similarly, the Y is a "he" and ascribed traditional masculine qualities—"macho," "active," "clever," "wily," "dominant," and also "degenerate," "lazy," and "hyperactive."[8]

Three common gendered tropes feature in popular and scientific writing on the sex chromosomes. This first is the portrayal of the X and Y as a heterosexual couple with traditionally gendered opposite or complementary roles and behaviors. For instance, Massachusetts Institute of Technology (MIT) geneticist David Page says, "The Y married up, the X married down. . . . The Y wants to maintain himself but doesn't know how. He's falling apart, like the guy who can't manage to get a doctor's appointment or can't clean up the house or apartment unless his wife does it."[9] Biologist and science writer David Bainbridge narrates the evolutionary history of the X and Y as a "sad divorce," set in motion when the "couple first stopped dancing," after which "they almost stopped communicating completely." The X is now an "estranged partner" of the Y, he writes, "having to resort to complex tricks."[10] Oxford University geneticist Bryan

Sykes similarly describes the X and Y as having a "once happy marriage" full of "intimate exchanges" now reduced to only an occasional "kiss on the cheek."[11] A 2006 article on X-X pairing in females in *Science* by Pennsylvania State University geneticist Laura Carrel is headlined "'X'-Rated Chromosomal Rendezvous."[12]

Second, sex chromosome biology is often conceptualized as a war of the sexes. In Matt Ridley's 1999 *Genome: The Autobiography of a Species in 23 Chapters*, the chapter on the X and Y chromosomes is titled "Conflict." It relates a story, straight from *Men Are from Mars, Women Are from Venus* (1992), of two chromosomes locked in antagonism and never able to understand each other.[13] A 2007 *ScienceNOW Daily News* article similarly insists on describing a finding about the Z chromosome in male birds (the equivalent of the X in humans) as demonstrating "A Genetic Battle of the Sexes," while Bainbridge describes the lack of a second X in males as a "divisive . . . discrepancy between boys and girls," a genetic basis for the supposed war of the sexes.[14]

Third, sex chromosome researchers promote the X and Y as symbols of maleness and femaleness with which individuals are expected to identify and in which they might take pride. Sykes offers the Y chromosome as a totem of male bonding, urging males to celebrate their unique Y chromosomes, and calling for them to join together to save the Y from extinction in his 2003 *Adam's Curse: A Future without Men*. Females are also encouraged to identify with their Xs. Natalie Angier urges that women "must take pride in [their] X chromosomes. . . . They define femaleness."[15] The "XX Factor" is a *Slate* column about women's work and life issues with the slogan "What Women Really Think" and is also the name of an annual competition for female video gamers.[16] The promotional video for the US Society for Women's Health Research, designed to convince the viewer of how very different men and women really are, is titled *What a Difference an X Makes!*[17]

The past decade has witnessed a wave of critical scholarship on the potential of the genome-sequencing projects to resurrect dangerous notions of human difference, but the focus has been more on race than gender.[18] With the painful history of racial science in the background, publicly funded genome projects have supported research by historians, philosophers, and bioethicists to study the impact of new genome science on marginalized, vulnerable, or underrepresented groups. The history of sexual science—an enterprise that famously admonished that higher education could impair women's ovaries, pronounced that women are ruled by their emotions and not their intellects, and asserted that women are developmentally arrested males halfway between men and children—

holds similar cautions of the dangers of uncritical scientific constructions of sex differences.[19] Yet the questions to which scholars have so carefully and urgently attended in the case of genetic research on race and ethnicity are not being asked of genetic research on sex and gender. With this book, I hope to open a conversation about the methods and models of sex difference research in a genomic age.

THE SEX CHROMOSOMES

Chromosomes, housed in the nucleus of each of our cells, are packages of DNA (see fig. 1.2). Humans have 23 pairs of them. Each pair is composed of a chromosome received from the egg and the sperm. The chromosomes contain tightly coiled strands of DNA unique to each chromosome. Twenty-two of the pairs are homologous, which means that other than the small differences in gene variants that we inherit from our parents, the two chromosomes are identical. The twenty-third pair is different. In the case of males, it comprises an X partnered with the much smaller Y. Males are thus said to be "XY." Females possess two Xs and are "XX."

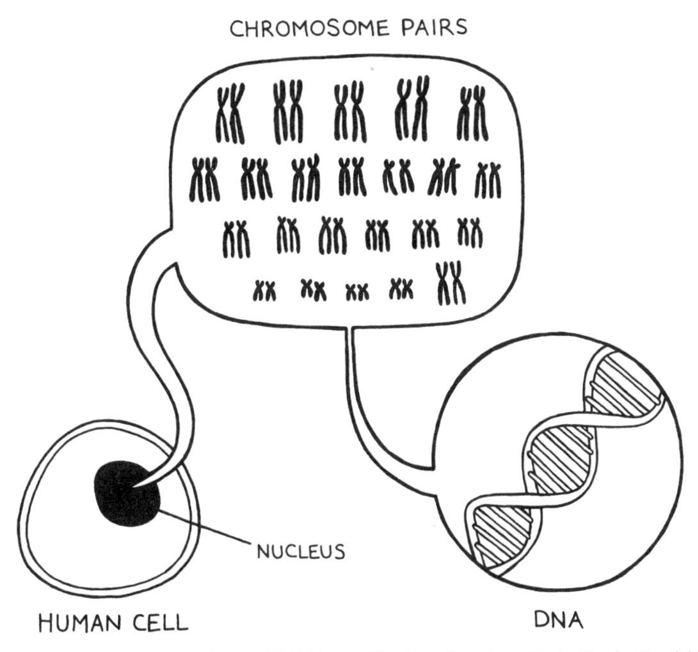

Figure 1.2. Chromosomes are packages of DNA housed in the cell nucleus. Illustration by Kendal Tull-Esterbrook; copyright Sarah S. Richardson.

The sex chromosomes are widely recognizable symbols of genetic science. Doctors use them to indicate the sex of a fetus, and school science textbooks feature memorable clinical anomalies such as the XXY Klinefelter's male, who has an extra X chromosome, and the XO Turner's female, who possesses one X instead of two. Classically, reproductively viable phenotypic sex in humans has been seen as a three-step process: chromosomal and genetic initiation of gonad determination (ovaries or testes), followed by the production of the correct ratio of sex steroids (androgens and estrogens), leading to the development and differentiation of the reproductive organs and secondary sexual characters. In males, testes determination begins during fetal development (see fig. 1.3). This involves a genetic pathway requiring the *SRY* gene on the Y chromosome and the *SOX9* gene on chromosome 17. As will be discussed in chapter 6, the genetics of female ovary determination has been less studied than that of the testes, but it too requires a sequence of genetic signals. The *WNT4* gene on chromosome 1 is emerging as a crucial genetic factor in ovarian development.[20]

Though often described as the "female" and "male" chromosomes, there is nothing essential about the X and Y in relation to femaleness and maleness. Chromosomes are only one form of sex-determining mechanism in the natural world. Birds have sex chromosomes, but the system is the reverse of mammals. In our avian cousins, males have the duplicate larger chromosome (called ZZ), while females are heterozygous (ZW), possessing one larger and one smaller chromosome. In fruit flies, sex is determined by the ratio of X chromosomes to autosomes, rather than the presence or absence of a Y chromosome, as in mammals. For turtles and many reptiles, sex depends on the temperature of the environment during early development, not sex chromosomes. Some species have just one sex, some have three or more, and some can change sexes during their lifetimes—and this can depend purely on the arrangement of sex chromosomes, wholly on exposure to environmental factors, or on a combination of the two.

Indeed, the ambiguous and indeterminate relationship between the X, the Y, and sex will animate much of the story in this book. In principle, any chromosome may contain genes relevant to sex differences. There are also other sexed processes in the human genome, such as maternal and paternal genetic imprinting, a process by which gene variants may be active or inactive depending on the parent of origin. Nonetheless, for the past century, the sex chromosomes have been the principal objects of analysis of genetic sex research, and today they continue to dominate the landscape of genomic reasoning about sex and gender.

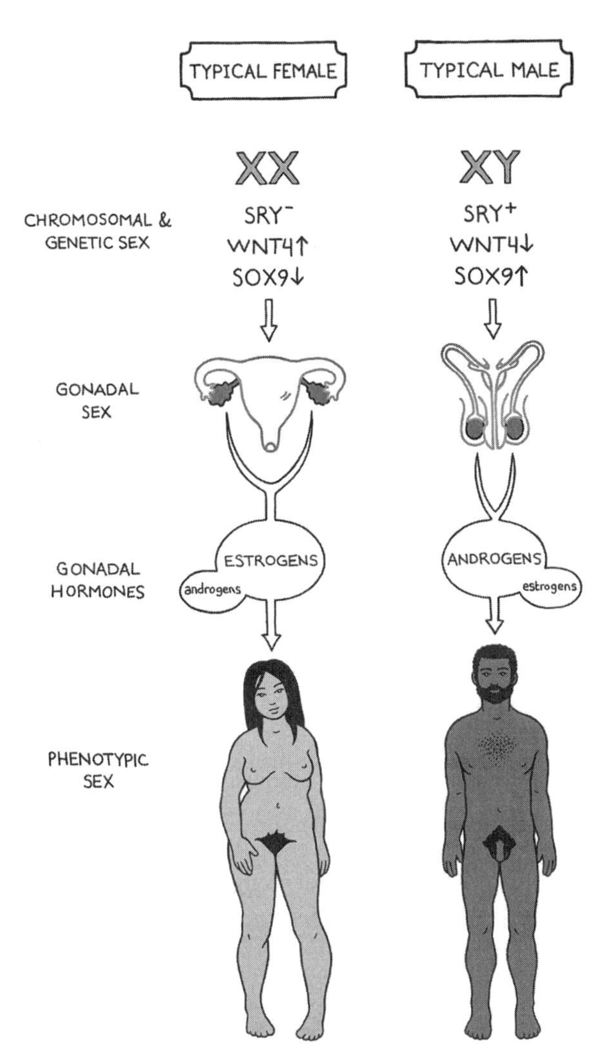

Figure 1.3. The relationship between sex chromosomes and sexual phenotype. Illustration by Kendal Tull-Esterbrook; copyright Sarah S. Richardson.

SEX ITSELF

In 1916, Morgan strenuously warned against thinking of the X and Y as "sex itself." He foresaw that the prominent dimorphism of the X and Y in the human genome would become imbued with special significance in the eyes of biologists. Indeed, the quest to know the *thing itself* or to access the *first cause* would become a central theme of twentieth-century molecular science. Within this new worldview, the anatomical markers of sex and the final expression of gender identity are not themselves "sex itself," but mere signifiers, traces, and elaborations of the genotypic dimorphism that underlies it all.

Biologists have never been under the illusion that genes and chromosomes are all there is to the biology of sex. Today, as in Morgan's time, researchers acknowledge that human biological "sex" is not diagnosed by any single factor, but is the result of a choreography of genes, hormones, gonads, genitals, and secondary sex characters. Today, academic sexologists typically distinguish between chromosomal sex, gonadal sex, hormonal sex, genital sex, and sexual identity. Some would add sexual preference, gender identity, morphological sex, fertility, and even brain sex to this list.

This layered conception of sex derives from clinical practice. It owes its origins to the diagnosis and treatment of disorders of sexual development and identity.[21] Consider an individual with the intersex condition androgen insensitivity syndrome (AIS). She may have "male" chromosomal sex (XY), "male" gonadal sex (internal undescended testes), and "male" hormonal sex (high levels of androgens), but "female" genitalia (breasts and vagina) and "female" identity. Similarly, a male-to-female transgender individual who has received hormonal and surgical therapy may have "male" chromosomal sex (XY), but "female" hormones, genitals, facial appearance, secondary sex characters, and identity. A unidimensional conception of sex falls apart in the face of these clinical encounters. Researchers and physicians have innovated accordingly, creating a conceptual apparatus for describing sex, gender, and sexuality in all of their social and biological dimensions.[22]

In the 1970s, feminist theorists such as Gayle Rubin helped to cement this layered and contextual understanding of sex.[23] Rubin distinguished between the biological category of "sex" (typically, male or female) and the social roles and expectations of "gender" (such as heterosexual masculinity and femininity). The sex/gender distinction analytically separates the anatomy and physiology of males and females (sex) from the behavioral and cultural expectations associated with the ideals of masculinity and femininity (gender). By design, the distinction underscores the idea that while sex may be fixed and given, gender is fluid and changeable.

Ironically, however, the sex/gender distinction served to harden the notion of the X and Y as "sex itself."[24] In the late twentieth century, Western society underwent a revolution in gender roles, signified by the widespread entry of women into public life. Lesbian, gay, bisexual, transgender, and intersex people asserted a political identity and became more socially visible. Hormone therapy and reconstructive surgery advanced to the point where doctors were able to alter the body chemistry, anatomy, and gendered appearance of a man or a woman.[25] Although the morphological, genital, gonadal, and hormonal body was rendered more fluid by these developments, chromosomal sex remained intact as the kernel or foundation of the biological sex concept. The X and Y came to represent the necessary alter ego of gender fluidity, signifying what nature intended the sexual fate of the infant to be.[26] Cynthia Kraus, writing on fly genetics, has memorably called this notion of genetic sex "naked sex," or "that part of sex which rebuffs the gender microscope."[27]

The X and Y emerged from this twentieth-century project of troubling our conceptions of sex and gender at the base of a hierarchy of conceptual layers of sex and gender that moves from "hard" to "soft": chromosomal or genotypic sex; then hormonal, gonadal, and morphological sex; and finally gender presentation and sexual orientation. In common parlance, the distillation of "chromosomal sex" even began to be substituted for the notion of "biological sex." Conceived as developmentally prior to hormones and culture, the X and Y chromosomes remain our closest approximation of "sex itself."

The visual binary of the XX and the XY signify that part of sex that is thought to be unchangeable and most fundamental. At the level of the chromosomes, the gender rainbow, it is thought, falls away. The binary is stark: XX is female and XY is male. Among all of the conceptions of "sex," chromosomal sex is considered the most elemental and ultimately diagnostic of one's "true" sex. In the biosciences as well as in the popular imagination, the X and Y are seen as a source of order and clarity. In the midst of gender chaos, some take them to represent the essence of maleness and femaleness and the ultimate naturalness, and hence rightness, of social customs and practices organized around the different roles, interests, and capacities of the two sexes. While hormones and culture help to shape gender, genetics alone, it is thought, can reveal "sex itself."

GENES AND GENDER

Feminist science studies scholars have developed powerful analyses of historical and contemporary scientific theories of sex differences, but they have left the gender dimensions of the sex chromosomes and genetic theories of sex difference virtually untouched. Only in the past few years has this subject been broached.[28] Perhaps this belated interest is, in part, due to the prevailing notion of the X and Y as simple, binary, and fixed, a notion often shared by feminist scholars themselves. From the perspective of the project of generating scholarship that challenges and complicates our conventional ideas about human sex and gender, scholars may have assumed that the X and Y were not as interesting, or not as yielding, as the malleable sex hormones or the soft tissue of genitals and gonads.[29] Moreover, unlike hormones and tissues, the sex chromosomes do not carry the gleam of a century of pharmaceutical investment or the popular intrigue of genital and gonadal surgery. Finally, it is only in the past few decades that genomics has been elevated to its now preeminent status as a widely shared framework and research platform for the human biomedical sciences.[30]

Feminist scholarship, however, does provide important resources for approaching the intersection of gender and genetics. Feminists have closely attended to the ideological implications of assertions that male and female differences in capacities, preferences, and behaviors are genetically hardwired. The tradition of feminist critiques of biological determinism dates back, in the United States, to nineteenth- and early twentieth-century feminist activists and science critics such as Antoinette Brown Blackwell, Eliza Burt Gamble, and Charlotte Perkins Gilman, and early feminist academics such as Helen Thompson Woolley and Ruth Herschberger.[31] In the 1970s and 1980s, Second Wave feminist scientists took on sociobiology, a controversial field that revived biological determinist approaches to human social behavior. Sociobiological claims in the 1970s professed to explain the biological foundations of male dominance, stereotypical masculine and feminine behaviors, and male and female gender roles, giving rise to misogynist pop science screeds such as Steven Goldberg's 1973 *The Inevitability of Patriarchy: Why the Biological Differences between Men and Women Always Produce Male Domination.*[32]

In 1977, a group of New York City–based feminist biologists formed to study and respond to these new claims. Ethel Tobach of the American Museum of Natural History, Department of Animal Behavior, and Betty Rosoff of the Department of Biology of the Stern College for Women organized the first "Genes and Gender" event. Jointly sponsored by the Committee for Women in Science Research of the New York Academy of Sciences, the New York chapter of the Association for Women in Science, and the Regional Women's Committee of the American Psychological Association, the symposium was held on January 29, 1977, at the American Museum of Natural History off of Central Park in New York City. The agenda for the symposium included a brief statement, titled "Why Genes and Gender?" The organizers wrote:

> Despite advances in our understanding of the processes underlying gender assignment in our society, there still persists the idea that "genes determine destiny." Sexism is at the core of many recent publications that attempt to interpret women's social roles as "biologically" or "genetically" determined. Behavior presumed to be sex-linked is frequently analyzed to disadvantage women. This symposium will present substantive material relating to the genetic processes, their expression in hormonal systems and in behavior, and the relationship of these phenomena to "differences" between women and men which affect their relationships.[33]

The registration forms for the 1977 conference, now archived at the Schlesinger Library at the Radcliffe Institute for Advanced Study, record penciled notes to the organizers from the registrants, a mix of biologists, psychologists, physicians, and activists: "GENES AND GENDER sounds interesting. Put me down for two tickets," wrote one. Wrote another, "I'm hoping its [*sic*] not going to be over my head. Genetics is a heavy subject!"[34] The symposium, with approximately 250 in attendance, was a hit. Over the next twenty years, the Genes and Gender Collective, as they came to be known, held lively regular conferences, producing seven volumes in the series "Genes and Gender: A Series on Hereditarianism and Women."[35] As Tobach wrote in a 1978 letter promoting the series' first book to feminist celebrities including Lillian Hellman, Susan Sontag, Jane Fonda, and Gloria Steinem, "this book . . . is concerned with exploding the myth, propagated in its most recent form by sociobiology, that women are somehow 'inferior,' or destined by their biological makeup to assume a second class role in our society."[36]

The Genes and Gender series featured contributions from leading scientists. They took issue with what they perceived as ideological claims about the biological inevitability of race, class, and sex hierarchies. Their critiques of biological determinism and its implications for women, alongside classic monographs by biologists from this period such as Ruth Bleier's *Science and Gender: A Critique of Biology and Its Theories on Women* and Richard Lewontin, Steven Rose, and Leon Kamin's *Not in Our Genes: Biology, Ideology, and Human Nature*, inform and help to motivate my approach to the history of sex chromosome genetics. Focusing on topics such as hormonal theories of women's behavior and evolutionary theories of rape, however, these writings never took up the basic genetic science of sex. Exploring the role of gender conceptions within the empirical claims, theories, and practices of genetic science itself is the concern of this book.

There also exists a significant stream of feminist research and activism on the implications of new reproductive technologies enabled by genetic science. Feminist bioethicists, science studies scholars, and medical commentators have attended closely to the social and ethical dimensions of "reprogenetics" technologies, such human genetic enhancement and genetic engineering, human cloning, genetics-based fertility treatments, human embryo selection, and prenatal genetic testing, as well as research practices such as the harvesting of human embryos in stem cell research.[37] This area of feminist scholarship focuses on gender and genetics in clinical practice and in public health policy and discourse. New genetic technologies, these works argue, are transforming human reproduction in ways that challenge human values. As such, vigilant scrutiny is required to protect women's health and reproductive autonomy. This book engages such scholarship by demonstrating that the myriad dimensions of the genes and gender question are not limited to the clinic. Beyond the clinic, in the realm of research on the basic genetics and biology of sex, scientists are investigating "sex itself." They are revisioning the conceptual terrain of human biological sex differences. The untouched—and seemingly untouchable—concept of "genomic" sex leaves us with a theoretical vacuum as we face the post–Human Genome era.[38]

GENDER AND SCIENCE

Sex Itself analyzes the operation of both traditional and alternative gender ideologies in the models, practices, and language of twentieth-century human sex chromosome research, attending carefully to historical context. Gender ideology is defined as widely held, socially inscribed beliefs about sex and gender difference. These include conceptions of maleness and femaleness, masculinity and femininity, sex and gender roles, and the nature of relations between the sexes. Traditional gender ideology comprises a set of stereotypes about gender difference that reflect and support dominant culture and institutional arrangements. Sociologist Michael Kimmel describes contemporary Western gender ideology as an "'interplanetary' theory of complete and universal gender difference," a reference to the popular relationship advice book *Men Are from Mars, Women Are from Venus*.[39] This view of gender has three principal components. First, male and female, and masculine and feminine, are complementary and binary. The gendered binaries of "public and private spheres," "active and passive," and "rational and emotional" represent this concept of gender difference. Second, conflict and antagonism characterize the relationship between male and female and masculine and feminine. Difference is conceptualized as a source of conflict. An example is the ubiquitous metaphor of a "War between the Sexes" in popular writing on men and women. Third, the biological assignment of sex and the social category of gender are completely elided. Differences between men and women are seen as residing in the individual and biologically determined, universal, and natural. Male is equated with masculine and female with feminine.

Gender is formative to the knowledge, practices, and institutional structure of the sciences. One the most sensitive guides to this territory is the historian and philosopher of science Evelyn Fox Keller.[40] Her studies of gendered language, metaphors, and discourse in science reveal how gender forms part of the deep fabric of scientific practices of reasoning and models of the world. Gender norms are, as Keller writes, "silent organizers of the mental and discursive maps of the social and natural worlds we simultaneously inhabit and construct."[41] "Masculinity" and "femininity" not only categorize men and women, they also schematize abstract concepts such as reason and emotion and mind and body. In this way, feminist approaches to science, Keller continues, "expose to radical critique a worldview that deploys categories of gender to rend the fabric of human life and thought along a multiplicity of mutually sanctioning, mutually supportive, and mutually defining binary oppositions."[42] This has consequences for the organization and valuation of forms of human knowledge, the perceived authority of differently embodied knowers, the practice of science, and the structure of scientific explanations of nature.[43]

The approach that I take in this book is most deeply inspired by feminist science scholarship that has documented, through historically and philosophically fine-grained case studies, how gender conceptions enter into the global theories, working models, and everyday interpretive practices of the biological sciences. The book *Sexing the Body* (2000), by feminist biologist and science studies scholar Anne Fausto-Sterling, is exemplary. Examining episodes in the twentieth-century science of sexuality, sex differences, and gender identity, Fausto-Sterling demonstrates how researchers repeatedly recourse to practices and frameworks that reinforce normative sex binaries, even when they do not fit the facts. In one chapter, "Sexing the Brain: How Biologists Make a Difference," Fausto-Sterling vividly shows how scientists, certain of the existence of sex differences in an important brain structure, the corpus collosum, relentlessly pursued a finding of sex differences, revising their measurement techniques and criteria each time evidence of a sex difference eluded them.[44]

In the path cleared by Fausto-Sterling, the past decade has seen a blossoming of high-impact studies of how gender ideology may distort our biological picture of sex. Elisabeth Lloyd's *The Case of the Female Orgasm: Bias in the Science of Evolution* (2005) offers a trenchant feminist critique of the gaps in evolutionary theory's accounts of female sexuality.[45] Joan Roughgarden's *Evolution's Rainbow: Diversity, Gender, and Sexuality in Nature and People* (2004) and her more recent *The Genial Gene: Deconstructing Darwinian Selfishness* (2009) present a sweeping critique and reconstruction of predominant evolutionary conceptions of sexuality and the theory of sexual selection, drawing on feminist approaches to science.[46] Most recently, Cordelia Fine's *Delusions of Gender: How Our Minds, Society, and Neurosexism Create Difference* (2010) and Rebecca Jordan-Young's *Brain Storm: The Flaws in the Science of Sex Differences* (2010) have launched rigorous feminist critiques of scientific theories of sex differences in the brain.[47]

The debate over gender in scientific knowledge has begun to enter the mainstream. Part of the context for this development is the rising urgency of the question of women's status in the science professions. In the past decade, funding agencies in the United States, United Kingdom, and European Union have undertaken significant investments in programs and social science research to advance women and girls in science.[48] These investments built on already seismic demographic shifts in the sciences. Whereas science was once largely the clubby domain of men of a shared class and cultural background, who perceived feminism as entirely external to scientific concerns, and even as a threat to science, we are now experiencing the entry of a generation of women and men, open to critical perspectives on traditional gender conceptions, into the senior ranks of many scientific fields.[49] This new generation of scientists has been trained in fully coeducational environments, may have had mothers who were professionals, and were likely exposed to gender studies in their college coursework. In my experience, for these scientists, the interrogation of how gender beliefs may influence science is not a threat but a personally compelling question and intellectually intriguing pursuit.

THE SOCIAL DIMENSIONS OF SCIENCE

At its heart, this book engages with one of the most fascinating and critical questions about science: how does the social and political context of science influence its cognitive content? Philosophers of science once averted their eyes from this question. They sealed science off from other human activity and posed it as an ideal form of knowing, free from ideology. One could politically "abuse" science, but, they asserted, performed correctly and used ethically, science is free from human politics. They characterized those who broached the matter of science and social values as "anti-science" and challenged them to explain why, if science was full of social and political assumptions, vaccines work and bridges stay up.

Few still hold that we must abandon a commitment to scientific realism and rationality, or give up on the idea that there is something distinctive about science, to acknowledge, also, the role of social context in the often contingent ways in which scientific knowledge comes about. Social context and social values are a part of good science as much as they are of bad science. To understand how science works, we need to understand its social dimensions. Today, the social dimensions of scientific knowledge are a rich and lively area of study, yielding new insights about scientific theory and practice.

I think of science as a set of historical and social, as well as knowledge-productive and cognitive, practices. This understanding of scientific practice was first vividly illuminated by Ludwig Fleck and Thomas Kuhn in the mid-twentieth century.[50] According to their picture of science, scientists, or communities of scientists, advance scientific hypotheses, models, and theories about their objects of investigation. These hypotheses are supported by a complex of facts of varying credibility; causal and mechanistic claims of varying solidity; and social forces, intuitions, and beliefs of varying degrees of transparency and visibility. Advocates of a scientific hypothesis apply the best technologies of the moment, as well as their skills of advocacy and argumentation, to make their case. In these debates, scientists appeal to pragmatic, explanatory, and moral virtues: parsimony, robustness, unification, reductionism, novelty, simplicity, efficiency, even beauty and ethics. In short, scientific hypotheses are not just collections of facts and data. They have empirical, theoretical, and pragmatic dimensions, too. Social context, beliefs, and values—including gender conceptions—may be relevant to any of these dimensions.

Following a method most prominently developed by Helen Longino to make visible the role of social values in science, I highlight cases in the history of human sex chromosome research in which different models or scientific theories vie for acceptance under conditions of incomplete empirical evidence. The history of sex chromosome science as I write it is not a chronology of the accumulation of more complete and accurate facts. It is, if one must simplify, a history of contestations. Active research models in the biosciences are often openly debated. Competing hypotheses are posed by two or more groups of scientists. Is the Y chromosome degenerating, or is it in a holding pattern? Does the *SRY* gene control male sex determination, or is it better conceived as one in a network of genes implicated in a convergent sex-determining pathway? Do males and females, like humans and chimpanzees, have different genomes? In time, these debates may be resolved clearly by empirical evidence. More often, however, the debate is not over the empirical evidence but, rather, how to interpret a set of suggestive findings within larger fields of knowledge. In these cases, we can see with special clarity the role that global models or scientific hypotheses, which may draw on cultural gender conceptions, play in shaping and motivating research programs.

MODELING GENDER IN SCIENCE

The episodes analyzed in this book show how gender conceptions may enter into and serve as a cognitive resource in the development of scientific theories. I call my theoretical approach to the history of sex chromosome science *modeling gender in science.* Close readings of the advancement of ideas in sex chromosome science with attention to how, in their time, these ideas were accepted as reasonable, plausible, or even true, reveal how gender beliefs shaped scientific knowledge in this field. Gender beliefs do not merely surround and accompany sex chromosome research; they are creatively and cognitively indispensable within the hypotheses, research programs, and theories of sex chromosome science.

In the past, analysts of gender in science have focused much attention on gender bias in science: cases in which the advancement of scientific knowledge may have been harmed by unreflectively incorporating narrow, ideological gender assumptions. Though I am also keenly interested in such cases, and indeed document them in great detail in this book, I employ a broader, more flexible, and, I believe, more fruitful framework for gender analysis of science.[51] *Modeling gender in science* is a methodological approach the historical, philosophical, and sociological analysis of gender in science that decenters bias as the central motivating question and is more neutral with respect to whether bias is good or bad for science. Rather than focusing on how ideas about gender have led to bias that has harmed science, this approach begins with the question: what work does gender do in this case?

Analyzing the case studies in this book and extending central insights from feminist philosophy of science, I argue that a comprehensive analytical approach to modeling gender in science requires going beyond the question of "gender bias" in science to also consider the constructive role of gender conceptions in the knowledge work of science. Through the varied case studies in the history of human sex chromosome science considered in this book, I seek to understand how ideas about gender enter into and influence scientific theories and in what contexts these ideas constitute unwelcome bias. I develop the concept of *gender valence* to refer to cases in which gender conceptions operate visibly and reflexively in ways that introduce productive partialities. In these cases, gender beliefs play a role in science that does not necessarily constitute harmful bias.

I believe that this approach yields a more sensitive and far-reaching analysis of the interrelation of science and gender ideologies than does one primarily framed by the question of gender bias in science. In particular, I argue that this approach is better suited to inviting scientists into transformative, interdisciplinary conversations about gender beliefs in the practice of science. I develop the concept of *gender criticality* to describe engaged reflective practices that might help to carry the insights of historical, philosophical, and sociological studies of gender and science into the practice of scientific research. I define gender criticality as the practice of making visible the operation of gender conceptions in knowledge. In this book, I describe and analyze examples of gender criticality in the history of sex chromosome science. I also demonstrate gender criticality through analyses of the scientific claims in play in particular episodes in the history of scientific research on the genetics of sex. The fruits of a historical-analytic method focused on modeling gender bias, gender valence, and gender criticality in science will become clear in the chapters to come, as we navigate the history of human sex chromosome science in dynamic relation to changing gender conceptions over time.

STRUCTURE OF THE BOOK

Sex Itself proceeds in ten chapters. My first task, in chapters 2, 3, and 4, is to unearth the origins of the concept of the sex chromosome and to locate the sex chromosomes within the context of twentieth-century genetics and sexual science. The theoretical debates that shaped the early concept of a "chromosome for sex" reveal the bold and controversial nature of the hypothesis of chromosomal sex determination and demonstrate how tightly the sex chromosomes are woven into twentieth-century genetic and genomic reasoning. The discovery and validation of the sex chromosomes fueled the development of the chromosomal theory of heredity, establishing the physical basis for Mendelian heredity and opening the door for the systematic experimental study of genetic mutation, linkage, and genome organization.[52]

Today we think of the notion of the "sex chromosome" as simply descriptive, but in reality it is the end result of a highly contingent course of events in the early decades of the twentieth century. Though scientists discovered the X and Y at the turn of the twentieth century, it took at least two decades for "sex chromosome" to stabilize as the technical term of choice for these odd chromosomes. Exploring terminological debates over what to call the X and Y, I uncover geneticists' early worries about the distorting effect of conceiving of these unusual chromosomes as the sex chromosomes. I show how these concerns were swept aside in the 1910s and 1920s as the "chromosome for sex" was caught up in the effort to evangelize the new chromosomal theory of heredity and in the new science of sex hormones.

Sex chromosomes began to emerge from the shadow of the sex hormones in the 1950s and 1960s, when the X and Y debuted in human genetics. These years saw the advent of enormous government investments in biomedical science. The revelation of the structure of DNA and new technologies for studying human genetic material prompted a series of breakthroughs in human chromosome studies.[53] Through widely publicized and sometimes sensational scientific discoveries about human sex chromosome anomalies, the X and Y moved to the fore as the central pillars of the human sex binary. It is here that the story of the gendering of the X and Y as objects of scientific knowledge takes off.

In chapters 5 and 6, I explore the intellectual origins of the stubbornly persistent identification of the Y with maleness and the X with femaleness. I focus on two cases: the notorious "XYY supermale syndrome" and X chromosome mosaicism theories of female biology and disease. Research on XYY, which flourished in the 1960s and 1970s, is the subject of chapter 5. The idea that XYY men are "supermales" prone to violence due to their extra dosage of Y genes flourished for a decade before large-scale studies in the late 1970s disproved any association between the Y chromosome and aggression.[54] I challenge received accounts of this classic case of scientific error, arguing that gender conceptions, not media hype, were the primary culprit. X mosaicism theories of female biology and disease, the subject of chapter 6, present a parallel case. Drawing strongly on resonances between traditional ideas of femininity and the concepts of chimerism, mixedness, and two-ness, human X mosaicism theories seek the origins of mysterious "female maladies" and female behavior in their doubled Xs. Analyzing X mosaicism theories of female biology and behavior and X-chromosomal theories of the higher incidence of autoimmunity in women, I show how gendered assumptions about the X chromosome operate to sustain and cohere hypotheses of dubious empirical merit in high-priority areas of women's health research.[55]

The advent of Second Wave feminism in the 1970s adds an intriguing layer to the history of genetic theories of sex and to our understanding of how gender conceptions work within science. In the 1970s, feminist scientists and science studies scholars began to analyze and critique biological theories of sex difference. Today, historians and philosophers of science may look back over several decades and examine the fascinating interaction between feminist ideas and scientific theories. The question of feminism's impact on scientific knowledge was first posed in Donna Haraway's trailblazing 1970s studies of feminist primatology.[56] Londa Schiebinger's concise 1999 book, *Has Feminism Changed Science?*, summarizes three decades of work on the question.[57] A substantial scholarship now documents the ways in which feminist criticism has led to improvements in science by exposing gaps in knowledge, identifying undefended assumptions, and generating alternative hypotheses. In chapters 7 and 8, I ask: has feminism—as a critique of scientific theories of sex and as a social movement—changed sex chromosome science?

The contextual salience of feminist critiques of science and changing gender norms simply leaps off the page when examining post-1980s scientific and popular literature on the genetics of sex. I offer two differently inflected examples of the interaction between feminism and the content of sex chromosome science during this period. The first, documented in chapter 7, is the case of sex determination genetics. This is an excellent example of an area in which feminist theories and critiques of science have contributed to advancements in science. Accustomed to thinking of the X as female specialized and the Y as male specialized, researchers in the 1980s and 1990s focused on the Y chromosome to locate the genes involved in male reproduction. Researchers assumed that the crucial sex-determining event was the initiation of testes development in the male fetus, and that the sex-determining gene would be located on the Y chromosome.[58] This culminated in a race to locate the sex-determining locus on the human Y in the 1980s. In 1990, *SRY*, the putative sex-determining gene, was identified on the Y chromosome.[59] Yet this gene did not behave as expected. It did not control or activate a pathway, no gene target was evident for it in the human testes, and it was—uncomfortably for some observers—closely related to an important gene family on the X chromosome.[60] In the 1990s, feminist critiques and theories of gender contributed to the development of a new model of sex determination. The

SRY gene was reconceived as one of many mammalian sex-determining factors in sexual development, and researchers at last began to study the genetics of ovarian determination as well as that of the testes. In a close reading of diverse primary source and ethnographic materials from this episode, I show how feminist gender-critical perspectives played a role in this transformation in models of sex determination in the 1990s.

Chapter 8 turns to a different kind of interaction between feminism and sex chromosome science in recent decades, the case of theories of Y chromosome degeneration. Bringing about sweeping transformations in cultural gender roles and expectations, the feminist movement has also changed the backdrop of conventional ideas about gender. For example, the movement of women into traditionally male realms has forced the refiguring of traditional conceptions of masculinity. This cultural change now registers in emerging debates in sex chromosome genetics. The hypothesis that the human Y chromosome is "degenerating" offers up a fascinating example of this phenomenon. During the 1990s and early 2000s geneticists mapped and sequenced the human Y chromosome revealing few genes on the male-specific region of the Y. Some scientists suggested that the Y chromosome is losing its genes, a process that may eventually lead to its extinction. This hypothesis gave rise to polarized debates over human Y chromosome structure, function, and evolution, with distinctly gendered dimensions.[61] In scientific and popular literature, the debate over whether the Y chromosome is "degenerating" became richly entangled with anxieties about the decline of men in a postfeminist age.[62]

Last, I arrive at the genomic age, the period since the release of the complete sequence of the human genome in 2001. I worry that we are presently on track to reinscribing, with little reflection, old and problematic frameworks for understanding sex and gender in the new and authoritative language of genomics. In chapters 9 and 10, I marshal the book's preceding historical discussion to motivate a critical discussion of this new research and to offer analytical tools and frameworks for thinking about its implications and potential pitfalls. I argue that we should resist today's push to carve a binary understanding of sex differences into human genome science.

My deepest concern is with how we choose to conceptualize sex differences in the genome as we enter this new era of research. In chapter 9, I analyze the widely circulated 2005 claim that males and females differ genetically by "2 percent," "greater than the difference between humans and chimpanzees," and that males and females should be thought of as having "different genomes." I show why these claims are unsound. More interesting, however, is what these claims reveal about the simplistic conceptual thinking about sex that remains prevalent in the molecular genetic biosciences today. Sexes are not like species, and the differences between them cannot be genomically conceptualized in the same way as species differences. There is not a "female genome" and a "male genome." In place of this distortive and harmful binary model of sex differences in the human genome, I offer a model of genomic sex differences as conceptually unique in biological ontology. Sex is a *dynamic dyadic kind*: "male" and "female" are dyadic, dynamic, and relational classes within a species. Hewing to this conceptual picture of sex, I argue, would yield a more empirically adequate picture of sex difference in the human genome and reflect our best understanding of the biology of sex.

Current trends in sex differences research and in the postgenomic biosciences show a dramatic uptick in molecular sex difference research in the past decade.[63] As I show in chapter 10, "sex-based biology" is emerging as a distinct discipline, with new conferences, journals, and private and public funding. The University of California, Los Angeles, geneticist Arthur Arnold recently coined the term "sexome" to describe the field's integrative focus on a comprehensive investigation of sex differences in the genome.[64] Both within the discipline of sex-based biology and beyond it, genomic research on sex and gender is proliferating at a stunning pace. Outlining some of the common methodological pitfalls and potential harms of sex difference research, I argue in this chapter that as we enter the genomic age there is a need for ongoing interdisciplinary dialogue about the practices of contemporary genomic research on sex. It is my hope, and an animating objective of this book, that the history of scientists' pursuit of "sex itself" in the X and Y chromosomes will help to build a more gender-critical practice of genomic science today.

◇

New science is never a blank slate but proceeds in a loud echoing hall of old and new ideas, data, and scientific frameworks. As *Sex Itself* shows, the sexual science of the past continues to resonate, in ways both subtle and explicit, in modern day research on the genetics of sex and gender. One need not look far in emerging genomic research on sex to find similar pathways of reasoning toward concepts of fixed, discrete, and biologically ordained sex differences, and an enduring zeal to draw out the implications of biology—in this case the genome—for gender arrangements in society. From the perspective of the history of gender and science, reading twentieth- and twenty-first-century genetic research on sex provides the historian's thrill of seeing older, once debunked, theories of sex difference revived decades later in the language of molecular genetics. Even in a postfeminist age, popular and biological theories of sex and gender, and the attitudes, inclinations, and intuitions that bred them, are remarkably resilient. They are part of the intellectual web of sex difference reasoning in biology.

With the sequencing of the human genome and the advent of the genomic age, the X and Y chromosomes, and the genome as a whole, have come under intense scrutiny in efforts to resolve persistent questions about human sex differences.[65] Genetic analysis of the sex chromosomes, some contend, will at last elucidate the extent and nature of biological sex differences and provide the clear-cut, objective measure of human sex differences that hormonal, morphological, and other measures have so far failed to yield. As we enter an era in which whole-genome technologies are a ubiquitous platform for biomedical research, I hope that this book will help in clarifying the place of genetics in our overall understanding of human sex and gender difference, and also contribute to building a more gender-critical practice of genetics for the genomic age.

CHAPTER 2

The Odd Chromosomes

Edmund Wilson coined the term "sex-chromosome" in 1906. It may be "convenient," he wrote, to call the X and Y the "sex-chromosomes or 'gonochromosomes.'"[1] Wilson's term, however, did not become widely used until the 1920s. Alternative terms predominated. Strikingly, "sex chromosome" and "gonochromosome" (referencing the gonads) would be the only terms proposed for the X and Y that spotlighted their role in sex determination, as markers of sex, or as carriers of traits linked to sex.

The X and Y chromosomes, first called the "odd chromosomes," were discovered in 1890 and 1905. Their discovery is, in large part, an American story. Five scientists at the forefront of an emerging generation of American experimental biology will serve as our guides: Clarence McClung, Thomas Montgomery, Thomas Hunt Morgan, Nettie Stevens, and Edmund Wilson. While research institutions in Germany, France, and Britain were the center of chromosome science in the late nineteenth century, the establishment of American universities funded by post–Civil War American wealth led to a blossoming of the experimental life sciences in the United States in the first years of the twentieth century.[2] These embryologists, zoologists, and cell biologists were part of a network of early career scientists who interacted through the major East Coast institutions of American biology: Johns Hopkins University in Baltimore, the University of Pennsylvania and Bryn Mawr College in Philadelphia, Columbia University in New York City, and summer stints at the Marine

Biological Laboratory at Woods Hole, Massachusetts. During the 1900s and 1910s, these American scientists forged the empirical foundations of the chromosomal theory of sex determination.

Contrary to what one might expect, the discovery of a link between sex and the X and Y chromosomes did not initiate a radical new understanding of the biology of sex determination—at least at first. Scientists struggled to understand the significance of these "odd chromosomes" to sex. They sought to fit the newly found chromosomes into preexisting theories of sex determination. As we shall see, they resisted any notion that "femaleness" and "maleness" were discrete genetic characters represented by the X and Y chromosomes.

SEXUAL SCIENCE AT THE TURN OF THE TWENTIETH CENTURY

Today the Victorian era conjures the rigid gender roles of the ideology of separate domestic and public spheres, embodied in the sexual prudishness of Victorian womens' high-necked, floor-length black dresses. In striking contrast to this image, late nineteenth-century cell biologists and embryologists understood sex as a complicated, spectrum-like, and highly variable phenomenon. They were fascinated by the diversity of forms of sexual dimorphism and intersexuality in nature. Cases of hermaphrodites (possessing both male and female reproductive organs), freemartins (male-female twins in which the female has been androgenized in utero), and gynandromorphs (variants, often in insect species, that exhibit typical morphological features of both sexes) appeared regularly in the scientific literature and were presented as holding the key to unraveling the biology of sex (see fig. 2.1). In the late nineteenth century, the scientific problem of sex was also broadly defined. "Sex" covered such diverse phenomena as sex determination, sexual dimorphism, the role of gametes in fertilization, the clinical and agricultural control and prediction of sex, explanations for the varying sex ratios in different species, and the existence of what we today call "intersexes."[3] Researchers had not yet drawn a definitive line between processes of sex *determination* (the initial cause of sex) and sexual *development* (the ensuing processes and systems of sexual development over the organism's life course), and they believed that sex was a continuous (spectrum) rather than a discontinuous (binary) trait.[4]

Developmental and externalist theories of sex reigned.[5] Embryologists asserted that the embryo is highly plastic and that external factors may alter its development. Characters such as sex were not thought to be established at fertilization. Rather, sex was seen as flexible and open

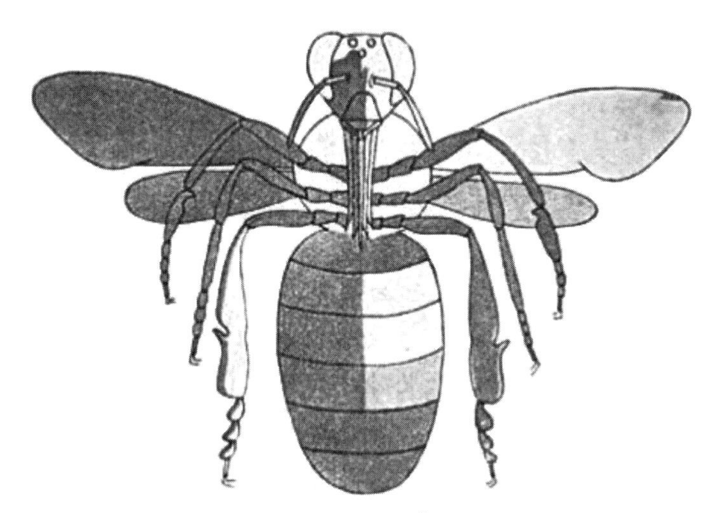

Figure 2.1. Gynandromorph. Reprinted from Elsa Mehling, *Über Die Gynandromorphen Bienen Des Eugsterschen Stockes* (Würzberg: C. Kabitzsch, 1915).

to influence from external cues in the environment. These environmental factors might include nutrition, temperature, time of fertilization, number of sperm penetrating the egg, age or vigor of the parent, ripeness of egg, and right or left ovary.[6] Agricultural breeders echoed this consensus. In the early 1900s, British and US agricultural breeders and statisticians produced numerous studies of fluctuations in species-typical sex ratios that seemed to support a role for environmental factors in determining sex during gestation.[7]

Concurrently, British physiologists were forging dramatic breakthroughs in the study of gonadal secretions (what today we call sex hormones). Their early studies, in birds and rodents, suggested that simple changes in hormones could lead male-bodied animals to behave like females, and vice versa. These experiments confirmed the going view that sex was highly malleable and could be modified well after fertilization. British physiologist Ernest Starling named the blood-borne "internal secretions" of the endocrine glands "hormones" in 1905. Continued experimentation with sex-reversed castrated chickens and ovary-transplanted guinea pigs would shortly lead British and German scientists to the "sex hormone" and a dazzling new model of sexual development and differentiation, a subject to which I return in chapter 4.

In sum, in the early 1900s, prevailing theories of sex determination stressed the environment, both external and internal to the egg, in determining the ultimate sexual fate of the organism. They also embraced a view of sexual development that assumed significant plasticity in response to time-sensitive and contingent physiological events and exposures, such as hormones. Cell biologists who were looking at the chromosomes joined this consensus. They believed that the egg cytoplasm was the mechanism by which cues from the mother's body and the wider environment directed development, and they emphasized the cytoplasm's influence on the sexual fate of the zygote. By the late nineteenth century, this picture of sex determination was codified into the "metabolic theory of sex," eloquently elaborated by British biologists Patrick Geddes and J. Arthur Thomson in their influential 1889 book *The Evolution of Sex*.

THE METABOLIC THEORY OF SEX

The first book-length treatment of the question of sex determination by modern sexologists, Geddes and Thomson's *The Evolution of Sex* "put sex on the scientific agenda."[8] "Leaving mere hypotheses behind, as well as theories based on insufficient statistics," the authors promised a theory of sex determination based solely on "induction from experimental evidence."[9] The book presented detailed critiques of popular theories of sex and argued vigorously for bringing the question of sex into experimental biology and under the mechanistic laws of physiology.

At the center of Geddes and Thomson's theory was *anisogamy*: the observation that females have large gametes (eggs) and males have little ones (sperm). Anisogamy remains central to our biological definition of sex today. The subclass of a species that produces the larger gamete is always considered female, regardless of its chromosome complement or its mating and offspring-rearing behaviors. In the late nineteenth century, Geddes and Thomson elaborated anisogamy to argue that maleness and femaleness results from different metabolic demands. They asserted that females' "more serious reproductive sacrifice" required "much greater flow of blood to the ovaries than to testes." Males, "of smaller size, more active habit, higher temperature, shorter life," have lower metabolic demands.[10] Metabolic rate, which Geddes and Thomson believed was established early in development through environmental mediators in the cell cytoplasm, determined sexual fate. Sex determination, according to this account, was a matter of the construction of male or female metabolic pathways early in development, subject to environmental influence and mediated by the cell's cytoplasm.

As evidence for their model of sex, Geddes and Thomson pointed to observations that while male sperm are numerous and highly active, and show a wide range of variation, female eggs are sparse, passive, and identical. They also cited sex-ratio studies purportedly showing that males flourish in "adverse circumstances" while females thrive in times of abundant nutrition and good conditions. They hypothesized that adverse conditions led to the production of males, with their small sperm and hardier metabolic constitutions. Good conditions, such as higher nutrition, led to females, with their larger eggs and demanding reproductive systems.

With its seemingly extensive docket of empirical evidence, rigorous theoretical foundation in developmental biology, plausible physiological mechanism, and route for environmental and cytoplasmic influences on sex determination, Geddes and Thomson's metabolic model of sex represented the culmination of nineteenth-century investigations. At the turn of the century, Geddes and Thomson's was the most prominent, sophisticated, and mechanistically plausible theory of sex. From the 1890s to the 1920s, Geddes and Thomson's model provided the central theoretical framework for sex determination research. Indeed, the metabolic theory of sex would become the sourcebook for the new chromosomal account of sex determination.

CELLS, CHROMOSOMES, AND HEREDITY

Chromosome studies originated in the field of cytology, the study of cells. Cytology began in the 1600s, following the first good glimpse of plant fibers, insect limbs, and microbes through a microscope in the late sixteenth century. By the early 1800s, cytologists had shown that distinctive cell structures characterize different tissues. Cytology's modern era, however, began in the 1850s when Rudolf Virchow asserted that all tissues arise from a single cell. The "cell doctrine" posited the cell as the basis of life.[11] It was complemented shortly by Charles Darwin's *On the Origin of Species* and the subsequent invigoration of biological research on heredity.[12]

In the second half of the nineteenth century, cytologists raced to locate the hereditary units in the cell. In sexual species, including mammals, heredity refers to how traits are transmitted and recombined through sexual reproduction. Thus, reproductive biology quickly became the site of investigation of these matters. Of special importance was the study of gametes—particularly the plentiful and easy-to-obtain spermatozoa.

By the late nineteenth century, improved microscope lenses and tissue-staining techniques allowed researchers to see nuclear bodies within the cell. Scientists could now observe the process of cell division in somatic cells (mitosis) and gametes (meiosis). The German cytologist Walther Flemming described and named the process of mitosis in 1878. He observed and carefully documented the separation and copying of "threads" in the nucleus during cell division.[13] In 1888, German anatomist Heinrich Waldeyer called these threads "*chromo*somes"—"for the ease with which they stain, and hence their tendency to stand out in microscopic preparations of the cell nucleus."[14]

Experimental studies in the 1880s confirmed that the hereditary material must be located in the cell nucleus. In the 1890s, German physiologist August Weismann integrated nineteenth-century cytology with Darwin's theory of evolution, generating a unified theory of heredity. Weismann asserted that gametes (which he called the germ plasm) are distinct from somatic cells (somatoplasm), so that acquired characteristics in the body are not transmitted to offspring. In turn, meiosis, permitting the recombination of germ plasm, produces the variation that fuels evolution by natural selection.[15] With this theoretical framework in place, all that was needed was to locate the physical basis of this process in the cell, a problem that had flummoxed even luminaries such as Darwin.[16] Chromosomes, remarkably consistent in size, shape, and number from species to species and transmitted in precisely the manner predicted by Darwin, were leading candidates for the physical basis of heredity. By the turn of the century, the reigning question in the life sciences was whether these "chromosomes" were actually the carriers of the hereditary elements.[17]

THE X-ELEMENT

In 1891, the German cytologist Hermann Henking found an atypical element in the sperm of *Pyrrhocoris apternis*, the fire wasp. Henking observed what appeared to be a spare chromosome, which entered just one of two resulting sperm during meiosis. He termed this body, unsure whether it was a chromosome, a "peculiar chromatin-element" or the "X element."[18] A variety of theories of the "peculiar" element's origin and function appeared in the decade following Henking's 1891 report. Thomas H. Montgomery, chair of zoology at the University of Pennsylvania and an accomplished young cytologist, dismissed the X as a "cast-off" chromosome.[19] Such chromosomes, he speculated, are "degenerate" and "no longer carry on exactly the same activities as the ordinary chromosomes." Smaller and darker-staining, they are "metabolically different," representing "the last perceptible stage in their history."[20] The X did not again pique interest until 1899, when Clarence E. McClung, another young cytologist, then pursuing his PhD at the University of Kansas, reported in the *Zoological Bulletin* a "peculiar nuclear element" in the sperm of locusts. He dubbed his finding the "accessory chromosome."[21] Noting similarities with Henking's mysterious X-element, he drew attention to "striking features" of the X's structure and behavior.

In 1902, McClung went further. He asserted that his "nuclear element" was indeed a chromosome and proposed a link between the accessory chromosome and sex. He reported "two kinds of spermatozoa: those with the accessory chromosome and those without."[22] This finding corresponded with the occurrence of two kinds of individuals in sexual species. As he wrote, "a careful consideration will suggest that nothing but sexual characters thus divides the members of a species into two well-defined groups, and we are logically forced to the conclusion that the peculiar chromosome has some bearing upon this arrangement."[23]

Simple and tantalizing—a chromosomal dimorphism corresponding to sex dimorphism. Yet at that time, linking a chromosome to a trait as profound and prominent as sex was a radical claim. McClung knew this. Citing his "considerable reluctance" to go "further afield than I should desire," McClung presented the hypothesis that the X chromosome is "the bearer of those qualities which pertain to the male organism."[24]

THE RACE FOR THE SEX CHROMOSOMES

Cytologists Nettie M. Stevens at Bryn Mawr College and Edmund B. Wilson at Columbia University in 1902 separately began intensive studies of the X—then called the accessory chromosome. They used insect sperm, featuring the *hemiptera*—grasshoppers, bed bugs, waterbugs, and stinkbugs. Stevens and Wilson kicked chromosome research on sex into high gear and within a few years yielded impressive evidence for the chromosomal basis of sex.

McClung, while examining the accessory chromosome, had been a visiting doctoral student in Wilson's laboratory at Columbia, and he may have stimulated Wilson's interest in the subject. Trained in embryology at Johns Hopkins, the leading medical and life sciences institution in the United States, Wilson had begun his career at Bryn Mawr College. He was recruited to Columbia in 1891, where he built a major cytology laboratory. Along with his colleague, Morgan, he would soon come to represent the new American eminence in biology. Wilson's monumental text, *The Cell in Development and Inheritance,* had been just published in 1896. It would be subsequently revised and reissued, remaining in wide use into the 1930s, becoming the principal advanced biology textbook of the day, and recording decades of vibrant developments in American cell biology and genetics.[25]

Stevens, one of a small but growing cohort of women who obtained advanced degrees in the sciences at the turn of the twentieth century, casts a very different profile than Wilson. It is worth briefly visiting her story (see fig. 2.2). Stevens obtained her undergraduate and master's degrees at Stanford University in California in 1900 and then moved to Bryn Mawr, where she completed her doctoral studies under Morgan in 1903. With funding from the Carnegie Institution, she became a postdoctoral researcher at Bryn Mawr. Though she later became a research associate and instructor there, she was never offered a full faculty position. Stevens's achievements and her persistence in the face of few opportunities for women are extraordinary. Particularly moving is her 1903 letter of inquiry to the Carnegie Institution in which she noted that "college positions for women in Biology this year, seem to be very few" and stated that she would prefer research to teaching, "but I am dependent on my own exertions for a living."[26] In the letter, she wrote of her dream to be part of "a School for research work in which members were to receive salaries and give their time to investigation. That is exactly what I should like."[27] For her Carnegie application, she assembled stunning letters of recommendation from Morgan and Wilson, among others. None failed to note her brilliance—for a woman. One letter from M. Carey Thomas, president of Bryn Mawr, stated, "Miss Stevens is one of the few women I know who seems to me to possess original power of a high order."[28]

Stevens made three major contributions to the chromosomal theory of sex. First, she provided incontrovertible empirical confirmation of McClung's initial conjecture. In a series of detailed studies of the X chromosome in a wide class of insects during the years 1903 to 1905, she demonstrated that the presence of the accessory chromosome correlates with sex. Stevens's "Studies in Spermatogenesis," published in the Carnegie Monograph Series in 1905, provided the definitive data demonstrating McClung's proposition; however, her findings reversed McClung's assumption that the extra X determines maleness. Stevens confidently demonstrated that the accessory X correlates with femaleness, concluding that "here . . . it is perfectly clear that an egg fertilized by . . . a spermatozoan containing the larger heterochromosome develops into a female."[29]

In the course of this research, Stevens uncovered a genuine surprise—a small chromosome in males that pairs with the X. Stevens had discovered the Y chromosome—a second major contribution. Her early work on

Figure 2.2. Nettie Maria Stevens. Courtesy of University Archives, Columbia University in the City of New York.

insects did not detect a Y, but when Stevens turned to *Tenebrio molitor*, the common mealworm, she was able to discern a small chromosome in the male. Tenebrio revealed two sperm types—one with ten large chromosomes and one with nine large and one small chromosome. In somatic cells, females had twenty large chromosomes, and males had nineteen large plus one small chromosome. Stevens was thus able to surmise that females in this species are XX and males XY. As she wrote of Tenebrio, "this seems to be a clear case of sex-determination . . . by a definite difference in the character of the elements of one pair of chromosomes of the spermatocytes which contain the small chromosome determining the male sex, while those that contain 10 chromosomes of equal size determine the female sex."[30] Stevens quickly confirmed the presence of a Y in the male fruit fly and other organisms.

Wilson was simultaneously advancing on the sex chromosomes. In 1905, he published findings of two differently sized chromosomes—what he termed "idiochromosomes"—in hemipteran bugs. He found that the male produced two types of sperm, one containing the large idiochromosome, the other a small one, while the female produced one type of egg. Historian Stephen Brush has argued, based on correspondence from the Carnegie Institution's archives and early drafts of Wilson's paper, that "Wilson probably did not arrive at his conclusion on sex determination until after he had seen Stevens' results."[31] In a later 1905 paper, now citing Stevens's corroborating "important direct evidence," Wilson also disputed McClung's hypothesis, agreeing with Stevens that the X chromosome is the female sex determinant.[32] In 1906, Wilson concluded, like Stevens, that the presence of an extra X determined femaleness, and an absence of it led to maleness, writing:

> The facts leave no doubt that both forms of spermatozoa are functional; that all of the eggs possess the same number of chromosomes; that all contain the homologue, or maternal mate, of the accessory . . . chromosome of the male; and that fertilization by spermatozoa that possess this chromosome produces females, while males are produced upon fertilization by spermatozoa that do not possess it.[33]

While Wilson and Stevens both insisted on the significance of the odd chromosomes for the theory of heredity, they parted ways on whether chromosomal sex determination could be brought in line with Mendelian genetics. Wilson was a skeptic of Mendelism, the new theory of heredity that asserted that genetic traits are inherited as pairs of discrete unit characters, one from each parent, with qualities of dominance and recessiveness.[34] He believed that sex was determined quantitatively. Sex was a whole-chromosome effect: one X kept things tilted toward maleness, while two X's pushed the balance in favor of femaleness. As we shall shortly see, Wilson's view was the most favorably received at the time. Stevens was a staunch Mendelian who saw sex as determined by unit characters: a factor or factors on a chromosome rather than by the whole chromosome. In her 1906 paper, Stevens wrote that the chromosome theory, "which brings the sex determination question under Mendel's Law in a modified form . . . makes one hopeful that in the near future it may be possible to formulate a general theory of sex determination."[35]

Demonstrating the consistency of the XX/XY model of sex with the Mendelian theory of heredity became Stevens's central theoretical interest between 1906 and her untimely early death in 1912, and should be considered a third major contribution of Stevens to the developing sex chromosome theory. Wrote Stevens with confident pique in 1906: "there would hardly seem to be any basis for Wilson's attempt to associate the difference in development of male and female germ cells with activity or inactivity of chromosomes."[36] Stevens's role as a bold contestant in theoretical debates over the mechanism of sex determination during the first decade of the twentieth century, despite her meager platform as a research associate at a suburban women's college, is in my view the most overlooked and impressive dimension of Stevens's contribution to the chromosomal theory of sex.

Stevens has only recently gained recognition for her contributions to the chromosomal theory of sex.[37] This recognition, however, has been uneven, and often overlaid with the superficiality brought about by a perceived imperative to include female figures in the historical vignettes of science textbooks. While historian of genetics Garland Allen refers diplomatically to the "independent work of E. B. Wilson and Nettie M. Stevens on the accessory chromosome," others are not so charitable.[38] Historian Peter Bowler, in his book *The Mendelian Revolution*, portrays Wilson as the innovator of the theory of chromosomal sex determination—with Stevens providing the follow-up data. Writes Bowler, "[Wilson] began to argue that, in insects at least, sex is determined by an accessory chromosome which had been discovered in the sperm-producing cells of the males. Crucial work *supporting this view* was performed by Nettie M. Stevens in 1905. . . . Wilson now became a leading exponent of the view that chromosomes are responsible for transmitting hereditary characters."[39] Similarly, historian Elof Carlson's *Mendel's Legacy: The Origins of Classical Genetics* dubs the XX/XY theory "Wilson's Solution."[40] In his inexplicably unbalanced account of the discovery of the sex chromosomes, Carlson describes Wilson as developing the chromosomal theory of sex in "brilliant papers," while writing that Stevens was "confused." He portrays her as a bumbling cytologist whose contributions were redundant and minor compared to the more far-reaching and theoretically sophisticated efforts of Wilson.[41] Most egregious is historian John Farley's account of Stevens in his touchstone work on the history of modern biological theories of sex, *Gametes and Spores*. Farley dismisses evidence that Stevens was the first to discern the significance of the X and Y to the Mendelian account of sex, contriving that Stevens's hypothesis was merely bolder than Wilson's—an overconfident lucky guess, more the result of her lack of experience than her scientific acumen. "Whereas, of course, Wilson was far more aware of the difficulties inherent in the Mendelian scheme," Farley writes.[42]

By 1906, Stevens and Wilson had laid down the fundamentals of the chromosomal theory of sex, and between 1906 and 1912 they each confirmed and extended the hypothesis of chromosomal sex determination in a wide range of species. During these years, Stevens published more than a dozen papers summoning evidence against Wilson's hypothesis and arguing vigorously that sex is determined by unit characters, not a whole chromosome. In retrospect, neither Wilson's nor Stevens's position on the precise mechanism of sex determination was scientifically vindicated. Yet both made indisputably valuable intellectual contributions. To both Stevens and Wilson we may credit the careful observations that empirically grounded the correlation between the odd chromosomes and sex and the theoretical debates that developed and honed the early theory of chromosomal sex determination.

BUTTERFLIES AND OTHER PUZZLES

Despite Wilson's and Stevens's persuasive data in insects and worms, the chromosome science of the early twentieth century was not yet able to produce transparent, consistently replicable results. From our perspective today, the evidence on which the hypothesis of chromosomal determination of sex rested in its early days was astonishingly limited, uneven, and contradictory.[43] Two issues particularly confounded attempts to validate a causal relation between the sex chromosomes and sex determination. The first was underdeveloped techniques and methods for precise descriptive characterization of chromosomes. The second was conflicting findings in different species. As a result, as Morgan records in his 1912 memoriam for Stevens in *Nature*, "many cytologists assumed a skeptical or even antagonistic attitude for several years toward the new discovery."[44]

Unsettled technical practices in the new field of chromosome cytology plagued the formalization of descriptive convention around the X and Y chromosomes in the early 1900s and helped to generate doubt as to whether reported cases of "odd chromosomes" were credible. Researchers suspected, probably rightly, that many observations of odd chromosomes were spurious. Distinguishing the X and Y chromosomes from the autosomes was no small task. It required fresh material, proper fixation of the chromosomes during cell division, and precise staining and counting practices. Racing to identify chromosomes with unique structures or behaviors, many claimed to find extra or "odd" chromosomes in their samples. This led to such a mess of data that Wilson sent up alarms in 1913 about "the danger of confusing with chromosomes other compact and deeply staining bodies that may lie near or among them, or at the spindle poles—extruded nucleoli or nucleolar fragments, chromatoid bodies, 'acrosomes,' yolk-granules, or the like." These persistent problems threatened the credibility of chromosome studies and the chromosomal theory of sex. As Wilson continued, "one can never avoid the suspicion that some of the existing contradictions in the literature may have arisen from such a source. . . . So deceptive is this resemblance that any observer without careful study might readily conclude the body in question is in each case an accessory [X] chromosome."[45]

Technical problems were compounded by stubbornly variable findings among species. Many expected that a true "sex chromosome" would provide a universal key to sex determination, but the search for odd chromosomes in different organisms revealed that they did not always correlate with sex. Odd chromosomes did not obey regular laws from species to species, nor were they adequate to explain sex determination in many species. This confused and destabilized the hypothesis of a relationship between the X and sex.

Insect chromosomes vary considerably from one species to another. Scientists pursuing the odd chromosomes could not always clearly distinguish them from "normal" chromosomes in the cells of a particular bug. Additionally, among those species with odd chromosomes, the males in some possessed just a single large X—notated "XO"—while in others they carried an X and a Y—"XY." It was unclear whether these systems were related (though Wilson argued forcefully, and ultimately correctly, that they were). [46] Species yielding an even—as opposed to odd—number of chromosomes in the sperm led to much confusion. Wilson wrote, "this question has been complicated in a most unfortunate way by errors in counting the spermatogonial chromosomes. . . . Regarding the forms that possess an accessory or heterotropic [X] chromosome the existing accounts are . . . in conflict in giving sometimes an even number, and sometimes an odd one."[47]

Weird animals also proved to be theory wreckers. "Parthenogenesis" is the occurrence of occasional asexual reproduction in sexual species. Aphids alternate between parthenogenesis and sexual reproduction, and Stevens, Morgan, and Wilson invested considerable resources to test the sex chromosome theory against this phenomenon. Most believed that environmental factors determined whether aphids reproduce sexually or parthenogenically. After years of effort, Stevens ultimately found a single unpaired odd chromosome in aphid males. The question of how precisely the aphid alternated between sexual forms, however, remained unresolved.[48]

Other species with odd chromosomes showed what appeared to be a confusing reverse pattern, known today as "female heterozygosity." In early X-Y studies in insects, females were found to be XX and males XY or XO. Later it was discovered that in some species females were heterozygous XY while males were XX. By 1910, females had been found to be the heterogametic sex, rather than males, in sea urchins, birds, moths, and butterflies. Yet females of these species still produced eggs, not sperm. As Wilson wrote in 1914, "the results of genetic experiments on *Lepidoptera* moths and on birds lead us to expect the existence in these forms of a different cytological type, in which the eggs, instead of the spermatozoa, are of two different classes; but the cytological facts have not yet become sufficiently clear to warrant any definite conclusion. In the case of birds, indeed, a conspicuous contradiction still appears between the cytological and the genetic results."[49] Were species with XY females a refutation of the chromosomal theory of sex determination, vindicating the view of critics that the X and Y chromosomes were really spurious chromosomal elements with no sex specificity? Or were they merely a different adaptation of chromosomal determination of sex?

Such findings led to a realization that the odd chromosomes, at best, explain a subclass of sex determination in the natural world. Study of sex determination systems in multiple phyla quickly suggested that no single, unified theory of sex—or of chromosomes in relation to sex—should be expected. As Morgan wrote, "it seems not improbable that this regulation is different in different species, and that therefore it is futile to search for any principle of sex determination that is universal for all species with separate sexes."[50]

THE METABOLIC MODEL OF CHROMOSOMAL SEX DETERMINATION

The critical question, of course, was how—by what mechanism—the odd chromosomes might determine sex. As cytologists crafted models of the role of chromosomes in sex determination, environmental and metabolic theories of sex provided an intellectual framework. Cytologists looked to Geddes and Thomson's theory of sex, in particular, as they worked to model the physiological role of the odd chromosomes in maleness and femaleness.

Exemplary is the very first formulation of the hypothesis of a chromosomal link to sex, by McClung in 1902. McClung proposed a metabolic mechanism for chromosomal determination of sex. Citing *The Evolution of Sex*, McClung suggested that the "extra" odd chromosome determines sex by accelerating metabolism. Interestingly, he interpreted Geddes and Thomson's theory to imply that males need something "extra" to develop. Thus, he speculated that the extra, odd chromosome is male determining. As he wrote, "it would seem most natural that the determinant should be for the purpose of carrying the transformation beyond the production of ova to spermatozoa."[51] Notably, like the predominant theories of the day, this early model of X-chromosomal sex determination embraced environmental and cytological factors. The X chromosome, McClung believed, acted as a selective mechanism that ultimately decides a fertilized egg's sex "in response to environmental necessities." This allowed the egg the flexibility to produce "the sex most needed by the species." The X chromosome functioned, he hypothesized, to help the egg discriminate between male- and female-producing sperm, selecting "either the spermatozoa containing the accessory chromosome or those from which it is absent."[52]

By 1906, Stevens and Wilson had reversed McClung's theory: it was females, not males, who carried the "extra" X. Researchers now concluded that the X is female determining. The metabolic model of sex again filled in the details. Citing Geddes and Thomson's "bigger gamete" theory of constitutional femaleness, Wilson argued that a bigger gamete needed more chromosomal material—thus, the female-determining properties of the X. Wilson proposed that the X was a growth factor or stimulant that mediates sex. Metabolism was Wilson's mechanism for the sex-differentiating action of the X and Y chromosomes: as he wrote, "the primary factor in the differentiation of the germ cells may, therefore, be a matter of metabolism, perhaps one of growth." The X acts to up-regulate the "degree or intensity" of metabolism—with more X, or higher X intensity, leading to femaleness.[53] Wilson called this a "quantitative" model of chromosomal sex determination. As he described it, his was a model of genetic sex determination "based on the quantitative relations of the 'sex-chromosomes' without assuming alternative male and female genes."[54]

Like McClung's model, Wilson's quantitative-metabolic account of the odd chromosomes' physiological action allowed a role for cytoplasmic and environmental factors. Prior to the discovery of the X and Y chromosomes, Wilson had been among the strongest proponents of externalist explanations of sex. In the 1896 first edition of his textbook, *The Cell in Development and Heredity*, Wilson asserted "it certain that sex as such is not inherited" and that sex is produced "not by inheritance, but by the combined effect of external conditions."[55] In his earliest development of his sex chromosome theory, he wrote, "I have wished to indicate that a hypothesis of sex-production which recognizes in some cases a fixed predetermination in the chromosome-groups of the fertilized egg is not inconsistent with the control of sex-production in other cases by conditions external to the nucleus." A genetic account of sex that permits sex to be pushed in one direction or another by a balance of X-chromosomal factors, Wilson believed, provided a possible mechanism for the influence of external factors on sex. As he wrote, "in some cases the chromosome-combination, established at fertilization, may be in something like a balanced state that is capable of modification by conditions external to the nucleus."[56]

This theory also preserved the commonplace conviction that sex is a continuous and multifactorial trait. Like many of his contemporaries, Wilson was convinced by observations of intersex organisms and so-called gynandromorphic insects that maleness and femaleness exist on a seamless spectrum. His quantitative-metabolic model allowed that "the same kind of activity that produces a male will if reinforced or intensified produce a female. . . . In these cases the decisive factor may be a merely quantitative difference of chromatin between the two sexes." A quantitative-metabolic model of chromosome action would explain, for instance, "the latency of female characters in the male, and the development of such secondary female characters as may be regarded as an exaggeration or intensification of corresponding characters in the male."[57]

Geddes and Thomson's theories influenced Wilson as he developed his chromosomal account of sex in the first decades of the twentieth century.[58] As the larger size of the egg compared to the sperm was the starting point of Geddes and Thomson's theory of sex, so it was for Wilson's chromosomal model. "This growth of the oocyte involves the production of a mass of protoplasm . . . thousands of times the bulk of the spermatocyte" began Wilson.[59] "This suggestion," Wilson notes, "recalls the theory developed by Geddes and Thomson, in their well known work on the 'Evolution of Sex,' that 'the female is the outcome and expression of relatively preponderant anabolism, and the male of relatively preponderant katabolism.'"[60] He continued:

> The thought cannot be avoided that there is a definite causal connection between the greater activity of these chromosomes in the oocytes and the great preponderance of constructive activity in these cells; and it is especially this coincidence that leads me to the general surmise that one of the important physiological differences . . . between the chromosome-groups of the two sexes, may be one of constructive activity. . . . Perhaps a direct causal relation here exists.[61]

Incorporating Geddes and Thomson's distinction between the "constructive activity" of the more "conservative" anabolic female and the relatively "destructive" activity of the katabolic male, and importing Geddes and Thomson's assertion that "the final physiological explanation [of sex] is, and must be, in terms of protoplasmic metabolism," Wilson presented the odd chromosomes as a final puzzle piece in the picture of sex determination established by the metabolic theory.[62]

Wilson is not an isolated case. Scientists were simply not prepared to abandon the existing developmental, externalist science of sex in favor of a purely chromosomal one. A quantitative, metabolic understanding of sex underpinned all of the authoritative accounts of the mechanism of chromosomal determination of sex in the first decades of the twentieth century, from those of McClung and Wilson to discourses by Thomas Hunt Morgan, Arthur Darbishire, Joseph Cunningham, and even F. A. E. Crew as late as 1927. These intellectual architects of sex determination theory in the early twentieth century favorably cited and discussed the metabolic theory of Geddes and Thomson as a framework for conceptualizing the role of odd chromosomes in sex determination. When Geddes and Thomson updated and reiterated the metabolic theory of sex in their 1914 text *Sex*, the intervening discovery of the odd chromosomes required no modification of their original 1889 metabolic theory. The X and Y chromosomes appear only as a side note. The odd chromosomes, wrote Geddes and Thomson, now citing Wilson, are "not at all incompatible with a physiological interpretation, for it may be that this mysterious sex-determinant acts as an accelerant or depressor of the rate of metabolism."[63]

In sum, the prevailing early twentieth-century conception of sex as a complex, environmentally sensitive trait made researchers wary of postulating a chromosome as the genetic determinant of sex. Moreover, while the role of the odd chromosomes in sex was the subject of much theoretical speculation, the science of the time did not allow ascertainment of their precise mechanistic role in sex determination. The leading early theories assumed a model of sex determination that was labile, open to many influences, and nondeterminate. Scientists worked to fit the odd chromosomes into preexisting accounts of the development of sexual characters.

◇

"As to what the X-element really signifies in connection with sex," wrote the University of Cincinnati cytologist Michael Guyer in 1911, "at least four possibilities have been suggested":

1. that it is an actual qualitative sex-determinant;
2. that sex is determined by purely quantitative conditions of the chromatin;
3. that the X-element is merely sex-accompanying and not sex-producing;
4. that sex is the resultant of several essential factors and is not established unless all work together, the X-element being the *decisive* factor.[64]

A decade after the discovery of a link between the X, Y, and sex, the chromosomal theory of sex remained deeply unsettled. The precise role of the X and Y in sex determination would remain opaque until the 1950s and 1960s—and beyond. Yet, as we shall see, within a few short decades after Guyer's assessment, the X and Y had become the "sex chromosomes"—and soon after they hardened into the "female" and "male" chromosomes.

CHAPTER 3

How the X and Y Became the Sex Chromosomes

In 1960, an international body of geneticists assembled in Denver, Colorado, to agree on a system for naming and classifying the human chromosomes. The X and Y presented a point of debate and conceptual dissonance. The inherited system, devised in the first decades of the century, postulated two kinds of chromosomes: the "sex" chromosomes and the "ordinary" chromosomes, or autosomes. The scientists wondered: is there justification to continue this central distinction between sex chromosomes and the others? Or should the X and Y simply be classified by size and structure, like autosomes? Why are the X and Y called the "sex chromosomes," after all?[1]

The distinction between "ordinary" and "other" chromosomes was honed in the early twentieth century. Debates over cytological nomenclature raged on the pages of science journals. The question of the "chromosome for sex" took center stage. By 1910 the X and Y had "received a great variety of names," wrote Thomas Montgomery.[2] The most common terms for the X and Y chromosomes, and their first traceable appearance in the early literature, include odd chromosomes (1890s), accessory chromosomes (1899), heterochromosomes (1904), idiochromosomes (1905), heterotropic chromosomes (1905), monosomes/diplosomes (1906), gonochromosomes (1906), sex chromosomes (1906), and supernumerary chromosomes (1908).[3] Three terms, "accessory chromosomes," "heterochromosomes," and "idiochromosomes," dominated literature on the

X and Y from the first hypothesis of a link between the X and sex until the 1920s, when "sex chromosome" began to come into general use.[4]

The story of how the X and Y ultimately became the "sex chromosomes" offers a window into how contextual and contingent factors shaped the conceptual terrain of genomic sex from its earliest days. "Sex chromosome" was neither the first nor the most preferred name for the X and Y. Several alternative terms for the X and Y, which did not point to their link to sex, flourished into the 1920s. Many prominent biologists openly denounced the concept of a "sex chromosome." So vituperous was the climate that the geneticist Thomas Hunt Morgan called the concept of the sex chromosome a "travesty." To understand why "sex chromosome" triumphed, despite strong objections, we must visit the opportunistic role of the concept of the sex chromosome in consolidating two great biological efforts of the first quarter of the twentieth century: the chromosomal theory of inheritance (this chapter) and later the new biology of sex heralded by the discovery of the "sex hormones" (chapter 4).

THE TERMINOLOGY OF ABERRANT CHROMOSOMES

Each of the popular terminological contenders for the X and Y—"accessory," "heterochromosome," "idiochromosome," and "sex chromosome"—captured different features of the X and Y and reflected a different view of the empirical and theoretical significance of these unusual chromosomes.

Coined in 1899 by Clarence McClung, the term "accessory chromosome" originates in observations of "XO" insect species, in which half of the sperm appeared to have an extra, or accessory, chromosome. The term was first applied to the X chromosome. It would later be used generically to refer to the X and Y—as in the "accessory elements."[5] "Accessory chromosome" was used far more widely than "sex chromosome" and remained popular into the 1920s. McClung insisted on the superiority of "accessory chromosome" to "competitive terms with improper associations."[6] Compared to the alternatives, the term "accessory chromosome" laid special emphasis on the existence of an "extra" chromosome in one sex.

Montgomery's "heterochromosome" joined the lexicon in 1904. Initially, Montgomery intended the prefix *hetero-* to distinguish the X, which he believed was a nonautonomous, "reformulated" chromosome, from the ordinary *auto*-somes, which were independent cell bodies that maintain their form. But "heterochromosome" quickly became the preferred term for the X and Y for other reasons—it seemed to describe precisely their status as a heterogeneous pairing unit. Montgomery argued that the pairing behavior of chromosomes, not the number of chromosomes, should take center stage in chromosome naming. He pressed "the value of chromosomal relations as a taxonomic character" as opposed to the "less constant" "number of the chromosomes."[7] The status of the X as an "extra" or accessory chromosome was not as central, he argued, as its status as a nonpairing or hetero-pairing chromosome.

Nettie Stevens favored "heterochromosome" for the XX/XY type. The term "heterochromosome" offered the additional advantage, over the term "accessory chromosome," of highlighting the commonly observed size differential between the X and Y. She first used it in 1905 to distinguish XY (heterochromosome) from XO (accessory chromosome) forms of sex determination. Describing the two sizes of chromosomes she found in the worm *Tenebrio*, for example, she wrote: "This seems to be a clear case of sex-determination, *not by an accessory chromosome*, but by a definite difference in the character of the elements of one pair of chromosomes."[8] Rather than an accessory chromosome, Stevens described *Tenebrio* as possessing "smaller" and "larger" heterochromosomes.[9]

Wilson proposed an alternative term, "idiochromosome," in 1905 to describe cases of two paired unequal-sized chromosomes. He called the X the "heterotropic chromosome" and saw the X as one of a *pair*, either present or ancestral, of sex-determining chromosomes. Wilson specifically disfavored the term "accessory chromosome." Defending his alternative, Wilson wrote:

> Since there is no reason for considering the "accessory chromosome" as in any sense accessory to the others, it appears to me that McClung's term might well be abandoned in favor of a less compromising one. I suggest that until their physiological significance is positively determined chromosomes of this type may provisionally be called *heterotropic* chromosomes (in allusion to the fact that they pass to one pole only of the spindle in one of the maturation-divisions) in contradistinction to *amphitropic* chromosomes, the products of which pass to both poles in both divisions.[10]

Wilson's system of nomenclature stressed chromosome pairing behavior during cell division. Males had a pair of idiochromosomes, large and small (XY); females had a single heterotropic chromosome (XO).

The term "idiochromosome" emphasized the sportive, nonhomologous features of the X and Y chromosomes and, like Montgomery's term, focused on the fact that the X and Y are a pair of "unequal" chromosomes. As Wilson defined them, idiochromosomes are those "left without a fellow of like size."[11] "The *purely descriptive* term 'idiochromosome' (peculiar or distinctive chromosomes)," wrote Wilson, "will be applied to the two chromosomes, usually unequal in size, which . . . undergo a very late conjugation and subsequent asymmetrical distribution to the spermatid-nuclei."[12] It was the distinctive, idiosyncratic pairing behavior of the X and Y, not their link to sex, that merited a special term for them. As Wilson put it, "were it not for their failure to unite to form a bivalent body until the end of the first mitosis we should find *no ground* in this case for designating these chromosomes by a special name."[13]

In a 1906 contribution to *Science* titled, in part, "The Terminology of Aberrant Chromosomes," Montgomery, who had recently coined the term "autosome" for the "ordinary chromosomes," decried the proliferation of nomenclature for the X and Y and asserted "a pressing need for a conciser and more uniform nomenclature." "I call upon fellow workers to discard their previous names," Montgomery enjoined.[14] While "autosome" quickly became the term of choice for homologous chromosomes, the question of what to call the X and Y would remain unsettled. In a 1913 *Science* editorial, McClung, a stickler for terminology who had now been elevated to the chair of zoology at the University of Pennsylvania, charged that cytological nomenclature had "fallen into lamentable confusion." This, he argued, was due to the scattershot naming of odd chromosomes, including the X and Y, by a *function* or *character* rather than by an accurate pointer to physical identification.[15]

The term "sex chromosome" was indeed inconsistent with the predominant descriptive practices in chromosome studies. A demanding ocular art with an elusive and phasic object of study, chromosome science in its early days required a precise and rich descriptive convention. Chromosomes were named, characterized, and classified strictly by pairing behavior during meiosis, and by their size, shape, and structure. The term "sex chromosome" transgressed this classification system. It specified the chromosome instead by the *trait* putatively linked to it. As such, researchers worried that the concept of a sex chromosome might introduce distortions into reasoning about the relationship between chromosomes and traits. Specifically, the term "sex chromosome" raised suspicion of confused and overly deterministic reasoning about the relation between chromosomes and sex. Sensitive to this, even Wilson, the originator and eventual champion of the term, warned that it was meant only as a form of shorthand: "sex chromosome" was a designation, he wrote, to be used "whenever it is desirable to avoid circumlocution."[16]

IN THE CROSSHAIRS OF MENDELISM

Below the surface of these terminological debates over the sex chromosome roiled a deep philosophical rift in turn-of-the-century biology. The sex chromosome arrived just as Mendel's new genetics was transforming and polarizing the world of biology. It became a flashpoint in these debates.

In 1900, three investigators rediscovered Mendel's laws of inheritance, which had been lying in obscurity since his plant hybrid experiments were first published in 1866.[17] Mendel crossed plants that differed in discrete characters, such as his famous round and wrinkle-seeded peas. These experiments yielded a large body of data about the frequency of trait inheritance. This data showed that traits must be determined by pairs of factors, which today we call gene variants, or alleles, and that these factors can influence phenotype differently depending on whether they are dominant or recessive. Mendel's findings also predicted that these gene pairs segregate from one another independently during gamete formation and randomly recombine to produce genetic diversity. These results are known as Mendel's laws of *segregation* and *independent assortment* (see fig. 3.1).

By 1903, the British geneticist William Bateson had confirmed Mendel's laws of heredity, as well as the ideas of dominant and recessive traits and recombination. Mendelian genetics operated on simple, abstract quantitative principles. The research program could theoretically proceed without regard to the nature, location, or mechanism of the genetic material itself. Nonetheless a structure that accounted for the physical basis of Mendelism would provide the definitive evidence needed to establish Mendelism more broadly in biology. In this way, Mendelism became intimately tied up with the science of chromosomes already in progress.

In the first years of the twentieth century, German Theodor Boveri and American Walter Sutton demonstrated that chromosomes were autonomous and distinct bodies—good candidates to be the carriers of the material that assures genetic continuity. Boveri showed that each species is characterized by a distinctive chromosome set, that each chromosome is a discrete individual, and that homologous chromosomes of maternal and paternal origin pair during cell division. On this basis, Boveri believed that chromosomes specialized in different functions, carrying genetic factors for biological traits.[18] Sutton asserted, furthermore, that chromosomes were the carriers of Mendelian alleles. Chromosomes come in homologous pairs—one inherited from each parent—that segregate and assort independently during cell division. Sutton explicitly con-

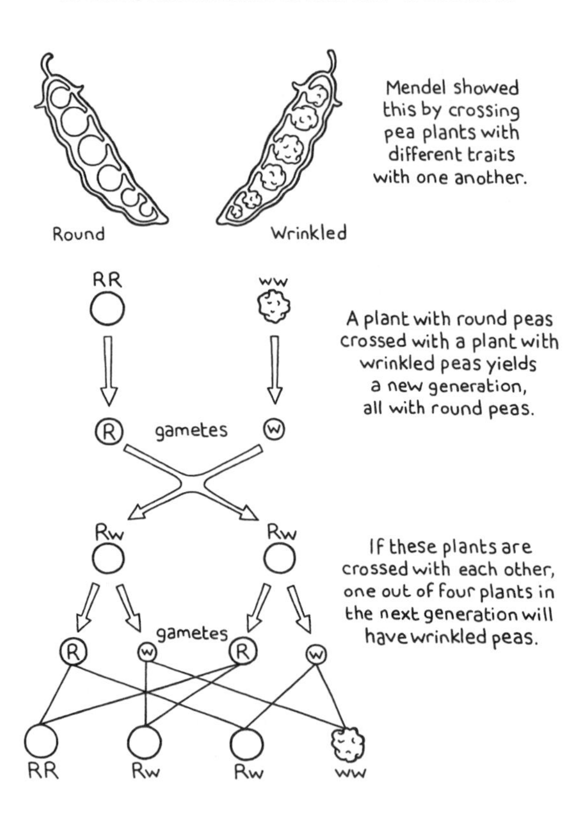

nected the meiotic division and recombination of the chromosomes to the segregation of Mendelian hereditary factors. Over the next two decades, what we know today as the chromosomal theory of heredity took shape. The so-called sex chromosome would become the controversial vehicle of the chromosome theory and the new Mendelism.

Very early on, Mendelism became associated both with the X and Y chromosomes and with a conception of sex as an either-or binary. Bateson, the principal figure in the reinvigoration of Mendelism at the turn of the century, argued in 1900 that sex is a binary, discontinuous variation in nature and thus a prime candidate for Mendelian explanation.[19]

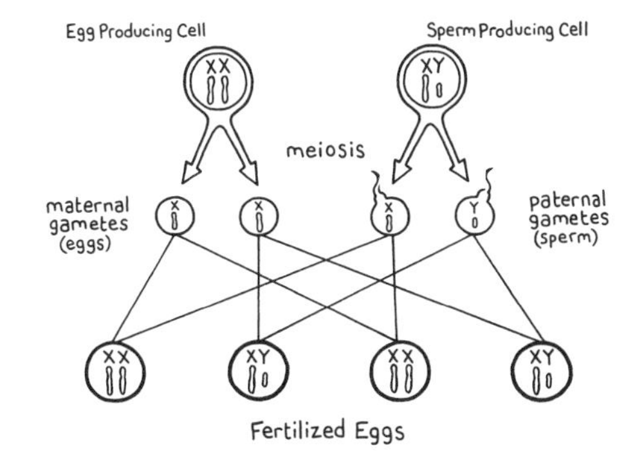

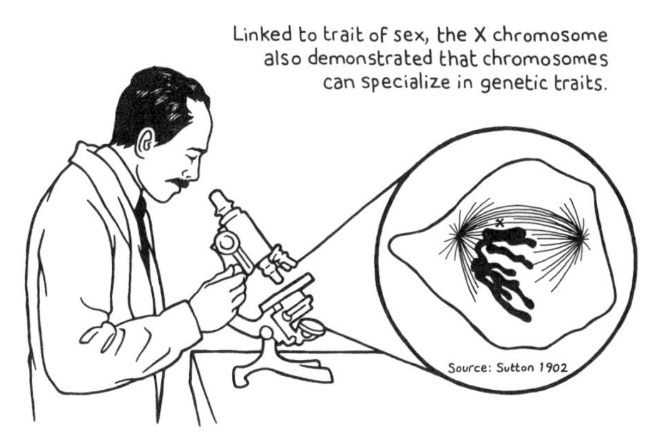

Figure 3.1. The chromosomal theory of inheritance. Illustration by Kendal Tull-Esterbrook; copyright Sarah S. Richardson.

A year following McClung's 1902 accessory chromosome hypothesis, Harvard geneticist William Castle, "the chief advocate of Mendelism in the United States,"[20] suggested that the X chromosome could provide the physical basis for a fully Mendelian genetic theory of sex and offered an elaborate model of how it might work.[21]

These Mendelian-chromosomal models of sex determination produced enormous irritation among cytologists and embryologists who were skeptical of Mendelism's broad claims about the nature of heredity. They vigorously rejected the plausibility of a Mendelian account of sex, and along with it, the notion of a "sex chromosome." They used the so-called sex chromosome to tar the Mendelian view that sex was a "unit character" as naïve and deterministic. French neo-Lamarckians such as Maurice Caullery joined British biometricians such as Karl Pearson and other skeptics to raise a chorus of criticism against the "sex chromosome." In America, the leading critics included Thomas Montgomery and Thomas Hunt Morgan.

A chromosome expert at the University of Pennsylvania, Montgomery had earned his PhD in Berlin and was deeply influenced by the German cytological school. Montgomery's opposition to the chromosomal theory of sex was philosophically grounded in his anti-Mendelism and his skepticism about Mendelism's corollary, the chromosome theory of heredity. He derisively cited the notion of the sex chromosome as proof that "modern Mendelian explanations represent a determinant theory far more rigid and complex than that of Weismann" (who held that hereditary characters were transmitted by the germ plasm).[22]

To Montgomery, the sex chromosome hypothesis represented an absurd and simplistic overextension of the chromosome theory of heredity. He maintained that the evidence showed only that the X chromosome and sex were "coincident," not causally connected.[23] As he wrote in 1910, "there is no valid reason to interpret sex as an immutable unit character resident in or presided over by particular chromosomes, and sorted out and distributed by Mendelian segregation."[24] He continued:

> This nuclear element is not an independent unit, but only a part, even if it be the most important part, of the cell whole. Thus the idea is erroneous to speak of the chromosomes as automatic units, for they are but parts of the cell or cell complex. . . . Now to assume that particular chromosomes alone are sex-determinants is to disregard this complex inter-activity.[25]

Moreover, the chromosomal theory of sex seemed to imply that maleness and femaleness are distinct, alternate developmental routes. In contrast, Montgomery argued that sex is "a labile process which may be changed by a variety of influences" and that maleness and femaleness are but "two modes of one process . . . not radically different conditions."[26]

Mendelism, critics such as Montgomery charged, required shoehorning complex traits into discrete unit characters. The chromosomal theory of sex appeared symptomatic of this central anti-intellectualism of Mendelism. Wrote Montgomery:

> He would be rash who would venture to claim that a particular chromosome determines excretion, another determines locomotion; yet these processes are relatively simple compared with that of sexuality, which some have contended may be controlled by a particular chromosome. The hypothesis is too naïve, it assumes too great simplicity of the cell, it tastes too strong of rigid predetermination.[27]

The chromosome theory of sex, Montgomery believed, showed the true colors of the naïvely atomistic and deterministic view of heredity promoted by the Mendelians.[28]

Morgan's condemnation of the "sex chromosome" was even more heated than Montgomery's. At the turn of the century, the young Morgan was emerging as a leading figure in American embryology. A 1900 trip abroad occasioned a meeting with the Dutch botanist and experimental geneticist Hugo de Vries. The visit turned Morgan's interest to questions of heredity and sex determination. Like Wilson, Morgan was a product of American biological training, having received his doctoral education in embryology at Johns Hopkins. Graduating in 1890, he became a professor at Bryn Mawr College, where he worked with Stevens and Wilson, and in 1904 he followed Wilson to Columbia. He would later develop the fundamentals of classical genetics and chromosome theory in his famous *Drosophila* labs with his students Alfred Sturtevant, Calvin Bridges, and Hermann Muller.

To Morgan's mind, Mendelism was a narrow and reductive theory that could not possibly account for the richness, complexity, and wondrous diversity of the natural world. His 1910 article, "Chromosomes and Heredity," famously castigated Mendelians for offering "intellectual security" at the expense of "restlessness of spirit," "adventurous character," and "the modern spirit of scientific theory."[29] The theory of inherited factors residing on chromosomes was repugnant to Morgan's training as an embryologist and his inclination toward developmentalist approaches to biology. As he wrote, "Mendelian principles seem to imply that hereditary elements are preformed in the germ cells, and hence exist as small units corresponding to given adult characters." To Morgan, this preformationism seemed to "ignore all that we know about the physiology of development."[30]

The sex chromosome took the brunt of Morgan's attack. Like Montgomery, Morgan believed that sex is a spectrum trait, exemplified by the cases of gynandromorphs and hermaphrodites.[31] As Morgan wrote, the model of genetic sex determination proposed by Mendelians such as Bateson, R. C. Punnett, and William Castle (he did not mention Stevens) presumed an overly binary conception of sex: "[It] assumes a *unit character* that determines femaleness, but for this very reason it introduces an interpretation of sex that is extremely hypothetical. For, the unit character is no longer simply a quantitative factor but a special element that has the power of turning maleness into femaleness. It is an entirely imaginary factor and lacks observational evidence in its support."[32]

The matter of the sex chromosome so troubled Morgan that even after he later assented to Mendelism and the chromosome theory, he remained irked by the excesses of the notion of a chromosome for sex. He spent a hefty five-page digression of the otherwise concise 1915 text, *Mechanism of Mendelian Heredity*, his powerful initial statement of Mendelian genetics and the chromosome theory, formulating an alternative terminology. Avoiding the term "sex chromosome," he referred instead to the "sex factors on the chromosomes." He proposed that instead of notating sex as XX and XY—implying that sex is determined by the whole chromosome—researchers ought to employ an alternative notation representing dominant and recessive factors for sex: FF/FO or FmFm/Fmfm (F standing for female-determining factor). Use of the X and Y to represent sex, he argued, should always be qualified so as not to imply chromosomal determination of sex. "[It is only acceptable] to use the symbols for the sex chromosomes as the symbols for the sex factors also, if it is at the same time recognized that the whole chromosome is not involved in determining sex."[33] The term "sex chromosome," Morgan wrote, "prejudices" thinking:

> There is no need to assume that X is the *sex chromosome* in the sense of carrying sex. The use of this term, I fear, prejudices the situation by the very aptness of the application. It may be that X only means more X, and that this is a factor in sex determination. . . . I should also object . . . on the general grounds that it refers a particular character to a single chromosome.[34]

Morgan raised the notion that the X is "female determining" as an example. Since males also have an X, as Morgan correctly pointed out, "a 'female-producing' X-sperm will give rise to a male if it enters an egg without an X." Sperm that "are sometimes loosely referred to as male and female," Morgan noted, "are properly only male-producing and female-producing in certain *combinations*."[35] (Morgan was prescient: chapter 6 of this book explores the persistence of the assumption that the X is the "female chromosome" even today.)

Morgan warned that the so-called sex chromosomes are only "the *best known* or *most visual* factor-difference among any number of possible theoretical ones." He bemoaned that "the identification of sex as a character with the factor for sex determination has led to needless confusion."[36] Maleness and femaleness, Morgan argued, involve many genetic and developmental factors, and "there are factors in the sex chromosomes that affect many parts of the body."[37] He concluded with a flourish: "To assume that all the factors that are shown by the male or by the female must be carried by a sex chromosome of some kind, if carried at all by the chromosomes, is a travesty."[38]

THE TRIUMPH OF THE SEX CHROMOSOME

Today, "sex chromosome" is the textbook term for the X and Y, thought of as entirely uncontroversial. How did the "sex chromosome" move from being the derided and unsettled term it once was to the consensus term of today?

One view is that the diverse alternative names for the X and Y—"heterochromosome," "accessory chromosome," and "idiochromosome" among them—were fuzzy terms representing early ignorance about the significance and function of the X and Y. Once the chromosomal determination of sex was proved *true*, leading scientists were forced to concede the aptness of the term "sex chromosome." However, historical evidence does not support this explanation. The eventual acceptance of the notion of a sex chromosome did not follow from empirical confirmation of the theory of the chromosomal determination of sex. While sex chromosomes were shown to *mark* sex and to segregate differently by sex during *gamete formation*, they were never proven to *determine* sex. Well into the 1950s, and some might say even up to today, the precise causal relation between the X, Y, and sex in mammals remained indeterminate. Moreover, it was not obvious that "sex chromosome" was the best technical term for the X and the Y. Alternatives such as "accessory chromosome" and "heterochromosome" were no less precise, generalizable, or descriptively rich than "sex chromosome." As we have seen, competing terms picked out different features of X and Y structure, behavior, and function that are just as significant as their connection to sex—features that would be sidelined by the eventual triumph of the sex chromosome concept.

To see how the sex chromosome triumphed despite these objections, we must first momentarily step back in our chronology, to the beginnings of the chromosomal theory of inheritance. When the X and Y were first discovered at the turn of the twentieth century, the question of sex determination was of marginal interest to chromosome scientists. These scientists were seeking to understand the basics of cell biology. What were the chromosomes? How does cell division occur? Where is the hereditary material housed in the cell? The X and Y intrigued cell biologists principally because of their ease of identification and their curious behavior during cell division. In the early years of the twentieth century, these special features helped to reveal a new and broadly explanatory physical theory of heredity known today as the chromosomal theory of inheritance. Sex was both central and incidental to this effort. "Maleness" and "femaleness," like Mendel's smooth and wrinkled peas, were salient biological characters used to test, confirm, and evangelize a new theory of inheritance, physically instantiated in the squiggly, paired bodies in the nucleus of the cell. The "sex chromosomes," like Mendel's peas, became the mnemonic touchstones of the new chromosome theory. As we shall shortly see, the X chromosome uniquely demonstrated that chromosomes were autonomous units that maintained their constancy in the organism—a necessary feature for any physical carrier of Mendelian traits. Key elements of the chromosomal theory of inheritance emerged from observations of the X and Y chromosomes in the first decades of the twentieth century, including that chromosomes are individual units, pair as homologues, carry hereditary material, and specialize in different genetic factors. These advances brought Mendel's laws into line with cytology, embryology, and developmental biology, laying the theoretical groundwork for classical genetics.[39]

THE CHROMOSOMAL THEORY OF INHERITANCE

Walter Sutton, a graduate student of Wilson's at Columbia, seized on McClung's demonstration of the individuality and constancy of the X chromosome in 1899 and 1902 to argue forcefully that chromosomes are the units or carriers of heredity (see fig. 3.1).[40] The X, he argued, showed that chromosomes have characteristic sizes and structures, that they maintain their individuality throughout cell division, and that they carry particular hereditary traits. As Sutton wrote:

> Perhaps the most important thing to be gained at present from the knowledge of the behavior of the accessory chromosome . . . is the light which it throws upon the question of the individuality of the chromosomes. . . . Now, if it be admitted that the body *is* a chro-

> mosome, inspection quickly shows us that it maintains throughout the spermatogonial divisions, as well as in those that follow, an indubitable independence. . . . Having, then, one of the chromosomes which preserves its individuality in this way, and seeing the other chromosomes enclosed for a part of their development in similar individual vesicles . . . have we not a right to suppose that at one time they too enjoyed the same independence as their more exclusive mate? . . . If this be granted, then we have at least one more ground for belief in the individuality of the chromosomes. . . .[41]

In Sutton's 1902 and 1903 work establishing the chromosomal theory of inheritance, the X chromosome featured as the "most unequivocal evidence of chromosomic individuality."[42] The clincher, of course, was that the X was also associated with the dramatic biological trait of sex. As Sutton wrote, "this particular chromosome" possesses "the power of impressing on the containing cell the stamp of maleness, in accordance with McClung's hypotheses."[43] The elegant meaning of Sutton's conclusion, even as it reiterated the erroneous idea that the extra X must be male-determining, could not be missed. In the X chromosome, Sutton had located every piece of evidence required to demonstrate that the chromosomes carry the hereditary material, behave as Mendel predicted, and carry discrete traits with stunning consequences for organismic development.

The role played by the X and Y in the development of the chromosomal theory of inheritance illuminates the "sex chromosomes" not merely as agents of sex determination but as conceptual innovations carved into the fibers and foundations of modern genetics. Historians of biology have not ignored the X chromosome's pivotal role in the development and dissemination of the chromosome theory of heredity. The flip side, however, remains to be considered. What was the role of the chromosome theory of heredity in shaping the "sex chromosome"?

One answer to this question can be found in the remarkable record left by Wilson. In 1905, Wilson recounted how observations of the X and Y ultimately persuaded him of the chromosome theory's validity. As he wrote:

> I must frankly confess that until I had followed step by step the behavior of the idiochromosomes . . . I did not appreciate how cogent is the argument . . . that synapsis involves an actual conjugation of chromosomes two by two, and that the chromosomes thus uniting are the paternal and maternal homologues.[44]

Wilson's account points to how consolidating a chromosomal link to the trait of sex served key evangelistic purposes within biology during the first quarter of the twentieth century. Well into the 1910s, much later than is generally appreciated, the sex chromosomes constituted the principal direct evidence for the chromosomal theory of inheritance. The chromosome theory had not gained the adherence of all geneticists—let alone biologists in general—and the sex chromosome remained a central selling point for chromosome theory proponents. As Wilson observed in 1914, "some eminent students of genetics are still reluctant to accept this theory."[45] He continued, "the conjugation of chromosomes, to say nothing of paternal and maternal homologues, has been obstinately contested; it must be admitted that the proof is still far from complete for the chromosomes generally," however, "in the case of the sex-chromosomes . . . the probability becomes a certainty."[46]

Wilson's 1914 lecture to the Royal Society of London, "The Bearing of Cytological Research on Heredity," trotted out the sex chromosomes to illustrate every major insight of chromosome research in the post-Mendelian era, including independent assortment, conjugation, and linkage. Wilson wrote:

> To the cytologist the *interest of the phenomena extends far beyond the special problem of sex*. Nature has here performed a series of experiments which gives a crucial test of many of our earlier conclusions, provides a secure basis for further advances, and at the same time brings vividly before us the connection of the chromosomes with heredity.[47]

He continued:

> . . . The cytological phenomena of sex-production lend strong support to the theory of the genetic continuity of chromosomes. They give unquestionable proof, in the case of a particular chromosome pair (XY), of the conjugation and subsequent disjunction of corresponding maternal and paternal chromosomes. . . . They give the first direct evidence of a difference of nuclear constitution between the homozygous and the heterozygous conditions, and of corresponding gametic differences. And finally . . . they fully substantiate the general cytological explanation that has been offered of Mendel's law.[48]

The sex chromosome, therefore, was not only an object of intense interest and debate, from a broad range of angles within genetic theory, during the critical period of the development of classical genetics. It was also a simple explanatory hook with which one could evangelize for the chromosomal theory of inheritance. It certainly did not hurt that the chromosomal trait was the dramatic and tangible phenotypic trait of *sex*, an object of popular fascination and a long-standing target of biological investigation.

MORGAN'S SEX-LINKED TRAITS

The story of the role that the chromosomal theory of inheritance played in establishing the "sex chromosome" continues in Morgan's fly lab in the years 1910 to 1920. Most flies have red eyes, but in 1910 Morgan found a rare mutant white-eyed male. When crossed with a red-eyed female, white-eyed males disappeared for a generation but then showed up in the subsequent one. When a white-eyed female was crossed with a red-eyed male, all of the male progeny had white eyes. The only explanation for this pattern of inheritance, now familiar to us from the case of human hemophilia, was that the white-eyed variant was recessive and carried on the X chromosome. It was "sex linked."[49]

Sex linkage allowed normal phenotypic traits to be mapped to the X chromosome and opened up a world of chromosomal linkage mapping. Morgan subsequently developed sex linkage into an experimental system and chromosome-mapping industry for genetics. All of the first genetic mutations, and many of the most important principles of genetic inheritance, were found on the X chromosome. Over the next ten years, with the aid of the X chromosome and the fruit fly's simple biology, short lifespan, and four plump chromosome pairs, Morgan and his students elaborated the chromosomal basis of inheritance (see fig. 3.2). Morgan's detailed studies of sex linkage in *Drosophila* ultimately provided the first definitive experimental evidence for the alignment of Mendelism with the chromosome theory of heredity.

Morgan, we recall, had vigorously objected to the notion of a "chromosome for sex." He railed at length against its imprecision and the view of heredity, development, and life that it seemed to represent. He was an antichromosome theory partisan, and the sex chromosomes represented everything that was wrong with the theory. The concept of the sex chromosome, he wrote, was a "travesty." Yet in 1910, after a decade of

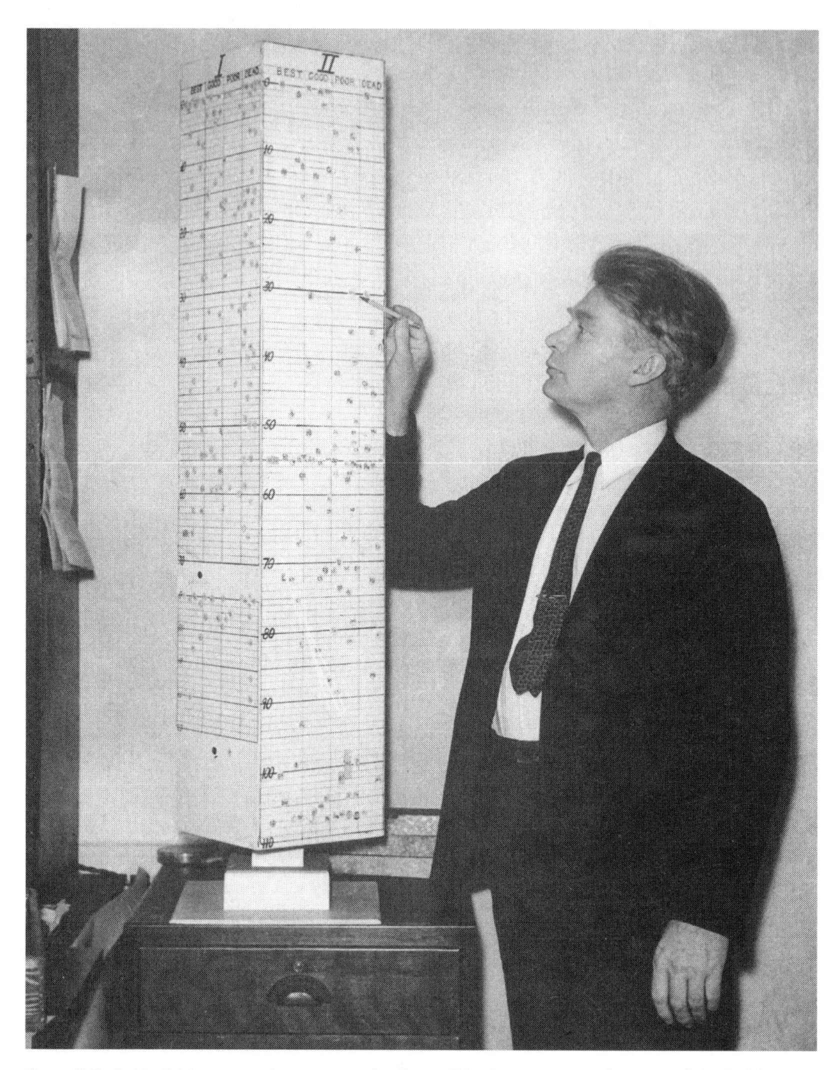

Figure 3.2. Calvin Bridges mapping genes to the *Drosophila* chromosomes. Courtesy of the Archives, California Institute of Technology.

principled resistance, Morgan began to relent. In short order, he became the principal champion of Mendelism and the chromosome theory. By 1916, he had even adopted the term “sex chromosome” in his own writing. His shift had everything to do with the special properties of the X and Y chromosomes.

Morgan’s change of view began with his acceptance, in 1910, of evidence from Nettie Stevens and Edmund Wilson, about the role of the

X and Y in sex determination: that there are "two classes of sperm, differentiated on the basis of the accessory chromosome," that in meiosis a separation of male and female elements does occur, and that Wilson's and Stevens's model held for higher animals as well as plants and insects.[50] This was persuasive evidence, indeed, that chromosomes might carry hereditary elements. The most important factor in Morgan's shift in perspective, though, was his own finding of his famous white-eyed male in *Drosophila*: the first sex-linked genetic mutation.

As historian of science Sharon Kingsland writes, Morgan's 1913 *Heredity and Sex* "cast the entire problem of sex determination in a Mendelian framework."[51] Morgan now championed the chromosome theory against those voicing his own previous doubts. Whereas in 1910 Morgan decried the chromosome theory as "violating the modern spirit of scientific theory," in his 1915 *The Mechanism of Mendelian Heredity* he wrote, "why then, we are often asked, do you drag in the chromosomes? Our answer is that since the chromosomes furnish exactly the kind of mechanism that the Mendelian laws call for; and since there is an ever-increasing body of information that points clearly to the chromosomes as bearers of the Mendelian factors, it would be folly to close one's eyes to so patent a relation."[52] Whereas in 1915 Morgan was still denouncing the misleading nature of the term "sex chromosome," in his 1916 *Sex-Linked Inheritance in Drosophila*, Morgan acceded, writing with Bridges that, "following Wilson's nomenclature, we speak of both X and Y as sex chromosomes."[53]

◇

Critics had attacked the sex chromosome for its simplistic binary conception of sex, for its failure to explain sex in all species, and for lacking a mechanism for its role in determining sex. The notion of a chromosome for sex survived these attacks, in part due to its intimate and indispensable role in building the case for the new chromosomal theory of heredity. The explanatory success of the chromosomal theory of heredity ultimately overwhelmed critics' worries about the sex chromosome. These critics silenced their objections, not because they were satisfactorily addressed (even after he accepted Mendelism, Morgan still did not feel the matter of the sex chromosome was resolved, as we can see from his 1915 discourse), but because the broader explanatory utility of the chromosomal theory simply swept them aside. In the face of the chromosome theory's explanatory success, critiques of the sex chromosome became minor and distracting terminological quibbles. The priority of generating a consensus around the chromosome theory overtook the sex chromosome dispute. The case of Morgan vividly demonstrates this point.

The term and the concept of the "sex chromosome" gathered steam throughout the 1920s, and by the 1930s it was the term of choice for the X and Y chromosomes. Alternative terms, such as "accessory chromosome" and "heterochromosome," drifted into disuse. Though the term "sex chromosome" had triumphed, few maintained that the X itself was *sex determining*. Wilson was an exception. He maintained that the X and Y were, in one way or another, the material hereditary basis for maleness and femaleness. Wrote Wilson in 1925: "That these chromosomes bear the differential factors of sex has been proved to be the case by an irresistible body of more detailed and concrete evidence drawn from both genetics and cytology. The chromosomes in question may, therefore, appropriately be called the *sex-chromosomes*."[54]

Others resolved that the X is the "sex chromosome" only in the sense that it serves as a histological *marker* of sex. For instance, as the Edinburgh geneticist F. A. E. Crew wrote in a text on the genetics of sex in 1927, "since the tissues from male and female differ chromosomally only in this respect, these chromosomes are referred to as the sex-chromosomes." Notably, this formulation altogether avoids the matter of the causal relation between the X and the sex.[55]

Most commonly, the "sex chromosome" construct derived from the phenomenon of *sex linkage*, which lay at the heart of the new genetic research program produced by Morgan and his students. This "sex chromosome" concept entirely sidestepped the question of the X's role in sex determination. *Drosophila* geneticists were not primarily interested in the question of sex determination, but rather in sex linkage as a clue to traits especially amenable to genetic study. As Morgan and Bridges wrote in the 1916 monograph, *Sex-Linked Inheritance in Drosophila*:

> This term (sex-linked) is intended to mean that such characters are carried by the X chromosome. It has been objected that this use of the term implies a knowledge of a factor for sex in the X chromosome to which the other factors in that chromosome are linked . . . [but] we are unable as yet to locate the sex factor or factors in the X chromosome.[56]

In this formulation, the X was a "sex chromosome" not because it was the sex determiner, but because it carried at least one critical factor corresponding to sex, causing traits on the X to be reliably sex linked.

Perhaps unknown to the scientists at the 1960 Denver conference, then, their questions about the reasoning behind the inherited terminology of the "sex chromosome" were not new. Doubts about the precision and clarity of the concept of the "sex chromosome" have been a persistent theme since the discovery of the X and Y. The X and Y debuted in a charged moment in the history of biology when the fundamentals of genetics and heredity fell into place. They became a centerpiece of this effort. At the same time, they were ambivalently and instrumentally joined to sex, opening a speculative space that carves the terrain of this book.

In the first years of the twentieth century, cytologists asserted that chromosomes carry the hereditary material and that they provide the mechanism for the generation and recombination of novel hereditary characters. McClung's, Stevens's, and Wilson's evidence for an association between chromosomal complement and sex, and the visually compelling case that the X and Y made for the individuality and constancy of the chromosomes, offered the principal early evidence for the chromosome theory. In the 1910s, this theory was further confirmed and extended by studies of X-linked traits. Sex linkage became the principal system for early studies of genetic mutations. The link between the X and sex became the crucial fact in these scientific research programs, and in its powerful simplicity the so-called sex chromosome became the prized plume of the new genetics.

Terms such as "heterochromosome" and "accessory chromosome," while more consistent with terminological convention, would not do. They did not point to the critical relationship between a chromosome and an obvious phenotypic trait. The clear way in which the X and Y marked sex difference became important: *rhetorically*, in the recruitment of the biological community to the new theory, and *scientifically*, as the central technology of the "sex-linkage" research program that confirmed the chromosomal basis of Mendelian traits. For these aims, terms that were perhaps more descriptive, capturing the structural and behavioral features of the X and Y, went by the wayside. Wilson's efficient shorthand of the "sex chromosome" won the day.

CHAPTER 4

A New Molecular Science of Sex

Sadie Salome was a sex chromosome,
Thru her first mitosis she was starting to roam,
Leaving her home,
Causing this pome'
Anaphases, metaphases, telophases, all the crazes known,
Sadie went the paces, heeding not her parents' groan,
Hear them moan!
Till at last she met a man upon her ruin bent,
Who gave her picric acid to hasten her descent.
Alcohol, xylol, and all kinds of goo
Filled her with frenzy in each tiny mu,
And she lept into the xylol paraffine,
Ain't it fine, ain't it fine!

Oh, that cytology jag, that microbiology rag.
Sadie soon went slipping down that keen microtome,
Section after section of that sex chromosome.
Oh, you chrom, chromosome on that sliding microtome.
Stained in haematoxylin
Mounted on a slide,
Naughty Sadie cannot hide
Any of her chromatin! What a sin! See them grin!
Oh, that cytology, that microbiology rag!

The "Cytology Rag," relating the travails of Naughty Sadie, an X chromosome who meets "a man upon her ruin bent" in a microbiology laboratory, loosely parodies the Irving Berlin hit "Sadie Salome Go Home," about a girl gone astray in the world of burlesque theater. A photocopy of the anonymously authored tune was slipped, probably in the 1920s, into a privately owned edition of Thomas Hunt Morgan's 1915 *The Mechanisms of Mendelian Heredity.* At some point, the University of California acquired the book, stowing it away on a dusty bookshelf. In 2007 a librarian scanned and uploaded it, loose-leaf ephemera and all, as a publically available digital copy of Morgan's masterpiece—thus preserving the visage of Sadie the sex chromosome slipping and sliding, exposed under a microscope, for our delight and critical reflection.[1]

In the mid-twentieth century, the sex chromosomes indeed began to roam. The X and Y moved into popular culture and gender politics as they traversed new interest in the biology of sex differences. In the 1920s, the X and Y were swept into a new science of sex, heralded by the discovery of the sex hormone. With the arrival of the "sex hormone," the once-contested concept of the "sex chromosome" gained firmer foundations and wider legibility as it was folded into a broad modern account of the molecular science of sex. The new molecular science of sex, with hormones at the center, solidified, legitimized, and elevated the X and Y as biological agents of sex by generating a commanding model of sex determination and differentiation in which the chromosomes occupied a distinct and necessary role. As the concept of the sex chromosome hardened and the science of genetics moved to the center of modern biology and medicine in the mid-twentieth century, scientists and the public increasingly turned to the X and Y to divine the essence of femininity and masculinity.

HORMONES: THE NEW SCIENCE OF SEX

While scientific, philosophical, and medical scholars had, of course, speculated on the origin, nature, and physiology of sex for centuries, the new science of sex that emerged in the early twentieth century was the culmination of a particular vision of a modern, empirical, comprehensive biology of sex.[2] Fundamental scientific investigations of sex had quickened in the late nineteenth century as questions of women's proper role in society, their fertility, and their suitability for education came to the fore.[3] Reproductive sciences, in particular, flourished in the heat of these social battles. The eugenic, birth control, and population control movements shared a faith that the new sciences of experimental biology and genetics would build a better society. They hoped that the biological sciences would soon yield technologies for the control and medical direction of human reproduction. This objective stirred the imagination of social reformers and prompted generous public and private investment in scientific studies of sex determination, sexuality, and reproduction.

By the 1890s, as the biological sciences were becoming increasingly specialized and professionalized, empirical sexual science coalesced as the academic field of sexology. British researchers such as Havelock Ellis, Patrick Geddes, and J. Arthur Thomson and German researchers such as Magnus Hirschfeld developed sexology into a field with regular conferences, academic positions, and journals—with the *Journal of Sexology* debuting in 1908. The field embraced a broad interdisciplinary study of sex, including in its purview biology; genetics; reproductive physiology and gynecology; psychology; the study of venereal diseases and sexual "deviancy"; and social, historical, and anthropological research on sex across cultures and throughout history. The sex chromosomes were discovered very much outside of this milieu, as the chance byproduct of fundamental investigations into cytology and heredity. In contrast, the sex hormones emanated from the center of scientific investigations into sex shaped by the practical concerns of agricultural breeding, reproductive health, and public health policy.

Hormones are chemical agents that transmit signals from one cell to another, acting at a distance to create metabolic changes in cells and tissues. Gonadal transplantation and castration experiments in the late nineteenth and early twentieth centuries first established the role of so-called gonadal secretions in fertility and stereotypical secondary sexual behaviors and traits (see fig. 4.1).[4] In 1905, these secretions became known as the "sex hormones." Rapid developments in the 1910s propelled the sex hormone from the small world of European hormone research, largely in a clinical context, into the American biosciences, in the realm of basic research and academic biochemistry. The University of Chicago embryologist Frank R. Lillie, with his famous freemartin research in 1916 and 1917 (more on this shortly), lifted the sex hormone from its relatively marginal status as a pharmaceutical research target and chemist's curiosity to the centerpiece of biological models of sexual development.[5]

Between 1915 and 1930, the scientific study of sex underwent rapid consolidation, professionalization, and expansion. In the footprint forged by sexology, a robust new field emerged, with the gleam of the chemist's lab and the authoritative language of hormones, genes, and chromosomes.

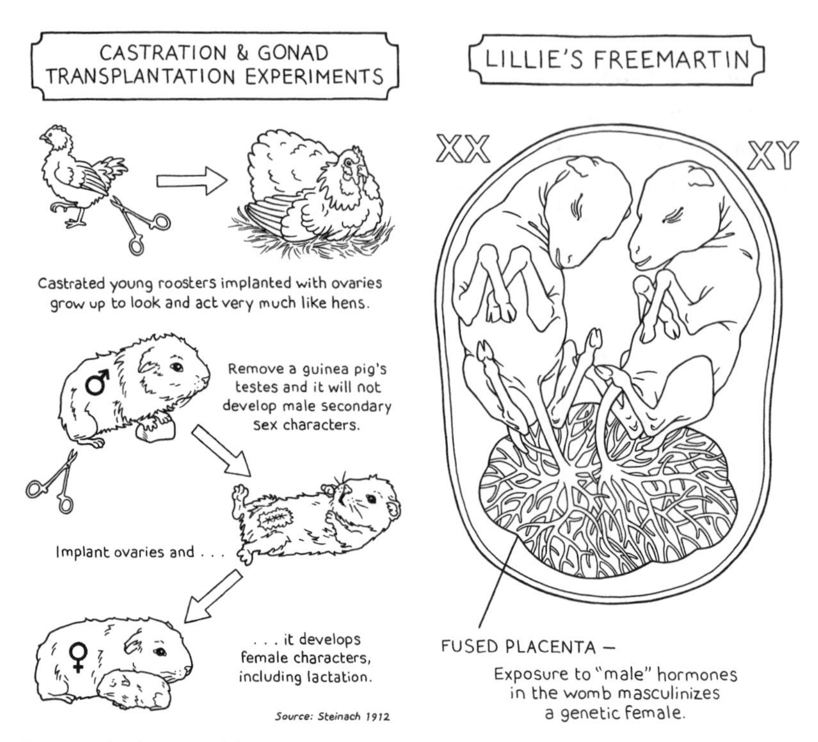

Figure 4.1. The discovery of the hormonal model of sexual development and differentiation. Illustration by Kendal Tull-Esterbrook; copyright Sarah S. Richardson.

The Association for the Study of Internal Secretions was established in New York City in 1917, the same year as the founding of the journal *Endocrinology*. By the 1920s, biochemists had entered sex hormone research *en force*, and research shifted from castration and transplantation methods to extraction, purification, and synthesis of hormones. This intensive work yielded the clinical isolation of estrogen and testosterone by 1923 and 1927, respectively.[6]

By the mid-1920s, hormones had become, like genes today, the most prominent object of biomedical, pharmaceutical, and popular interest to emerge from modern biology. Sex hormones seized the public imagination and became a node through which ideas were exchanged between scientific theory and cultural norms, ideologies, and expectations. Scientists promoted the view that the sex glands were "the 'master glands' of the endocrine system."[7] As Chandak Sengoopta writes, "scientists and the general public came to regard the testicles and the ovaries not merely as sources of gametes but also as secretors of potent chemicals determining not only the narrowly sexual aspects of our existence but also our very beings and lives."[8] Hormones entered popular culture. Writes Julia Rechter: "everyone knew about hormones; they joked about 'monkey' and 'goat' glands; and they read in their daily papers about the miracles of hormonal magic."[9] Pharmaceutical hormone therapies promised new fertility aids and offered the prospect of a simple, highly effective means of birth control. Many also believed hormones would permit the correction of modernity's gender deviants—feminist spinsters, homosexuals, impotent males, and frigid wives.[10] The endocrinology pioneer Eugen Steinach promoted testicular transplantation as a medical cure for homosexuality and a "rejuvenation" therapy for low virility and listlessness in elderly men.[11] During this period, as Anne Fausto-Sterling has influentially argued, a new, molecular notion of sex differences began to cement: "Hormones, represented on paper as neutral chemical formulae, became major players in modern gender politics."[12]

A concrete institutional result of the growth of sex research was the 1921 founding of the National Research Council Committee for Research in Problems of Sex (CRPS), a body that due to its risqué subject would have been inconceivable—and politically unviable—only a decade earlier. Lillie, "one of the great entrepreneurs of early-twentieth-century biological sciences,"[13] played a key role in mobilizing the effort, drawing together birth control advocates, eugenicists, gynecologists, social science researchers, endocrinologists, geneticists, and agricultural scientists to build what Adele Clarke has called a "funding source and legitimation device extraordinaire" for sex research.[14]

Lillie's CRPS provided an institutional platform, newfound status, and resources for research on the genetics of sex. In 1922, in the first annual report of the Committee, Lillie placed genetic sex determination and its control at the top of the CRPS's research priorities. He wrote that the CRPS's work

> should be directed specifically towards the fundamental problems which are these: (1) How is sex determined? Can sex determination be controlled? How? (2) What are the factors active in the development of sexual characteristics whether anatomical, physiological, or psychological? Can sex development be quantitatively controlled? (3) the problem of sex relationships; their nature, their control.[15]

Accordingly, the CRPS actively involved and funded scientists investigating the genetics of sex. F. A. E. Crew, author of the definitive book on the genetics of sex determination in the 1920s, was a consultant to the Committee, and famous geneticists such as Harvard's Edward East were associates of Margaret Sanger's CRPS-funded birth control research institute. As hormones transformed predominant theories of sex, the sex chromosomes rode their coattails, gaining a new concreteness as profoundly important molecular determinants of sex as a result. "Sex chromosomes" came to constitute the genetic contribution to the simple, coherent, and remarkably explanatory new molecular biology of sex.

SEX DETERMINATION AND SEX DIFFERENTIATION

The alliance between sex hormones and sex chromosomes was not merely institutional: it was built into the theory of the sex hormone. Lillie erected modern hormone theory on a foundational distinction between sex determination, which would be genetic, and sex differentiation, governed by the hormones. This distinction, laid out in his 1916–17 articles on the freemartin, defined sex determination as the narrow question of sex determination at fertilization and delegated this function to the chromosomes. In doing so, Lillie alleviated concerns about the imprecision, overbroadness, and apparent empirical contradictions and weaknesses of the notion of a "sex chromosome." Lillie's hormone theory carved out a limited but certain function for the chromosomes and supplied hormones as the answer to apparent explanatory gaps in the chromosomal theory of sex.

The freemartin, a curiosity known to sheep and cattle breeders, is a zygotically female but morphologically sex-ambiguous member of a male-female twin pair with fused placentas (see fig. 4.1). Previous theories held that the freemartin is "really a male." Lillie demonstrated that the freemartin is instead a genetic female masculinized by circulating male hormones. As the female members of dizygotic male-female twins fused by vascular tissue, freemartins share blood-borne hormones, including sex hormones.[16]

The significance of this finding among atypical male-female twin births in farm animals may seem opaque today. But, as Lillie argued, "nature has . . . performed here a perfectly controlled experiment."[17] Despite the research amassed by the early twentieth century on hermaphrodites, sex-reversed castrated and gonad-transplanted animals, parthenogenesis, and other oddities of sexual development, the freemartin demonstrated two critical facts not previously appreciated. First, it showed that hormonal factors play an essential role in the *fetal* stage of sexual development. Fetal masculinization of a female in the case of the freemartin established, with conclusive embryological evidence, that chromosomes and genes play at best only a partial role in sexual development and differentiation, even in their very early stages. "The sex-characters that develop before birth are, like those arising after, dependent for the degree of their development upon sex-hormones," argued Lillie.[18] Second, and most important for our inquiry, because the freemartin is genetically female and developmentally male, it demonstrated a clear demarcation between primary chromosomal sex determination and secondary hormonal sexual differentiation. Wrote Lillie, "the sterile free-martin enables us to distinguish between the effects of the zygotic [genetic] sex-determining factor in mammals, and the hormonic sex-differentiating factors."[19]

Prior to Lillie's work, hormone scientists had not yet generalized their research findings on the gonadal hormones into a comprehensive theory of sexual differentiation. Lillie was the first to do this. He built a simple two-tiered model of sexual development, with genes as the initiators and sex hormones as the dominant agent of sexual differentiation. As Rechter writes, "to American scientists, Lillie offered an important theoretical and experimental framework with which to organize known sex phenomena. He drew together the major strains of research on sex—genetic, embryological and hormonal—in his experimental research on the development of cow-twins. In short, Lillie provided American sex researchers with a clear scientific paradigm."[20]

The new hormonal, biochemical model of sex displaced the previous metabolic model (see chapter 2) and helped to concretize the sex chromosome concept in two ways. First, hormone scientists relinquished initial "primary" sex determination wholly to chromosomal control, distinguishing it from the "secondary" development of sexual characters, which was governed by hormones. This step removed some major objections to the "sex chromosome"—notably, its inability to explain the plasticity and variability of sexual dimorphism and of sexual systems in nature. Second, building the sex hormone model on the foundation of the sex chromosomes also conferred on the chromosomal theory of sex a broad scientific legitimacy it had not previously enjoyed. Until this point, the sex chromosomes were a set piece for the chromosomal theory of heredity, and not at all a focus of the science of sex—a very different arena of scientific endeavor. When the sex hormone model leaned on a parallel chromosomal mechanism to complete its account, it magnified the relationship of the X and Y to sex. In drawing the chromosomes so prominently into its model of sex, hormone science helped to constitute and solidify the "sex chromosome."

SEX HORMONES AND SEX CHROMOSOMES

Lillie's theory inspired a new industry of investigations of the biology of sex. The generative space that the sex hormones helped to carve out for the sex chromosomes can be seen in various ways in the central writings of major theorists of the biology of sex during this period. In their theories, genes and hormones collaborated to determine a zygote's sexual fate. Here, I cite examples from Lillie, Morgan, and another principal expert on sex in the 1910s and 1920s, the German geneticist Richard Goldschmidt. In their writings, we can see how the sex chromosomes (and the genetic factors for sex that they presumably carried) were elevated as coequal molecular agents of sex as they were brought into the hormonal theory. Scientists joined endocrinological and genetic models, postulating complex interactions between hormonal and chromosomal "sex factors" in their endeavor to model the biology of sexual development. Hormones, rather than competing with genes for a theory of sex, enrolled and elevated them into a broader explanatory framework.

From the start, Lillie assumed a multifactorial model of sex in which genes and hormones acted in dynamic relation to one another. As Lillie wrote in his classic 1916 work on the freemartin, "the intentions of the genes must always be carried through by appropriate hormones developed in the gonad."[21] He advanced a quantitative multifactorial picture of how genes and hormones—"sex-factors" both—act in concert to advance development in the direction of either maleness or femaleness. In his 1917 "Sex-Determination and Sex-Differentiation in Mammals," he argued that initial zygotic sex is "a condition in which there is a quantitative superiority of one or the other tendency or set of factors. The advance of development progressively limits the possible operations of the inferior set of factors, so that, by both positive and negative limitations, reversal of the initial sex-index becomes increasingly more difficult."[22] Lillie predicted that in some organisms (such as insects) sex is determined wholly by genetics, in others (cattle and sheep) by sex hormones, and in others (humans) by a combination of the two. In Lillie's model, sex is not largely "determined" at the zygotic stage; rather, genes and hormones act in inhibitory or excitatory loops as development proceeds to canalize sexual development toward maleness or femaleness.

The geneticist Morgan was keenly aware of work by Lillie and others on the gonadal secretions, and throughout the 1910s and 1920s he mined the hormonal theory of sex to explain how genetic factors might play a role in sex determination. In his 1914 article "Sex-Limited and Sex-Linked Inheritance," Morgan referred to the "substance" produced by ovaries and testes as a key to how an extra X chromosome in females determines sex. "It only remains to point out some of the different ways in which a factor being present in duplex both in the male and in the female produces its effect only in the male. It some cases it has been shown that the ovary produces some substance that is inimical to the production of certain characters," he wrote.[23] Morgan also enthusiastically described experiments on the gonadal hormones in the course of his exposition of the chromosomal model of sex in his 1915 *The Mechanism of Mendelian Heredity* and in his 1919 *The Genetic and the Operative Evidence Relating to Secondary Sexual Characters*.[24] Morgan noted castration experiments in birds and mammals, including humans, which showed the impressive results of gonadal secretion deficiencies on the development of secondary sexual characters. He theorized that genetic and hormonal factors act both independently and together to produce the range of secondary sexual characters typical of a species: "some characters depend on the gonad. . . . Still other characters . . . are the direct product of the male or female genetic constitution. . . . [E]nough has been found out to indicate that we must be . . . prepared to admit that no one theory may be able to account for all of the secondary sexual differences that exist between the sexes."[25] Morgan concluded that both genetic and hormonal factors should be considered in explaining secondary sexual differences.

German geneticist and embryologist Richard Goldschmidt's ideas about genetic sex determination perfectly crystallize this intricate interweaving of hormonal and genetic factors into a holistic molecular account of sex determination and development during the 1910s and 1920s.[26] Goldschmidt was a renowned geneticist and the author of an internationally influential genetics textbook printed between 1911 and 1938. He was also a leading expert on the genetics of sexual differentiation. Goldschmidt was interested in developing a theory of gene action during development, an inquiry he termed "physiological genetics." He believed that genes were activated in response to physiologic constraints and threshold signals over the time course of organismic development. In the apparent interaction of hormones and genes in sexual development, Goldschmidt saw an ideal model for his physiological genetics. Examining cases of intersexes, he attempted to show how different degrees of maleness or femaleness could be achieved by manipulating the timing of exposure to sex factors. Lillie's case of the freemartin served as Goldschmidt's central example of this phenomenon. In Goldschmidt's account of sex, the "sex-differentiating substances"—sex hormones and genes—were elements of a complex regulatory signaling process that worked in a time-dependent fashion over the course of development (see fig. 4.2).[27]

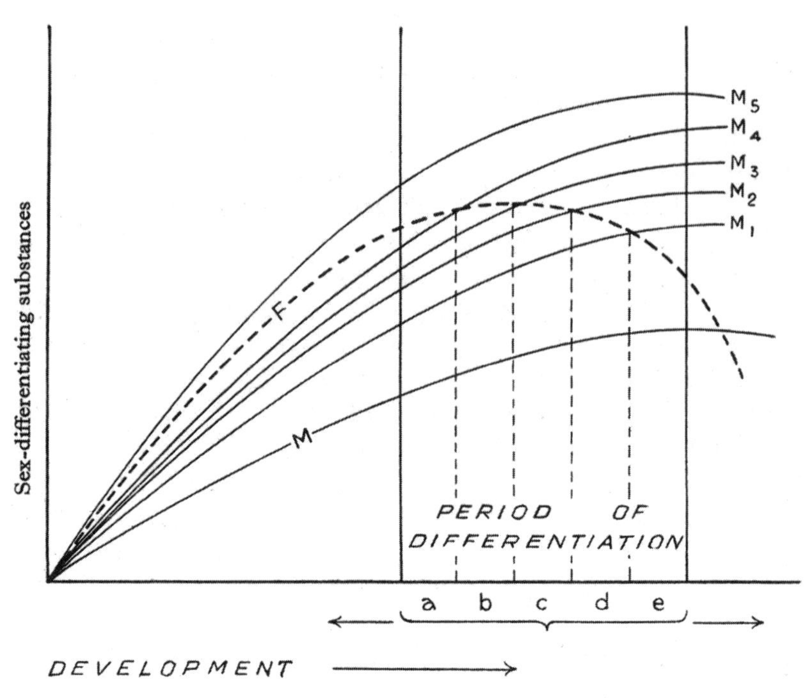

Figure 4.2. "The Sex-Differentiating Substances and the Time Law of Intersexuality, After Goldschmidt" Reprinted from F. A. E. Crew, *The Genetics of Sexuality in Animals* (Cambridge: Cambridge University Press, 1927).

These brief vignettes show how rise of the sex hormone concept during the years 1915 to 1930 drew genes and chromosomes into a powerful mutually reinforcing theoretical framework for the biology of sex. By the 1920s, the new science of sex, structured around the sex hormone, was firmly established, institutionally and intellectually. An expectation of a dynamically interacting two-tiered structure of explanation for sex at the genetic and hormonal levels fueled new biological models of sex and irresistibly pulled theorists toward the consistent and parallel terminology of "sex hormones" and "sex chromosomes."

In sum, a direct intellectual and institutional context for the development of the sex chromosome concept during the period from 1915 through 1930 was the concurrent and spectacular development of sex hormone research. Once the idea of "sex hormones" was established in the scientific world and the popular consciousness, it was easy to accept the terminology and concept of the "sex chromosome." With its rich resources, medical applications, and powerful institutional platform, the new molecular biology of sex drew the X and Y chromosomes into its center of gravity and folded them into the thick scientific account of sex it purported to offer. Through this process, the sex chromosomes began to take on firmer form in both scientific and popular understanding. Despite the apparent insistence of scientists such as Lillie, Morgan, and Goldschmidt on the complexity and fluidity of hormonal and genetic sex determination, today's notion of the X and Y as the molecular agents of sex and as the genetic homunculi underlying sexual dimorphism—as "sex itself"—is first recognizable in this moment.

GENDERING THE X AND Y

As the X and Y became the "sex chromosomes," they increasingly began to be perceived as the fixed essences of sex, bearing fundamental truths about male and female human nature. What does the Y chromosome do for males? Does the second X endow human females with something "extra," or is it more advantageous to have a single X chromosome? These questions, charged with gender politics, stalked the X and Y from their earliest appearance in the public and scientific consciousness.

Human genetics in the first half of the twentieth century existed in what historian Susan Lindee has called a "medical backwater."[28] Research on human inheritance took place largely under the umbrella of eugenics, a social and scientific movement for the improvement of the human gene pool by rational direction of human reproduction. In the United States, early twentieth-century eugenic fieldworkers fanned out across the land to collect family pedigrees. Eugenicists claimed to show the Mendelian inheritance of everything from eye color to boat-building skills to moral depravity. The human chromosomes, and indeed the cellular and biochemical basis of human inheritance, went largely unexamined.[29] For their part, cytogeneticists favored insects and plants. With their rapid generations, small sets of chromosomes, and simple binary traits, they were perceived as far more tractable for laboratory research than humans.

Nonetheless, scientists knew that humans had sex chromosomes, and they did not refrain from speculating about what the X and Y meant for human sex differences.[30] The sex chromosomes found their way into broader discourses around gender, discourses often framed by the expectation that the facts of biology would help to settle the matter of the hierarchy of the sexes once and for all. Most of the focus was on the implications of an extra X chromosome in females and a single X chromosome in males, but scientists also engaged spirited debate about the connection between the Y and maleness.

"Sex is quantitatively determined by the X chromosome. . . . [T]wo X's determine a female and one a male," wrote Morgan and Bridges in 1916, concisely summarizing the view of chromosomal sex determination that would stand until the 1950s. In the first half of the twentieth century, cytologists believed that the X determined sex and that the Y was a chromosomal fragment of little genetic importance. Wilson described the Y chromosome as merely a "supernumerary" "without significance in sex production." Guyer asserted that "the existence of the [Y-element] seems to be a wholly capricious one, at least as far as sex production is concerned." And Stevens wrote that "the indications are that the chromosome Y is of no hereditary value."[31]

Work on sex chromosomes in *Drosophila* by Morgan's student, Calvin Bridges, sealed the Y's fate in the 1920s. Bridges reported the existence of occasional XO males and XXY females.[32] These anomalies showed that in *Drosophila* sex depended not on the Y but on the ratio of Xs to sets of autosomes. This finding was taken to hold for humans as well and as definitive confirmation that the Y has no role in sex determination. As Crew concluded in his authoritative 1927 monograph on the genetics of sex determination, "the Y chromosome, a partner to the X-chromosome during gametogenesis, is not concerned in the determination of sex."[33]

Even as all agreed that the Y was not sex determining, however, there was persistent speculation in the first half of the twentieth century that the Y must be important, in some way, to maleness. William Castle was the first to propose the Y as the location of male-specific characters. In 1909, Castle queried, "if . . . the differential sex-character has its cytological basis in the 'X-element,' as Wilson designates it, it becomes an interesting question, what is the cytological basis of those numerous morphological characters possessed by the male, but wanting in the female?" While males were "a defect race, or regressive variation" in lacking a second X, Castle wrote that they were "progressive" in being able to acquire "characters purely male" on the Y chromosome. Because of the Y chromosome, he speculated, "the male may possess certain qualities not merely not manifested by the female, but even not possessed by it."

Specifically, Castle suggested that the Y provides a mechanism for the long-held view that evolutionary innovation happens in the male. "I would offer the suggestion that we have a mechanism suitable for the transmission of characters exclusively male in the Y-element," he wrote. The Y "would constitute a likely place in the germ-cell for new structures to find lodgment . . . and so would produce structures peculiar to the male, and unrepresented in the female." (Since the Y pairs with the

X, however, Castle felt compelled to concede that "striking" or advantageous traits, developed in males, may eventually be "shared with the female.")[34]

Wilson responded that Castle's view was a "tempting suggestion" but pointed out that there was as yet no evidence for it. He cited the fact that some species lack a Y in males but show the full range of male secondary sexual characters. Moreover, there were no documented cases of Y-linked inheritance in males. The unresolved relation of the Y to sex, Wilson worried, remained a gap in the chromosomal account of sex: "The Y-element still remains a puzzle; and until it has been satisfactorily accounted for our cytological view of the problem [of sex production] will remain defective."[35]

Morgan, for his part, found offensive the notion that the male is the result of absence of the female-determining X. This idea, he wrote, "ignores the Y element, and thereby makes the male condition the result of the absence of something which, if present, turns the embryo into a female. It seems to me that there is no warrant for considering the male in this sense a lacking female. The physiology and the biology of the males offer much to contradict such a view of his composition."[36] In 1926 Morgan again took up the cause of the role of the Y in maleness. While the Y has "less significance in sex determination than the X-chromosome," it should not "be ignored or treated as though it were 'empty,'" wrote Morgan. As Morgan noted, "a male *Drosophila* without a Y-chromosome is sterile."[37]

Despite the efforts of Castle and Morgan, speculation about the sex chromosomes and the biological origins of stereotypical male and female traits during the first half of the twentieth century largely focused on the X chromosome. By the 1920s, researchers had formulated genetic maleness as a result of the suppression or absence of the feminizing "extra" X chromosome. Genetic maleness was the outcome the absence of the second X—or, as in *Drosophila*, of the autosomes successfully overcoming the powers of the female-tending X to produce a fully functional male.

Columbia endocrinologist and notorious antifeminist Louis Berman, writing in 1921, interpreted the meaning of this picture of chromosomal sex as follows:

> It has been found that the twenty-two [*sic*] chromosome individual invariably develops into a female, and the twenty-one into a male. Therefore, femaleness is a positive quality, dependent upon the action of the X-chromosome, *and maleness an absence of femaleness*, due to lack of the extra, odd chromosome.[38]

In a refreshing reversal of common tropes about femininity, Berman here appears to see femaleness as a positive quality and, by implication, maleness as a negative one, an "absence." But in Berman's definition of maleness as the "absence of femaleness" and femaleness as marked by an "extra, odd chromosome," a door is opened to quite a different view, depending on the eye of the beholder. A 1922 *Chicago Daily Tribune* article, headlined "Stork to Take Orders for Boy or Girl Soon," captures this path of reasoning. In the article, prominent public scientist Julian Huxley opines, "among insects and some other animals the female possesses one more chromosome than the male. If, as is supposed, this chromosome is the carrier of sex, then it might be possible to eliminate the extra chromosome."[39] Apparently, the presumed use of the genetic technology would be to remove the "extra" X, that signal of femaleness, to turn the balance from female to male.

A related line of conjecture took up the specific advantages males gained by *not* having an "extra X." Researchers suggested that the presence of a single, unpaired X in males gave men an evolutionary edge. While the single X may subject males to damaging genes on the X, as in the case of color blindness and hemophilia, it also exposes them more wholly to advantageous genes. The risks that males take with their sole X are countered by rich potential rewards. While females enjoy the security of a second X, it dulls their potential for extraordinariness. Perhaps the X chromosome was the source of male genius. Scientists fastened on the X as a genetic basis for the "greater male variability" theory of male intellectual superiority.

Charles Darwin was among the most prominent early adherents of the concept of greater male variability, which framed research on cognitive differences between males and females from the 1870s to the 1930s. In *The Descent of Man*, he argued that males are the engine of evolution, accumulating variations that lead to species divergence and novel adaptive traits. He wrote that males are more "highly variable" than females and that "it is the male which has been chiefly modified, since the several races diverged from their common stock." Darwin related the greater variability of the male to "the distinctive characters of the male sex," among them that "man is more courageous, pugnacious, and energetic than woman, and has a more inventive genius."[40] In the late nineteenth century, sexologist Havelock Ellis and psychologist G. Stanley Hall influentially argued that males are more intelligent, on average, than females and that this is because men are more cognitively variable.[41] As evidence, they cited the long-observed predominance of males among residents of what were then known as institutions for the "feeble-minded" and, conversely, among the ranks of genius and the socially eminent.

The earliest geneticist to attach the X to male variability and female conservatism was McClung, who first conjectured the link between the X and sex in 1899.[42] In 1918, McClung wrote of the X chromosome that "it is possible that we have here the explanation of the greater variability of the male." He continued:

> There is a possibility that in the male, the sex [X] chromosome being unmated, or opposed by an inactive element, may be more free to react with the other chromosomes and in this way change their constitution, being in turn affected by the reaction. By the nature of its transmission it must, after this experience, pass into the female line where its relation to the complex is necessarily different. The contrast in these two conditions is obvious and the interpretation strongly suggested.[43]

In the mid-twentieth-century, observations of an excess of males among the intellectually disabled and documentation of a large number of mentally impairing X-linked conditions exclusive to males led scientists to take up McClung's suggestion. They began to speculate that the single X was a mechanism for the observed "greater variability" in male intellect, bestowing male intellectual superiority alongside higher rates of mental disability—and that the double X, in parallel to the dull brown of the female mallard duck in contrast to the male's brilliant feathers, was a source, in humans, of female dullness, tempering their intellectual capacity.[44] While the greater male variability theory remains influential in some circles today, it has been largely discredited. Studies show no significant differences between males and females in overall intelligence and demonstrate that, while men are more likely to be at the very low end of the IQ scale, they are not equally as likely to be at the high end.[45]

Not to be ignored, but certainly more marginal, was the contrary idea that double-X females are superior, advantaged, or special as a result of their "extra Xs." This was appropriated by women's advocates: "The ancient idea that the female is essentially an undeveloped male seems to be finally disproved by the fact that it requires more determiners—usually one more chromosome, or a larger sex chromosome—to produce a female than a male," pronounced the feminist psychologist Helen Thompson Woolley in 1914.[46] Even Berman conceded that biologists could no longer seek the source of female inferiority in the chromosomes: "For the time being, let the feminists glory in the fact that they have two more chromosomes [*sic*] to each cell than their opponents. Certainly there can be no talk here of a natural inferiority of women."[47]

In *The Natural Superiority of Women* (1953), the anthropologist and public intellectual Ashley Montagu began his case for women's superiority by citing female X chromosome advantage. Montagu asserted that females' extra X "lies at the base of practically all the differences between the sexes and the biological superiority of the female to the male."[48] In a chapter titled "'X' Doesn't Equal 'Y,'" Montagu argued that it is "to the presence of two well appointed, well furnished X-chromosomes that the female owes her biological superiority." Males, with their "X-chromosomal deficiency," fall prey to such diseases as hemophilia and colorblindness, and countless other speculated weaknesses, he wrote, while females, owing to an extra X, are "constitutionally stronger than the male."[49]

THE ADVENT OF HUMAN CHROMOSOME SCIENCE

After World War II, human genetics research rapidly expanded in the wake of massive postwar American investments in education, research, and medicine. The American Society of Human Genetics was founded in 1948, and its journal, the *American Journal of Human Genetics*, in 1949.[50] As Lindee argues in *Moments of Truth in Genetic Medicine*, the field's founders sought to distinguish themselves from the old eugenics. They developed a sweeping new vision for human genetics, one in which all disease was conceptualized as "a genetic phenomenon subject to technological control."[51] Genetics, they contended, would offer fundamental insights into human biology and the diagnosis and treatment of all human disease—not just rare hereditary and congenital disorders, but also cancer, infectious diseases, and basic processes such as aging.

In the 1950s and 1960s, fundamental breakthroughs in the structure and biochemistry of DNA arrived in rapid succession, facilitating the emergence of genetics as the backbone of modern biomedicine. James Watson and Francis Crick predicted the structure of DNA in 1953. By 1966, Marshall Nirenberg and Heinrich Matthaei had unraveled the genetic code, showing that triads of DNA bases precisely corresponded to amino acids, the building blocks of proteins. It was not molecular analysis of DNA, however, but the study of the structure, behavior, and function of chromosomes that led to the first insights into human heredity during this period. As in the early twentieth century, the sex chromosomes were at the fore.

Cytogeneticists received a jolt of postwar funding from the Atomic Energy Commission. They were charged with the task of assessing the long-term health and biological consequences of nuclear fallout.[52] They undertook large-scale studies of chromosomal variations in human populations, and a series of profound and triumphant discoveries followed. As Lindee writes, "human cytogenetics was a sleepy subspecialty of no interest to physicians in the late 1950s. By 1964 it was glamorous enough that practicing clinicians wanted to learn to work with chromosomes."[53] Among the cytogeneticists' victories was the startling 1956 revelation that humans possess forty-six chromosomes, rather than forty-eight, as had been widely believed.[54] There was also the surprise that an extra chromosome 21 causes Down's syndrome, a finding that spurred hope that geneticists were on the brink of determining the chromosomal underpinnings of many devastating congenital disorders. These dramatic discoveries were widely publicized, along with human karyotypes illustrating the chromosomal counts.

The sex chromosomes became one of the most flamboyant symbols of the new, seemingly determinate relationship between chromosome karyotype and human phenotype. At the height of human cytogenetic investigation in the 1950s and 1960s, sex chromosomes dominated the stream of exciting findings. The first significant postwar breakthrough for human sex chromosome research was the identification of a condensed body present only in female cells. Discovered in 1949, the "Barr body," an artifact of the presence of two X chromosomes, suddenly allowed nuclear sexing of any human cell.[55] Murray Barr described the revelation that the "nuclei bear a clear imprint of sex" as the "principle of nuclear sexual dimorphism."[56] The notion that every cell has a sex shifted the terms of human sex research and ushered sex difference into the molecular genetic age. Screening for the presence of a Barr body allowed sex chromosome aneuploidies (numerical errors), such as Turner's syndrome (XO) and Klinefelter's syndrome (XXY), as well as a host of rare finds, such as XXXs, XXXYs, and XXYYs, to be detected well before more detailed chromosome analysis and visualization techniques became available. These findings led to another spectacular development, the 1959 discovery that the Y, not the X, carries the crucial factor in male sex determination, reversing the long-standing consensus that the X is sex determining and the Y is genetically inert.[57]

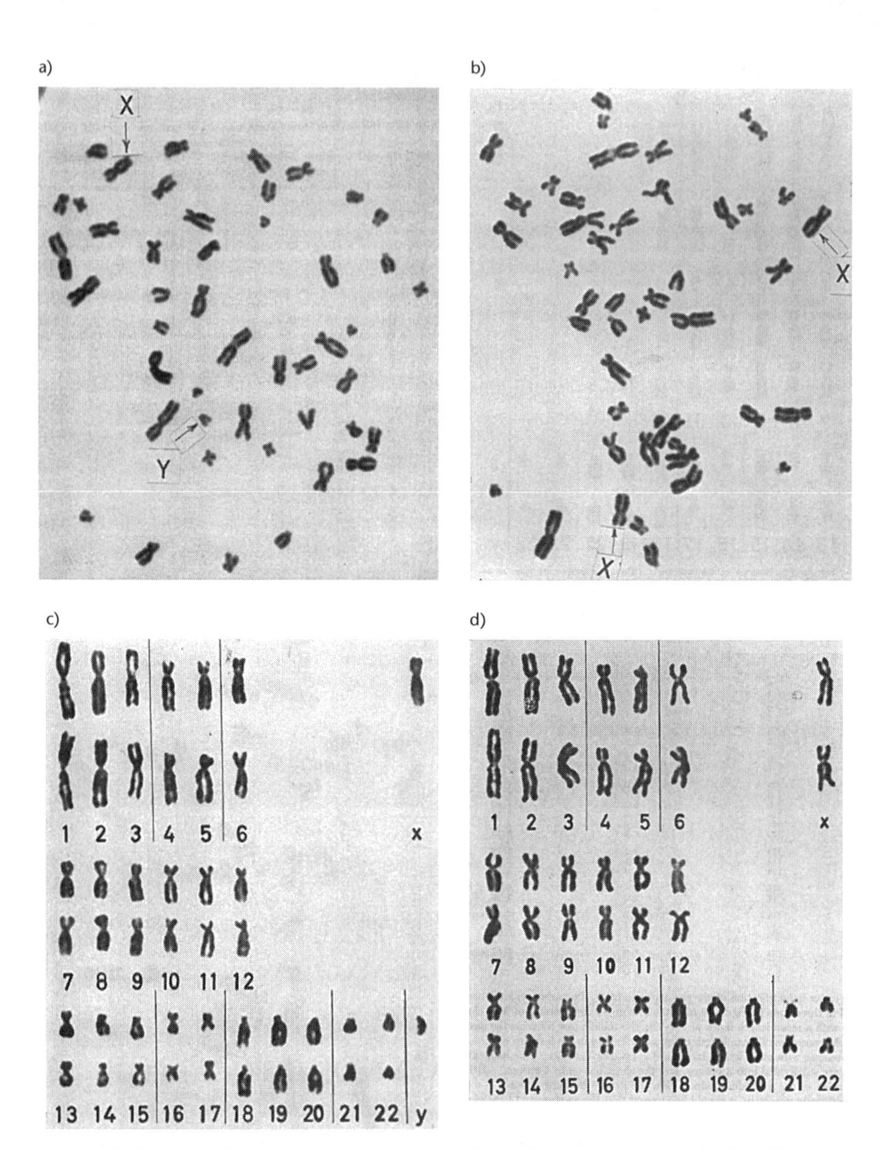

Figure 4.3. The human karyotype, icon of modern genetics in the 1960s. Reprinted from J. H. Tjio and T. T. Puck, "The Somatic Chromosomes of Man," *Proceedings of the National Academy of Sciences of the United States of America* 44, no. 12 (1958): 1229–37.

◇

By the 1960s, human sex chromosome errors and other chromosomal anomalies had become potent symbols of the sensational new genetics. The historian of midcentury genetics Soraya de Chadarevian argues that the representational schema of the human karyotype was the public icon of modern genetics in the 1950s and 1960s, before the double helix took its place. This human karyotype, with its forty-four autosomes and two outlying sex chromosomes was often presented in duplicate, a "male" and "female" karyotype printed side by side (see fig. 4.3). Through this striking visual binary, the freak-show novelty of sex chromosome aneuploidies, and the glamour of the new genetic age, the notion of the X and Y as the molecular pillars of biological femaleness and maleness was cemented in the public and scientific consciousness.[58]

Interestingly, the 1960 Denver commission on human chromosome nomenclature briefly considered dropping the separate "sex chromosomes" terminology and numbering all of the chromosomes strictly by their size and structural features. In the alternative schema, the X would be clustered with chromosomes 6 through 12 and the Y with chromosomes 21 and 22. In the end, however, the commission decreed that the old distinction would continue. The autosomes would be ordered together by size. The X and Y, to remain the "sex chromosomes," would be depicted separately, in the stark whitespace of the right margin. Today the Denver "Standard System on Nomenclature of Human Mitotic Chromosomes" remains the template for visual and terminological representation of the mammalian karyotype.

As genetics grew into the authoritative science of the post–World War II period, the sex chromosomes increasingly became flashpoints for speculation about the nature of male and female, as we shall see. Early theoretical debates over the binary picture of sex seemingly implied by the "sex chromosome" concept were pushed aside, and the X and Y hardened into the "female" and "male" chromosomes. This leaves us to wonder, as scientists such as Morgan once did, how might the story have differed had the X and Y remained "heterochromosomes," "accessory chromosomes," or even the "odd chromosomes" instead?

CHAPTER 5

A Chromosome for Maleness

In his popular 2003 book *Adam's Curse,* Oxford University geneticist Bryan Sykes describes his first glimpse of "the DNA that had made me a man":

> This is my Y-chromosome, the bearer of my maleness and the token passed unaltered down from a long line of fathers. This is the chromosome I have come to see. I see it in my own father, as he leads his RAF squadron in the Second World War. I see it in my grandfather, fighting in the trenches and wounded in the battle of the Somme a generation earlier. . . . It is only my Y chromosome that now speaks with a single voice, one that has come to me from generations of men.[1]

In the book, Sykes animates the Y chromosome as the transcendent essence of maleness, shared by all men in genetic fraternity, and presses the Y into a nostalgic patrilineal narrative of family, blood, and ancestry.[2] Retelling history through the eye of the Y, Sykes depicts Genghis Khan's infamous violent temper and legendary promiscuity as driven by "the ambition of his Y chromosome."[3] According to Sykes, throughout human evolution, the Y compelled male sexual appetite, aggression, rape, and pursuit of wealth and power: "Forced by the relentless ambition of the Y chromosome to reproduce itself, women were reduced to a state of serial pregnancy, increasingly enslaved by dependence on men. . . . Men, driven on by the lash of their Y chromosomes, could copy their cattle and become the stud bulls of their own herd."[4]

Conceptions of the Y as male, and as the molecular agent of masculinity, are ubiquitous in popular and scientific writing on the sex chromosomes today. In his 2002 *Y: The Descent of Men*, geneticist Steve Jones dubs the Y the "vessel of manhood" and asserts that "the chromosome unique to men is a microscopic metaphor of those who bear it."[5] He continues: "Important as experience can be, the contents of a boy's skull are influenced by messages, both direct and indirect, from his insidious Y chromosome. From his earliest days his brain, like his body, is to some degree under its control."[6] Similarly, Craig Venter, the biotech entrepreneur who succeeded in sequencing the human genome in 2001, writes in his 2007 autobiography *A Life Decoded*, "I never fully got over what I viewed as my father's betrayal, which I considered even worse than having a gun pointed at me. I can blame it all on the Y." We also learn of his first sexual conquest: "My Y came into its own when Kim had a Sweet Sixteen party while her parents were away."[7]

Here I explore the origins of these associations between the Y, maleness, and masculinity and investigate their influence on human sex chromosome research in the twentieth century. One might imagine that the Y would be an uncontroversial case of a chromosome with a "sex." Since 1959, we have known that without a Y chromosome—or at least without the gene it carries that is needed for testes determination—the developing fetus will not become a male. Moreover, only males have a Y.[8] But outsized conceptions of the Y as the seat of biological maleness and the essence of masculinity are not fixed by these facts. These conceptions have a history and a context. They were developed and hardened within the particular dynamics of mid-twentieth-century cytogenetics, behavioral genetics, and sociobiology.

XYY "supermale" theories in the 1960s and 1970s, which asserted that males with an extra Y were bigger, taller, and more aggressive—more male—represent the first episode in the construction of the notion of the Y as the chromosome for maleness. Today scientists regard XYY as an embarrassing episode in the history of genetics. The studies were characterized by notorious empirical and methodological failings. In what follows, I show how a working model of the Y as the specialized agent of maleness, a model that draws on ideological conceptions of gender in the 1960s and 1970s, played a central and significant role in the development of the flawed hypothesis of an XYY supermale syndrome. The model of the Y chromosome that was advanced by XYY research—that the Y is the genetic substrate of "maleness"—would carry into the molecular era, and continues to influence both popular and scientific conceptions of the Y chromosome today.

A DOUBLE DOSE OF MALENESS

Research on the human Y chromosome, the genetic marker of male sex and the carrier of the testes-determining gene, began in the 1950s. First observed in insects in 1905, the Y's function in humans was disputed until 1959, when the first trait was linked to the human Y: the male sex-determining factor.[9] In 1959 and 1960, a series of clinical case reports showing XXY, XXXY, and even XXXXY genotypes leading to normal (if infertile) phenotypic males firmly established the Y—not the X, as previously believed—as the marker of male sex (see fig. 5.1).[10]

Phenotypic Sex and Sex Chromosome Constitution in Man

Sex chromosomes in: Male phenotype	Female phenotype
XY	XO
XYY	XX
XXY	XXX
XXYY	XXXX
XXXY	XXXXX
XXXYY	XO/XX
XXXXY	XO/XXX
XX/XXY	XO/XX/XXX
XY/XXY	XX/XXX
XY/XXXY	XO/XYY
XO/XY/XXY	XXX/XXXX
—/XXY/XX	XX/XXX/XXXX
XXXY/XXXXY	
XXXX/XXXXY	
XXXY/XXXXY/XXXXXY	

Figure 5.1. Sex chromosome aneuploidies demonstrating that the Y chromosome carries the male sex-determining factor. Reprinted from K. R. Dronamraju, "The Function of the Y-Chromosome in Man, Animals, and Plants," *Advances in Genetics* 13 (1965): 227–310, with permission from Elsevier.

Despite its link to male sex determination, the Y, an unusual chromosome that is one-twentieth the size of the X, was of little interest in the early 1960s. It was considered too skittish and small for cytogenetic analysis. Most geneticists assumed that it was genetically inert, carrying little other than the male testes-determining gene, until blood samples collected during routine physicals of 197 patients at a high-security psychiatric hospital in 1965 revealed an unexpectedly large number of males—3.5 percent—with an extra Y chromosome.[11]

Patricia Jacobs, the Edinburgh cytogeneticist famous for identifying the Klinefelter's (XXY) genotype in 1959 at the precocious age of twenty-three, conducted the study. In the 1965 *Nature* paper reporting the study's findings, Jacobs hypothesized that "an extra Y chromosome predisposes its carriers to unusually aggressive behavior" and that "we might expect an increased frequency of XYY males among those of a violent nature."[12] Jacobs's attribution of the larger representation of XYYs in security facilities to a double dose of Y-linked aggression framed the next fifteen years of research on the human Y chromosome.

In a 2006 interview, Jacobs recounted the path of reasoning that led her to hypothesize a link between the Y and aggression. In 1960, reading a Swedish study that showed an overrepresentation of extra-X, or XXY, individuals in institutions, Jacobs noticed an apparent overrepresentation of the very rare genotype XXYY. As Jacobs recalls, "a very significant proportion were XXYY . . . and I said to [my colleague]: 'That's far too many XXYYs . . .' and he said 'No, it's just chance.' I thought, I don't think it is."[13]

The arrival of the XYY male initiated the first serious cytogenetic studies of the Y chromosome. Research on the so-called supermale flourished in the United States, Britain, and Denmark during the 1960s and 1970s. The so-called XYY syndrome was a mainstream target of investigation in the most prestigious journals of biology, genetics, and cytogenetics. In 1970, the US National Institutes of Mental Health (NIMH) published an eighty-page "Report on the XYY Chromosome Abnormality," a move that conveyed a high-level sense of urgency about XYY males and fortified the impression that XYY was a veritable medical disorder and a serious factor in criminal sociopathy.[14] The link between XYY and aggression was hotly pursued. According to the Pubmed research database, by 1970, nearly two hundred papers on the link between XYY and aggression had appeared in the scientific literature. Between 1960 and 1970, XYY research comprised 82 percent of all published scientific studies on the human Y chromosome. It accounts for 28 percent of the entire body of

Y chromosome research generated in the quarter-century between 1960 and 1985.[15]

XYY research propelled the Y into the cultural lexicon as a molecular symbol of masculinity, similar to testosterone. Extensive print and television media images of XYY criminals tended to accentuate their height, muscularity, and sexuality. High school and college biology and psychiatry textbooks in the 1970s rapidly incorporated the idea of the XYY super-aggressive male, often featuring an image of 1970s serial killer Richard Speck (though Speck was not, in fact, XYY). Popular cultural forms reinforced this link between the Y and masculine behavior. As Jeremy Green, a British science studies scholar whose 1985 account of the XYY episode remains the principal secondary historical literature on it, records, "by the early 1970s, there had been at least two 'thriller' films in which the main character is a violent criminal driven by a chromosome abnormality, a series of crime novels with an XYY hero (who constantly wrestles with his inner compulsion to commit crimes), and as a spin-off from the novels, a TV series called 'The XYY Man'" (see fig. 5.2).[16]

Writers and cultural critics began using the term "Y chromosome" as a synecdoche for a man. The *Oxford English Dictionary* cites Peter Cave's 1974 *Dirtiest Picture Postcard* as the earliest English-language usage of the Y chromosome in a nonscientific text: "You've buttonholed me to give me long and boring lectures upon Germaine Greer, the faulty Y chromosome and the drudgeries of housework and child-bearing." In 1982, Edgar Berman wrote in *The Compleat Chauvinist*, "from that constantly on-the-make little tree mouse . . . to Mailer the magnificent, the DNA of the male Y chromosome has programmed us to lead our sisters."[17]

It took enormous resources and more than a decade of work by multiple research groups to reverse Jacobs's initial conjecture. Through large epidemiological studies published in 1976 and 1977, researchers proved that 97 percent of XYY individuals will never commit a crime, that the higher prevalence of XYYs in security institutions than the general population was due, if anything, to slightly lower median intelligence, that their offenses were actually less aggressive and violent than those of their fellow XY inmates, and that the only phenotype reliably associated with XYY was increased height.[18] This profile of the XYY male showed that any behavioral differences in XYYs were due to a generalized developmental effect, consistent with a chromosome imbalance, not a double dosage of Y-chromosomal genes specialized for aggression or violence. XYY vanished almost completely from the scientific literature by 1980.

Figure 5.2. XYY Man book cover. Reprinted from Kenneth Royce, *The XYY Man* (New York: McKay, 1970).

BIAS IN XYY RESEARCH

Today the XYY syndrome stands as an exemplar of discredited science and poor genetic reasoning. The overwhelming majority of interpreters agree that XYY research in the 1960s and 1970s was marked by bias, distortion, and gaps in reasoning. Methodological failings of these studies, many of which were raised at the time, included:

1. *Small sample size.* XYY individuals are very rare (about 1 in 1,000 males). In the small institutionalized populations studied, a single false positive would dramatically change the estimate of prevalence rates of XYY in prisons.
2. *Biased samples.* Researchers studied XYY only in incarcerated individuals in "high-security" institutions and did not study nonincarcerated men. This was, of course, a biased sample, including only individuals already deemed "criminal" or "aggressive."[19]
3. *Lack of double-blind protocol.* Most early studies, including the 1965 study by Jacobs, did not use double-blind procedures, but set out specifically to screen for XYYs. This created the likelihood that "wishful" false positives slipped into the data.
4. *Poorly defined phenotype.* The phenotype of "aggression," which was defined by the fact of subjects' incarceration rather than specific behaviors they exhibited, lacked rigor.
5. *Neglect of environmental variables.* XYY researchers failed to profile the family, social class, and medical background of their subjects, data essential for testing the viability and strength of any behavioral genetics claim.
6. *Lack of plausible mechanism; no experimental confirmation.* Researchers never offered a molecular or evolutionary mechanism for how an extra Y might carry the specific behavioral phenotypic consequence of criminal aggression, nor did they propose, or attempt to carry out, human or model organism experiments that might test this causal link.

The XYY theory also leaned on questionable biological assumptions and ignored contradictory evidence. Two issues in particular stand out. First, the XYY hypothesis defied the conventional framework for understanding how supernumerary chromosomes affect phenotype. As Green noted, researchers assumed that the Y chromosome "would make a *specific contribution to violence or aggression*, instead of having the generally impairing effects associated with an extra chromosome."[20] According to this double-dosage model, the extra Y doubled the contribution of a putative gene for "aggression" in a male, rather than producing complex, global developmental effects, as in the case of Down's syndrome (trisomy 21). Second, XYY researchers ignored a supremely problematic piece of evidence that contradicted the idea that a connection between the Y and aggression exists. Despite the finding that extra-X, or XXY, men were as prevalent as XYY individuals in security institutions, researchers associated aggression with the Y chromosome and not the X.

For all of these reasons, scholars today often hold up XYY studies as a classic example of poor reasoning in human behavioral genetics. Despite this, scientists invested nearly two decades of research efforts and funding in XYY studies. Even principal actors in the XYY controversy are at a loss to explain what went wrong in retrospect. Reflecting on the research blunders of XYY studies, Saleem Shah, the head of the NIMH juvenile crime unit and coauthor of several key publications on XYY, has said, "It is difficult to know what people's motives were for doing these studies. I have given up trying to figure out what the real facts are, because I do not know what they were. It depends, I suppose, on what you consider 'real.'"[21] In the aftermath of the XYY episode, geneticist Jonathan King, a critic of the XYY studies, was similarly flummoxed: "The interesting question is why the whole history of studies in this field were [*sic*] so full of biases and distortions. . . . Valid criticisms of those studies were made from the very beginning. They had much publicity and for various reasons were not widely circulated."[22]

The received view of XYY research among scientists today is as an "embarrassing" incident of mishandled public relations in the history of behavioral genetics.[23] In the decades following the XYY episode, many scientists disavowed it as a case of media sensationalism, portraying the idea of the XYY supermale as a marginal theory that was rapidly debunked and renounced by geneticists but that nevertheless lived on in the media.[24] For instance, James Watson's important textbook, *Recombinant DNA*, mentions XYY, but only as a cautionary tale of a poorly conceived line of research, which because of media distortion and amplification wrongly stigmatized XYY individuals.[25]

In his 1985 study, Green, a scholar in science studies and communications, analyzed media coverage of XYY research. Green showed that it was prominent scientists who hyped the XYY hypothesis, drawing on popular "expository strategies," while journalists were more willing than the scientists to raise questions and rein in incautious interpretations of the link between XYY and aggression.[26] Indeed, scientists used newspaper interviews "as an opportunity to present their more creative, and less empirical interpretations and speculations."[27] Green concluded that the media did not create a "myth" of the hyperaggressive XYY male, but accurately conveyed the already sensationalized endorsement that scientists gave to this image of XYY men.

Green attributes scientists' fever for the XYY hypothesis, despite its weak foundations, to two historical factors. He suggests, first, that an "upsurge of interest in crime . . . [and] the apparent failure of programs based on social intervention" in the 1960s and 1970s stimulated research on human aggression and encouraged researchers to overstate their claims.[28] Unfortunately, any evidence for the existence of a connection between crime waves in the 1960s and 1970s and XYY research is tenuous. While it is certainly plausible that XYY researchers imagined that genetic studies could help to solve crime problems, any such hope would have been in the broadest of generalities. Crime was not a matter of serious academic interest, nor of any demonstrative applied practice, among XYY researchers. Cytogeneticists with no specialty in crime led the research effort. Their use of men in prison populations as study subjects was a matter of happenstance and convenience of access, not interest in criminal behavior. Biologists, not criminologists, were the principal agents of XYY research. Indeed, biological theories of criminal behavior ran directly against the current of the field of criminology at the time.

In contrast, amongst US geneticists, the notion that genetics could explain individual and community variations in behavior was professionally and institutionally on the rise during these exact years. The first textbook on behavior genetics appeared in 1960.[29] The field entered the wider public and scientific consciousness with Arthur Jensen's infamous 1969 article asserting that the racial IQ gap was genetic and hereditary.[30] In 1970, the Behavior Genetics Association and the first scientific journal devoted to the subject were founded. Aggression was a topic of particular interest in this field. During the 1960s and 1970s, psychiatrists were beginning to conceptualize human aggression as a brain abnormality with an organic and likely a genetic basis.[31] As we shall shortly see, this newly intense focus on innate and inherited human behavioral traits also found traction in other fields, including ethology, physical anthropology, and sociobiology.

The spread of these developments to applied fields such as criminology would come later.[32] Mainstream criminology in the 1960s still associated crime primarily with mental health, youth, and social class, factors that were understood to impact males and females differentially because of the cultural influence of sex roles.[33] There were few sympathetic ears for the XYY hypothesis in the crime control community. While some geneticists, such as Richard Lewontin and Jonathan Beckwith, criticized XYY studies, aggression researchers from psychology, endocrinology, and criminology, more sensitive to the nuances and methodological requirements of behavior studies, led the way and were among the earliest and most vocal critics of XYY research.[34]

The second factor named by Green is the competitive race for chromosomal disease correlations during the 1960s heyday of cytogenetics. XYY surfaced at the height of cytogenetics, just after new techniques had given rise to a series of dramatic findings that linked chromosome abnormalities to human phenotype, including the unambiguous association of Down's syndrome with the presence of an extra chromosome 21. Like the gene-discovery race of the 1980s, in the late 1950s and early 1960s researchers competed to identify new chromosomal abnormalities, hyped as the key to much of human diversity as well as to the origins of debilitating diseases and birth defects.[35] Green suggests that researchers, swept up in the frenzy, prematurely and overconfidently linked dramatic behavioral effects with the XYY genotype. Yet despite this heady immediate scientific context of XYY research, there are not parallel cases of flagrant bias and runaway hypothesizing in the case of other non–sex chromosome abnormalities during this period. That is, something in particular about the notion of an extra *sex* chromosome led researchers astray. To account for the specificities of the methodological errors and biases that characterized XYY research, we must look to another central aspect of the development of the hypothesis: popular and biological ideas about gender in the 1960s and 1970s.

BIOLOGICAL THEORIES OF SEX DIFFERENCE IN THE 1960S AND 1970S

In what follows, I argue that prevailing biological theories of sex differences constituted the background explanatory framework within which XYY theories were received and understood in the 1960s and 1970s. XYY studies originated not among psychologists or criminologists, but among molecularly inclined biologists and those in the nascent field of behavioral genetics, who treated XYY research primarily as concrete evidence of a genetic basis of behavioral differences between the sexes. Specifically, the XYY supermale hypothesis was conceived, and received, within an associative framework of biological claims about male aggression, human nature, and sex and gender difference in the 1960s and 1970s.

During the mid-twentieth century, physical anthropologists developed a theory of human cultural evolution centered on the "man the hunter" and "woman the gatherer" gender roles. The idea that men are primarily adapted to be hunters—aggressive, competitive, physical, active—became the foothold for theories of human sex differences across many areas of biology. Man-the-hunter came to be understood as an explanation for why, as Lewontin, Rose, and Kamin quipped in their famous critique of this view of human nature, *Not in Our Genes*, "today men are executives and women are secretaries" and "why we sometimes behave like cavemen."[36] As Lewontin and colleagues noted, such arguments were ideologically convenient, entering a contentious wider societal debate over gender roles in the 1960s and 1970s. Just as women's liberation movements were challenging the traditional gender order, a new science emerged that suggested, wrote Lewontin et al., "that sexual divisions have emerged adaptively by natural selection, as a result of the different biological roles in reproduction of the two sexes, and have evolved to the maximal advantage of both; the inequalities are not merely inevitable, but functional, too."[37]

Texts publicizing these theories of the evolution of human sex differences, such as Desmond Morris's *The Naked Ape* (1967), Lionel Tiger's *The Imperial Animal* (1971), and David Barash's *The Whisperings Within* (1979), circulated widely during the 1960s and 1970s.[38] This literature, which would form the foundations of the new field of sociobiology, portrayed modern man as an animal honed by evolution for violence and conquest, a beast contained and restrained by the manners and restrictions of contemporary society. The claim that males are naturally more aggressive than females was arguably the central premise of sociobiological theories of biological sex roles.[39] Sociobiologists cited men's higher aggression to explain why they are better at math and abstract thought, more suited for leadership roles, prone to rape, attracted to pornography, and ill suited for monogamy. Men's innate aggression was even said to explain features of human culture and society such as capitalism, imperialism, and war.

In the context of this general approach to the evolutionary history of sex differences, the XYY male appeared as a well-timed confirming example. Within this associative framework, XYY offered a forceful piece of evidence that a certain level of male aggression—in the ordinary XY male—was both normal and natural. As Curt Stern, the eminent American geneticist and author of the authoritative genetics textbook of the 1960s, speculated in a 1968 *New York Times* article on XYY, "the female sex owes its gentleness to the absence of a Y chromosome and the normal male his moderate aggressiveness to the single Y."[40] Kurt Hirschhorn, chief of medical genetics at Mt. Sinai School of Medicine, explicitly connected XYY research to then-emerging sociobiological conceptions of masculinity and aggression:

> What if the genes for aggression and tallness do exist in the Y? They would have had survival value for the caveman, and there might have been an evolutionary selection for them. But civilized man has been breeding against aggressive genes. . . . The average male has just the single Y. Now, today, we find a man who gets a double dose of Y's; it's understandable that they might well be too much for him to handle.[41]

This logic appears in the actual research studies as well. As one XYY study stated, "the XYY genotype may be seen as highlighting the association between maleness and violence. . . . If the genes on a single Y chromosome of a given individual are such as to produce a tendency toward marked aggression, then a double dose of such a Y chromosome would be expected to lead to extreme aggression."[42] Consistent with sociobiological models of sex differences, here the Y chromosome is portrayed as encoding atavistic traits, once adaptive, that emerge when the Y dosage is upped by an accident of nature. Within the biologically determinist and traditional gender-ideological worldview of sociobiology, many received the association between the Y and aggression as confirmation of the naturalness of normal male aggression and as evidence that contemporary gender roles are based on innate biological sex differences.

ACCOUNTING FOR BIAS IN XYY RESEARCH

Understanding the influence of the emergent sociobiological model of sex differences in 1960s and 1970s biology helps make sense of the unstated assumptions and inferences that glued the XYY hypothesis together. To proceed with the hypothesis of Y-linked aggression, researchers had to ignore standard explanatory frameworks and data that clearly opposed it. The alignment between a model of the Y chromosome as specialized for male-stereotyped behavior and mainstream contemporary theories of sex differences made this possible. Compared to XX individuals, XY individuals were overrepresented among violent criminals, so researchers reasoned that XYY individuals should be expected to be even more overrepresented. As Jacobs recalls, "so I thought, that's very funny isn't it? Maybe the Y is affecting their behavior. Well that's quite likely isn't it? If you stop and think about it, ninety-eight percent or some such number of the prison population are males. So you can't say the Y has got nothing to do with behavior."[43] Friedrich Vogel and Arno Motulsky would later summarize the chain of reasoning in their 1979 genetics textbook: "Normal men are more aggressive than normal women; normal men have one Y chromosome, women do not. Hence, if someone has two Y chromosomes, he should be twice as aggressive as normal men."[44] The crucial point here is that Jacobs's hypothesis was not specifically about XYY males; rather, it was that we should expect the Y chromosome to contain the genes for stereotypic male traits, aggression foremost among them. That the Y was the obvious substrate of maleness went unquestioned, not just by Jacobs but by most geneticists at the time.

The XYY "supermale" theory flew in the face of customary ideas about extra chromosomes. Geneticist Park Gerald, leader of a major Harvard XYY study, called the XYY complement a "supermaleness syndrome" in a press conference, and the *New York Times* quoted geneticist Curt Stern referring to XYY as "double Y aggressiveness."[45] Yet, the prevailing hypothesis at the time was that chromosomal disorders, such as Down's syndrome (trisomy 21), produced phenotypic abnormalities by general chromosomal imbalance, rather than through the duplication of specific genetic material. An additional chromosome was thought to produce an imbalance of gene product and not a simple doubling of it. Explaining chromosomal aneuploidy in his 1965 textbook *An Introduction to Human Genetics*, Eldon Sutton, who would become president of the American Society of Human Genetics in 1979, nicely captured this prevailing view: "Since the genes on a particular chromosome may be quite unrelated in their primary action, trisomy would disturb many apparently unrelated functions."[46] Genes work together in complex regulatory relation. An extra chromosome can have very specific characteristic effects, but these effects are unlikely to simply be "more" of what the genes normally do. Rather, an extra chromosome results in a general, stochastic developmental effect of multiple mistimed or diverted genetic pathways. This was the consensus view of the phenotypic consequences of chromosome aneuploidy. Despite this, the leading explanation of the link between XYY and crime in the biology community assumed that the XYY simply provided a double dose of male aggression.

Gender conceptions played a critical role in shaping the conceptual framework within which researchers crafted the XYY theory and help to explain why researchers pursued a double-dosage explanation of supernumerary chromosome behavior in the case of XYY. The symbolic identification of the Y with masculinity, and the confluence of the double-dosage model of XYY with dominant proto-sociobiological theories that saw human males as possessing a "beast within," led researchers, in the case of XYY, to assert the unlikely double-dosage model over the more validated chromosome imbalance model.

By the mid-1970s it had dawned on many researchers that males with an extra X (XXY or Klinefelter's males) were also overrepresented in security institutions. A double-X chromosome complement in males looked to be as predictive of criminality as the XYY complement. Researchers did not know quite what to do with this bedeviling fact. Jarvik et al., among the earliest to raise the issue of XXYs, wrote in 1973: "It would seem . . . that patients in mental hospitals are more prone to have an extra sex chromosome than men in the general population, but that the tendency is nonspecific, the probability of an extra X chromosome being as great as that of an extra Y chromosome."[47] Jacobs conceded in 1975 that "the excess of males with an abnormal chromosome constitution in mental-penal settings is not confined to XYY individuals but also applies to XXY men."[48] In 1976, Witkin et al. ventured that an extra X or Y may lead to similar behavioral phenotypes with regard to crime: "The similarities between the XYYs and the XXYs suggest that . . . the consequences of an extra Y chromosome may not be specific to that chromosomal aberration but may result from an extra X chromosome as well."[49]

The profiles of XXYs and XYYs in high-security institutions were indistinguishable in terms of kinds of crimes committed and representation in institutions of incarceration. Nonetheless, no research program was initiated to study the correlations between double-X and aggression. The association of the XXY with crime had to be ignored or dismissed to sustain the hypothesis of a specific association between the Y and aggression. A gendered schema that localized maleness to the Y, along with a corresponding association of the X and femaleness, facilitated the dismissal of the problem that XXY posed to the XYY theory.[50] Attempting to hold together the logic of the Y-aggression association, researchers proposed that the presence of XXYs in high-security institutions might be due to different behavior than their XYY counterparts. The hyperfeminine manners of XXYs—their passivity, emotionality, and subpar intelligence—landed them in prisons, while the hypermasculinity of XYYs—their aggression and impulsiveness—led to their institutionalization.

In the 1970s, scientists brought a gendered schema for understanding the Y chromosome into their delicate hairsplitting to explain equal XXY and XYY representation in criminal institutions. Leading human population cytogeneticist Michael Court Brown, of the University of Edinburgh, for instance, attributed Klinefelter's criminality to low intelligence, asserting that XXYs specialized more in property offenses in the range "from larceny to child stealing."[51] In contrast, XYY criminality, he claimed, was due to an "aggressive," "disturbed pattern of behavior," "so violent and aggressive that they require to be kept in conditions of maximum security."[52] Geneticist Edward Novitski, author of one of the most widely used human genetics textbooks of the 1970s, solved the quandary by claiming that XXYs show "failure in accomplishing some task, or an inability to do so, a pattern different from that of an XYY."[53] In contrast, he wrote that XYYs show "behavioral problems related to the impulsive, exaggerated performance of a deed that is carried out without regard for the long-range consequences."[54] Reproducing a familiar gendered binary, here XXY criminality results from feminine ineffectiveness and passivity, while XYY criminality is a consequence of a masculine overactive personality. In a similar vein, in the 1980 version of his textbook, Sutton characterized the XXY's troubles as due to "a slightly increased risk of diminished intellect and of emotional instability," while XYY displays "destructive" sociopathy: "The expression is one of increased impulsiveness with little perception of future consequences."[55] In this case, XXY is feminized as overly emotive and XYY masculinized as cold and unfeeling.

Male identification of the Y, along with a corresponding female identification of the X, contributed to a distorted model of the genetics of male aggression and led XYY researchers in the 1960s and 1970s to neglect alternative models of the relationship between sex chromosome abnormalities and behavioral phenotype. Gender beliefs help to explain this stubborn identification of the Y chromosome with maleness and aggression, and the inability of midcentury XYY research to see, take up, or account for a similar criminal profile of multiple-X-chromosome individuals.

THE FOLDING OF XYY RESEARCH

By 1982, Patricia Jacobs, delivering an address to the American Society of Human Genetics, would reflect back on "the end of an era" of XYY research (and cytogenetics as a whole), now eclipsed by the "day and age of restriction enzyme polymorphisms, transposable genetic elements, introns and exons."[56] After the downfall of XYY, Jacobs left the high-status Edinburgh cytogenetics lab for Hawaii, "for an empty laboratory in a corner of the anatomy department," and shifted her attention away from the human Y chromosome.[57]

For most, the 1976 study published in *Science* by American psychologist Herman A. Witkin and colleagues put the XYY hypothesis to rest.[58] The study included all 4,558 Copenhagen-born males in the Danish national draft registry whose height was greater than 184 centimeters (a subgroup expected to contain a higher rate of both XYY and XXY males than the general population). Researchers visited the subjects' homes to take buccal smears and blood samples for chromosomal analysis, an exhaustive effort in which subjects not found at home "were subsequently revisited, up to a total of 14 times in the most extreme instance."[59] Examining claims of an association between XYY and criminality, the study found no difference in the rate of aggressive crimes between XYY and XY individuals. It showed that XYY individuals on average have slightly lower IQs and are a bit taller than the general population, which they speculated may account for their elevated rates of criminal detention. Finally, it showed that the rate of criminal detention for XXY and XYY males was identical. As the authors concluded, "no evidence has been found that men with either of these sex chromosome complements are especially aggressive. Because such men do not appear to contribute particularly to society's problem with aggressive crimes, their identification would not serve to ameliorate this problem."[60]

Suspicions of some underlying truth to the supermale syndrome, however, persisted among some even after the hypothesis was discredited in most geneticists' eyes. In the aftermath, some cytogeneticists believed that research on the XYY hypothesis had been snuffed out before the association between Y and aggression had been thoroughly explored. For some in the Edinburgh, Harvard, Johns Hopkins, and NIMH cytogenetics communities who had promoted the hypothesis, the perceived proximate cause of the demise of XYY research was public controversy, not an empirical dead end.

In 1975, a controversy erupted over genetic screening for XYY in newborn males at Harvard University's teaching hospital, Boston Women's Hospital. One of the largest and most ambitious studies of the phenotypic effects of XYY in young boys, the study described its broad aim as the elucidation of the genetic substrates of sex differences. Wrote Park Gerald, the study's lead investigator and founder of Harvard's first human genetics research and training program, "we began the study to understand the important connection, if any, between genetics and behavior differences between the sexes."[61] The specific goal of the study, funded by the NIMH Center for Juvenile Crime, was "a more precise estimation of the risk, if any, to XYY children likely to develop aggressive or psychosexual pathology."[62] Parents of boys identified as XYY would be informed that their child had a sex chromosome abnormality; XYY boys would be subsequently tracked over the years through parental interviews, psychological observations of play and classroom behavior, and other physical and mental measures. It was the largest and most ambitious study of XYY children to date.

Science for the People (SFP), a national group of American leftist scientists with a lively cohort at Harvard and MIT, took issue with the XYY study's biologically determinist model of human behavior and rallied university, legal, and media avenues to stop the study.[63] Initiated in 1968 at the height of antinuclear and antiwar activism, SFP first became widely known for its condemnation of Arthur Jensen's 1969 argument that genetic differences account for the IQ and educational achievement gap between whites and blacks in the United States. Later, SFP drew up forceful responses to E. O. Wilson's 1970s reinvigoration of neo-Darwinist sociobiology, which the SFP condemned as racist, sexist, and biologically determinist.[64] Taking on the Harvard XYY study in 1975, SFP charged the study's lead scientists with serious ethical violations and methodological oversights. The Harvard study allegedly acquired subjects by providing an open-ended consent form to expectant mothers while they were in labor. The study failed to include controls (non-XYY children would not be tracked alongside the XYY subjects). It also dropped the double-blind protocol for individuals identified as XYY. As SFP argued, this created the possible scenario of a self-fulfilling prophecy. Parents informed of a child's "high-risk" status may change their behavior toward the child; in turn, XYY boys might meet their parents' worries with aggressive and acting-out behavior. Parents would then report this behavior to researchers, creating a feedback loop of confirmation. If this were to occur, it would seriously undermine the study's findings by making it difficult, if not impossible, to untangle the effects of an extra Y from other factors. SFP argued that the Harvard study did not possess the rigor and scientific merit necessary to justify the study's risks. These risks included the stigmatization of XYY boys, who were too young to consent to the research themselves, by identifying them with the aggressive supermale stereotype. SFP's campaign, and the public uproar it created, brought the Harvard study to an end in 1975, cutting the last thread of the XYY theory's scientific credibility and effectively ending institutional support for future XYY research.

The history of research into the Y chromosome and aggression remains a sore subject for many who were involved. Jacobs perceived the end of XYY research as purely a case of censorship by the academic left. As Jacobs said in her 1982 address:

> [XYY] should have provided a useful, objective tool in the rational study of human behavior. Instead it aroused fierce and acrimonious controversy that resulted, at least in the United States, in the cessation of virtually all research aimed at objectively identifying sex chromosomally abnormal individuals and understanding the effect of the additional chromosome on their development and behavior. Why did this happen, and can we, with hindsight, try to insure that such an episode is never again allowed to sully human genetics?[65]

To this day, Jacobs remains convinced that there must be an association between the Y and aggression, stating in a 2006 interview: "And they try to tell me the Y has nothing to do with it. Absurd!"[66]

Some researchers concluded (and remain convinced to this day) that while an XYY syndrome had not been proven, it also had not been definitively disproved. The question could simply not be adequately studied, either because of present methodological limitations or the political climate. Exemplary is one of the last scientific commentaries on XYY, published in 1978. The authors reiterate the overrepresentation of XYYs in criminal institutions, concluding vaguely that "since it is probable that there is considerable variability in the phenotypic development of XYY males, it is inappropriate to allude to an XYY syndrome."[67] Rather than denying an association between XYY and aggression, the authors hedge, instead labeling the problem infeasible to study: "Since we lack an adequate system for classifying human aggressive behaviors and determining the base rates for different aggressive behaviors in the background population, *we cannot state definitely whether there is an increase in aggressive behavior associated with this genotype*."[68] The paper rebukes neither the crude model of genetic control of behavior nor the doubling hypothesis that underpinned the XYY theory. The view that there is still more to be explored in the relationship between the Y, aggression, and masculinity, was a not-uncommon understanding of the outcome of the XYY studies among members of the genetics community, even as XYY studies came to an end.

The XYY affair, along with Jensen's inflammatory race and IQ claims, left an indelible mark on the course of genetic science, derailing the emergent field of behavioral genetics. Many geneticists turned away from the new project of behavioral genetics in response to the poor science and unsavory overstatements of the XYY "supermale" syndrome and Jensen's race and IQ claims—two humiliating episodes in the history of behavioral genetics that remain forever paired in the minds of many. Cases such as the supposed XYY supermale so tarnished behavioral genetics' reputation that it would be decades before the field would emerge from the shadows. Chromosomal correlations with human behavior were off the table. As Robert Michels, professor of psychiatry at Cornell Medical School, stated of the XYY dust-up:

> the real cost of all this is that I will tell every young psychiatrist or physician or medical student wanting to do research on anything of social relevance to recognize that it is immensely expensive in terms of time and personal career because there are a lot of people in the world who are going to bother you if you do it. It would really make sense . . . to review whether there is something you can't do in bean metabolism or some kind of pigeon behavior because it is safer to work that way in our present society.[69]

Indeed, by the late 1970s, research was trending toward molecular approaches to genetics. Human behavior research on the Y chromosome did not present a testable, causal-mechanistic hypothesis and rigorous experimental system in the way that molecular approaches emerging in medical research or model organism biology could now offer. The comments of MIT biologist Jonathan King, one of the principal critics of the Harvard XYY study, make clear the role that this normative shift toward the molecular and the mechanistic played in the demise of XYY studies:

> Our objection was to the attempt to link the XYY chromosome to abnormal social behavior in very young children. . . . That was our critique. If someone says to me, I am interested in the way that the mitotic spindle forms in a cell that has an extra Y chromosome, fine. If someone says to me, I am interested in what happens to women's ovaries or men's seminiferous tubules that causes chromosome damage, fine. . . . If someone said to me I want to look at the level of histone biosynthesis in nuclei in XYY males, I might say yes. If someone said I want to look at the relationship between XYY and the frequency of temper tantrums, I do not know. I would probably suspect that it lacks a valid scientific or medical foundation.[70]

In the face of attacks and empirical dead ends, XYY researchers would follow King's lead. They would indeed retreat to "molecular analysis" of the Y chromosome rather than attempting to correlate karyotypes and human behaviors. Yet, as we shall see, even as the association between the Y and aggression was discredited, the model of the Y as the "chromosome for maleness" marched forward, unexamined.

FROM XYY TO THE Y CHROMOSOME

Today it is common for sex chromosome geneticists to distance themselves from XYY studies. In the 2003 *Nature* paper announcing the successful sequencing of the human Y chromosome, for example, project leader and MIT scientist David Page positioned the event as the reversal of a century of prescientific dogma about the Y chromosome.[71] According to Page, prior to molecular analysis, biologists assumed that the Y was a "genetic wasteland" involved in no processes other than testes determination. Beginning in the mid-1980s, Page said, geneticists reversed this view by bringing "recombinant DNA and genomic technologies to the Y chromosome, culminating in *molecularly based* conclusions about its genes."[72] The result, Page claimed, was a new model of the Y chromosome as gene-rich and specialized for maleness. This genealogy of Y chromosome research conspicuously erases the era of Y chromosome research in the 1960s and 1970s.

XYY studies were high-profile, mainstream cytogenetic research that represented the primary work on the human Y chromosome for two decades. XYY spurred interest in the gene content and natural history of the Y, technologically and model-theoretically prepared the ground for future Y chromosome research, and rallied researchers and resources to the study of the Y. XYY research also helped to cement a working model of the Y chromosome as the chromosome for maleness that, as chapters 7 and 8 will demonstrate, remained extremely influential in the coming decades.

The presence of XYY as a reference point and instigating research model for understanding the genetics of the Y chromosome is evident in the primary literature, even after the XYY supermale theory was eclipsed. A 1975 *Nature* article on the mouse Y chromosome, for example, referred to "the XYY karyotype in man" as part of "a growing body of evidence for the involvement of the Y chromosome in the inheritance of particular characteristics."[73] In 1976 in the same journal, researchers at Johns Hopkins reported the use of cells from XYY males in an experimental system to locate and validate the existence of Y-specific genes.[74] They argued that the Y chromosome is made up of male-specific sequences, as opposed to only material that also appears in duplicate on the X. The authors concluded that "correlating the presence of these sequences with the various phenotypes associated with particular qualitative abnormalities of the Y chromosome" may help reveal its male-specific function.[75]

XYY studies also led to the development of cytogenetic technologies for studying the Y chromosome. A 1970 *Nature* editorial noted the "fever of excitement" incited by a new staining test that "makes part of the human Y chromosome fluoresce particularly brightly." The editors wrote, "this is an important time saver for those engaged in screening human populations for XYY males, the possessors of an extra Y chromosome who tend, as adults, to be found in prisons slightly more often than it seems they should be."[76] These new techniques, crucial to late 1970s and early 1980s Y chromosome studies, were developed through intensive research on the XYY supermale.

XYY research contributed to substantive and lasting lines of research on the functional genetics of the Y chromosome. As Reed Pyeritz, a biologist at Johns Hopkins, would state in a 1980 public forum on XYY research:

> Molecular geneticists are devoting much effort these days to mapping human chromosomes. . . . The Y chromosome is a unique chromosome. It contains certain DNA sequences which are repetitious and which have been mapped to the Y chromosome; XYY cells have actually been used in this research to prove that the DNA sequences are Y-specific. In other words, men with the XYY karyotype have twice as much of this DNA. These are the sorts of molecular and biological applications of knowledge about XYY that have clinical applications and potential value.[77]

XYY studies are woven into the history of Y chromosome research in ways that are at once instrumental and grounded in a conceptual understanding of the Y as the chromosome for maleness.

As research shifted from XYY to genetic analysis of the "normal" Y, a model of the Y as gene-rich, rather than inert, and specialized for "maleness," not just the testes-determining gene, continued to motivate researchers. By the early 1980s, research on aggression in XYY males gave way to studies of Y chromosome polymorphisms and to structural analysis of the Y and, by the mid-1980s, to a search for sex-determining, fertility, and other male-specific genes on the Y. While the association between Y and aggression had taken a hit with the decline of the XYY hypothesis, the concept of the Y as the biological kernel of maleness survived.

In noting these connections between XYY research and later molecular and genomic studies of the Y chromosome, I do not mean to tar today's Y chromosome research with the same blatant methodological errors that characterized XYY studies. Rather, the aim is to document how assumptions about the Y as the chromosome for maleness were formed through, and in relation to, historically specific scientific research agendas and gender politics. Neglecting the role that the gendered assumption that the Y is the chromosome for maleness played in the XYY episode, researchers today run the risk of not fully embracing its cautionary lessons.

◇

The widespread idea of the Y as the chromosome for "maleness" owes its origins to XYY supermale theories of the 1960s and 1970s. While there had been speculation about the possibility of some male traits on the Y chromosome prior to the 1960s (see chapter 4), the XYY episode is when the conviction that the Y represents the genetic essence of maleness first entered human genetics. This conviction played a central part in the notorious methodological errors and biases that characterized XYY research. The notion of the Y as the "chromosome for maleness" remains strongly embedded in popular and scientific discourse today—evidence of the abiding appeal of the idea of a simple gender binary, writ molecular in the human genome. The lasting contribution of the scientific characterization of double-Y individuals as bigger, more aggressive, and more sexual males would be to cement and amplify the association between the single Y and maleness.

CHAPTER 6

Sexing the X

"Every daughter," Natalie Angier writes, ". . . is a walking mosaic of clamorous and quiet chromosomes, of fatherly sermons and maternal advice, while every son has but his mother's voice to guide him."[1] In her 1999 *Woman: An Intimate Geography,* Angier celebrates the female double-X as a privilege of womanhood and a source of special womanly qualities. The "mystical X," she writes, is a source of "female intuition." Women "have . . . with the mosaicism of our chromosomes, a potential for considerable brain complexity. . . . [A] woman's mind is truly a syncopated pulse of mother and father voices, each speaking through whichever X, maternal or paternal, happens to be active in a given brain cell."[2]

In his 2003 *The X in Sex: How the X Chromosome Controls Our Lives,* David Bainbridge writes that, like the Christian vision of Mary as "both virgin and mother," women "represent some intermediate hybrid state." This explains women's "unpredictable, capricious nature."[3] The female double-X, he continues, is a "natural reminder of just how deeply ingrained the mixed nature of women actually is. . . . Women are mixed creatures and men are not . . . in a way far deeper" than we ever thought.[4] As mosaics of their two X chromosomes, "each woman is one creature and yet two intermingled."[5]

Focusing on "X mosaicism" theories of female biology, health, and behavior, this chapter examines the long-standing and infrequently questioned association of the X with femaleness. The historical sweep of my analysis of the "female chromosome" goes from the discovery of the

X chromosome at the turn of the twentieth century all the way to the present day. Today, the conception of the X as the "female chromosome" remains a common assumption in twenty-first century genomics. Examining the empirical, conceptual, and historical dimensions of X mosaicism theories as represented in research on female autoimmunity and in Johns Hopkins biologist Barbara Migeon's *Females Are Mosaics* (2007), I show how the association between the X and femininity can become a source of bias, cohering and buoying otherwise highly speculative biological models of sex differences in areas urgently relevant to women's health.

HOW THE X BECAME THE FEMALE CHROMOSOME

Both males and females possess an X chromosome. Female sexual development is directed by hormones acting in concert with genes carried by many chromosomes and is not localized to the X. Indeed, the X is arguably more important to male biology, given the large number of X-linked diseases to which men are uniquely exposed. Despite the fact that this is well known, researchers across many scientific and medical fields attribute feminine behavior to the X (or double-X) and assume that female genes and traits are mediated by the X chromosome.

The "female X" has its roots in early sex chromosome science, which held for half a century that the X was female determining in humans. Historically contingent technical and material factors helped to brand the X as female. In the early days, chromosomes were studied almost exclusively in male gametes—the sperm.[6] Looking at sperm, which as reproductive cells possess only one member of each chromosome set, a perfect dichotomy appeared: half the sperm cells had the X, and half did not. This led to a hyperbinary view of the X and Y. The sperm with an X always produces a female and the X in the males' sperm is always inherited from the female parent. Failing to distinguish between the "sex" of the gamete and the sex of the organism, this partial perspective helped to prematurely assign the X to femaleness.

Sperm are plentiful, accessible, and easier to study than eggs or other human tissue. Thus, there are good reasons that male gametes were early chromosome researchers' tissue of choice. Nonetheless, the focus on sperm introduced a bias into early sex chromosome research. Had researchers looked at somatic tissue, the dichotomy would have been far less clear-cut: both males and females possess at least one X.

The dominance of studies of the fruit fly *Drosophila* in the first half-century of genetic research also played a central role in establishing the X as the female determiner. Unlike in mammals (as scientists would later learn), in *Drosophila* the X is female determining. This is a threshold effect, in which sex is determined by the ratio of X chromosomes to autosomes, with more Xs producing femaleness. In textbook explanations of sex chromosomes during the first quarter of the century, an ink drawing of *Drosophila* chromosomes was consistently used to illustrate the section on the chromosomal theory of sex.[7] So pervasive were *Drosophila*'s X and Y as the model for the sex chromosomes that the leading American geneticist Thomas Hunt Morgan simply termed the XX/XY chromosome constitution "*Drosophila* type," writing that "the genetic evidence so far gained has placed in the Drosophila type the following animal forms: Drosophila, man, cat; and the plants, *Lychnis* and *Bryonia*."[8] The *Drosophila* model suggested that in humans, as in flies, the X should be expected to determine femaleness. Thus, when in 1924, T. S. Painter, an American cytogeneticist at the University of Texas, first described the human sex chromosomes, he dubbed XX "the female chromosome complex," the X the "female-producing chromosome," and males as "heterozygous for sex" since they possess only one X.[9] The idea that the X is "female producing" or female tending focused theories of the biological determination of femaleness exclusively on the X well into the twentieth century.

The human cytogenetic research revolution of the late 1950s and 1960s, which revealed in 1959 that it is the Y that determines sex (an idea that we shall shortly complicate), marked the demise of the X-chromosomal model of human femaleness. The idea that the X was female determining was promptly discarded in light of the 1959 finding. The female or feminine resonance that had accumulated around the X chromosome, however, did not fall away. As Fiona Alice Miller notes with respect to the term "Mongolism" for trisomy 21 (Down's syndrome), "contrary to conventional beliefs about new, breakthrough technologies, the introduction of chromosome analysis in the late 1960s did not displace existing standards of interpretation and practice."[10] Similarly, old habits and the force of the idea of a molecular gender binary revealed in the X and Y were irresistible. As the Y would be the male chromosome, the X would continue to be the female one.

A brief visit to the case of Klinefelter's and Turner's syndromes demonstrates how the idea of the female-engendering X operated in the early years of human genetics. Both Klinefelter and Turner were well-documented syndromes of gonadal dysgenesis prior to human chromosome research. Physicians in the United States identified Turner in 1938 as a syndromic phenotype found exclusively in women. Traits included short stature, infertility, and neck webbing.[11] Physicians at Massachusetts General Hospital described Klinefelter in 1942 as a disorder of gonadal underdevelopment in males, resulting in hormonal deficiencies causing infertility and limited body hair.[12]

Barr body screening in the 1950s revealed that Turner females lack a second X, and Klinefelter males carry an extra X. Once associated with sex chromosome aneuploidy, the disorders were redescribed in more strongly sexed and gendered terms.[13] The infertility of the XO Turner woman was portrayed as evidence of her masculinity rather than a disorder of female sexual development (and of development in general). Scientists even claimed that Turner women displayed discomfort with female gender roles, possessed masculine cognitive traits such as facility with spatiality, and had defeminized body shapes. The eminent British geneticist Michael Polanyi proposed that XO females were "sex-reversed males."[14] Klinefelter males were portrayed as feminine, with much emphasis on their purportedly unmuscular body frame and female body-fat distribution, lack of body hair, and infertility. XXY Klinefelter males were consistently described in feminized terms during the 1960s and 1970s. (Recall chapter 5's discussion of scientists' contrasting explanations for criminal behavior in XXY and XYY males.) Eldon Sutton, author of a widely used 1965 genetics textbook, described Klinefelter individuals as "eunuchoid"; "genitalia are underdeveloped, and body hair is sparse."[15] In behavior, Klinefelter patients were stereotyped as "passive-aggressive, withdrawn, self-contented, and mother-dependent," with a reduced libido.[16] Cognitive tests were designed to assess whether Klinefelter males were, allegedly like females, more social and verbally oriented. Patricia Jacobs and John Anderson Strong described an XXY individual as "an apparent male . . . with poor facial hair-growth and a high-pitched voice."[17] They continued: "There are strong grounds, both observational and genetic, for believing that human beings with chromatin-positive nuclei are *genetic females* having two X chromosomes."[18] A 1967 *New York Times* article captured this mode of reasoning about the role of the X in femaleness. Headlined "If Her Chromosomes Add Up, a Woman Is Sure to Be Woman," it described XXY males as having "a few female traits."[19] Later, studies were even undertaken to determine whether Turner women show a tendency to lesbianism or Klinefelter men incline to homosexuality or cross-dressing.

In retrospect, we can see how these assumptions about the X as feminizing distorted understanding of these disorders, stigmatized individuals carrying them, and misdirected research and clinical care. Decades of intersex patient activism emphasizing that conditions such as XXY and XO are not "hermaphroditic" and that most people with these conditions live typical lives as male- or female-identified individuals helped to contribute to a transformation in the understanding of these conditions.[20] Developments in medical science reinforced activists' message, showing that nonsexual matters are the most urgent concern in the clinical management of Klinefelter and Turner. Now, clinicians specializing in Klinefelter and Turner stress that these are not diseases of gender confusion. Klinefelter patients are phenotypic males, and Klinefelter is not a syndrome of feminization. Today, Klinefelter is thought of as one of the most common genetic abnormalities and often has so few manifestations that many men live out their lives never knowing of their extra X. Writes Robert Bock in the US National Institutes of Health online guide for parents of XXY children, "for this reason, the term 'Klinefelter syndrome' has fallen out of favor with medical researchers. Most prefer to describe men and boys having the extra chromosome as 'XXY males.'"[21] Similarly, XOs are phenotypic females. Turner's syndrome, which has more profound and systemic phenotypic effects than XXY, is emphatically not a masculinizing condition. Physical deformities, heart trouble, hearing impairment, infertility, and autoimmune disorders are the principal concerns requiring lifelong medical management for Turner females.[22]

In short, researchers did not immediately give up the search for the relationship between the double-X and femaleness in the wake of the 1959 finding that the presence or absence of the Y determines sex. They continued to ask, what does the extra X do for females?

FEMALE X MOSAICISM

In 1961, British cytogeneticist Mary Lyon demonstrated that female mice are genetic "mosaics" for the X chromosome.[23] Lyon theorized that to equalize the expression of X-linked gene products in male and female mammals, one of the Xs in each somatic cell is inactivated early in female development. X inactivation, today known as Lyonization, results in approximately half of a female's cells expressing the maternal X chromosome and half the paternal X chromosome. Thus, females have two populations of cells, identical with respect to the twenty-two pairs of autosomes, but variable in X-chromosomal gene expression when females carry functionally different versions of an X-chromosomal allele. The distinctive patterning of calico cats, which are always female, is an effect of female "mosaicity" for the X-linked trait of coat color.

X CHROMOSOME MOSAICISM

X inactivation in females compensates for the presence of 2 X chromosomes.

The presence of a mosaic of active maternal and paternal X chromosomes usually shields females from X-linked diseases.

The single X in males can have health consequences. If Mom's X carries a genetic disease, the son will almost certainly develop it.

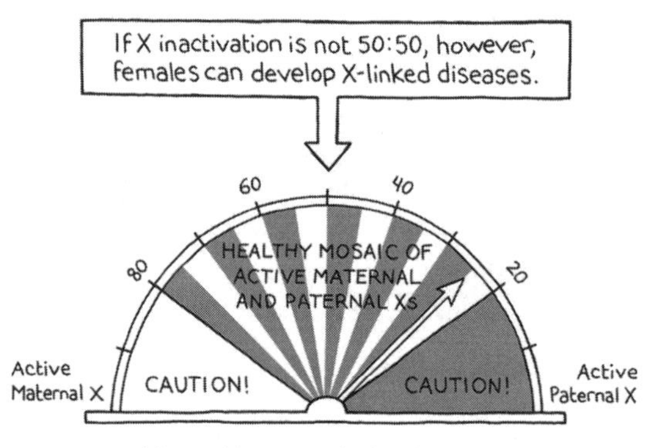

Figure 6.1. X chromosome mosaicism and its potential role in female pathology.Illustration by Kendal Tull-Esterbrook; copyright Sarah S. Richardson.

X mosaicism has some implications for human female biology (see fig. 6.1). Random X inactivation early in development leaves most women with a 50:50 ratio of cells expressing either their paternal or maternal Xs. As a result, females carrying a disease allele on one of their X chromosomes will generally not develop the disease since cells carrying the other X usually produce adequate amounts of the needed gene product to compensate for any dysfunction. For this reason, X mosaicism shields females from X-linked diseases. Classic X-linked diseases such as Duchenne's muscular dystrophy or hemophilia are infrequent in women and generally affect only men.

In rare cases, X mosaicism will begin to skew, resulting in tissues biased toward the maternal or paternal X chromosome. Tissues grow clonally, so skewing can happen randomly as a result of a bias in the cells from which the tissue grows. As we age, chromosomes fray, whither, and disappear due to the erosion of genetic repair mechanisms, making skewing more common. Usually, skewed X mosaicism has no phenotypic consequence and goes unnoticed. If a woman carries an X chromosome disease allele, however, extreme skewed X inactivation leading to dominance of the chromosome carrying the disease-causing allele can cause women to exhibit classic X-linked diseases generally restricted to men. Thus, the primary clinical implication of skewed X mosaicism for females is that it may leave them functionally monosomic for the X—like males—making them vulnerable to male-typical X-linked diseases.

FEMININITY AND THE DOUBLE-X

From its inception, the hypothesis that females are cellular mosaics for X-chromosomal genes has been received as confirmation of dominant cultural assumptions, both pejorative and beneficent, about women. The characterization of females as mosaics resonates with conceptions of women as more mysterious, contradictory, complicated, emotional, or changeable.[24] The future Nobel Laureate molecular biologist Joshua Lederberg wrote in 1966 that "the chimerical nature of woman has been a preoccupation of poets since the dawn of literature. Recent medical research has given unexpected scientific weight to this concept of femininity."[25] Reporting on the new finding in 1963, *Time* magazine asserted that "the cocktail-party bore who laces his chatter with the tiresome cliché about 'crazy, mixed-up women' has more medical science on his side than he knows. . . . Even normal women, it appears, are mixtures of two different types of cells, or what the researchers call 'genetic mosaics.'"[26]

Today, the notion of X mosaicism as scientific confirmation of traditional ideological conceptions of female instability, contradiction, mystery, complexity, and emotionality is thoroughly entrenched. As science writer Nicholas Wade was quoted saying in the *New York Times* in 2005, "women are mosaics, one could even say chimeras, in the sense that they are made up of two different kinds of cell. Whereas men are pure and uncomplicated, being made up of just a single kind of cell throughout."[27] A 2005 Penn State press release on a study by one of its scientists similarly announced that "for every man who thinks women are complex, there's new evidence they're correct; at least when it comes to their genes."[28]

These conceptions are not just the common trade of science journalists. They are also widely shared by present-day sex chromosome researchers. Duke University geneticist Huntington Willard, for instance, is quoted as saying that "genetically speaking, if you've met one man, you've met them all. We are, I hate to say it, predictable. You can't say that about women"; and MIT geneticist David Page says, "women's chromosomes have more complexity, which men view as unpredictability."[29] British sex chromosome geneticist Robin Lovell-Badge has similarly said that "10% [of genes on the X] are sometimes inactivated and sometimes not, giving a mechanism to make women much more genetically variable than men. I always thought they were more interesting!"[30]

There are, of course, long-standing associations between femaleness and the connotations of mosaicism articulated by scientists and science writers. Representations of females as crazy, mixed-up, chimerical, mysterious, contradictory, changeable, complicated, disordered, unreliable, monstrous, complex, unstable, emotional, coy, duplicitous, and impure crowd Western art, film, and literature. As feminist theorist and phenomenological philosopher Elizabeth Grosz has written, "the metaphorics of uncontrollability, the ambivalence between desperate, fatal attraction and strong revulsion, the deep-seated fear of absorption, the association of femininity with contagion and disorder, the undecidability of the limits of the female body . . . are all common themes in literary and cultural representations of women."[31]

In Greek mythology, a chimera is a female monster comprised of incongruous parts. The Medusa, once a beautiful maiden, is a jealous snake-haired woman whose hideous laughing face will turn a man to stone. Women as seductresses who turn into murderous or possessive nightmares are a common theme in Western literature and art. World War II anti–venereal disease posters depicted beautiful young women as two-faced, carrying gruesome diseases that threatened the security of the nation behind their lithe bodies and made-up masks. Many horror and noir films hinge on the audience's shock and dread at a beautiful, feminine woman turned suddenly grotesque, possessed, animalistic, or bipolar. In the classic film *Cat People*, for example, the chilling female main character is haunted by the threat of transforming into a panther if aroused to passion, anger, or jealousy.[32]

Within biology and medicine in particular, these tropes carry special potency. In 1916, the Harvard geneticist William Castle wrote of the double-X female that "if one were inclined to be facetious, he might say that . . . *duplicity* is synonymous with femaleness, *simplicity* with maleness!"[33] Conceptions of females as "mosaics" or "chimeras," and hence as mysterious, contradictory, complicated, or changeable, have been interlaced with biological theories of sex difference for centuries. Biomedical ideas of female chimerism and changeability arise especially around pregnancy, puberty, and menstruation, which are characterized by physical transformation, permeability, and leakage.[34]

In the history of psychology and psychiatry, women have long been conceived as essentially bipolar and unstable, subject to identity split, and on the edge of hysteria. The classic women's malady, hysteria was a female-specific disease of a woman at odds with her wandering womb or libido. One of the principal characteristics of the hysteric was pathological lying—two-facedness. In the late nineteenth century, hysterics were considered cases of normal femininity gone out of control. The portrayal of hysterics as rapidly cycling between different moods and physical states is particularly rich with frightening chimeric, shape-shifting imagery.[35]

Hormone science, too, is filled with notions of women as unpredictable, dangerous, emotional, and nymphomaniac. Louis Berman, in his 1921 *The Glands Regulating Personality*, asserted that women are "liable to rapid and pendulum-like fluctuations," "which constitute the essence of the life story of woman," while "man is relatively free of these liabilities, and so remains man by his freedom."[36] Today, the figure of the suddenly monstrous or violent hormonal woman remains a pervasive cultural stereotype.[37]

The conviction that the X chromosome underlies female biology, femaleness, and femininity, and the concordance between X mosaicism and cultural notions of females as complex, contradictory, and changeable, remain palpable influences in contemporary biomedical research on women's health.

X MOSAICISM AND FEMALE MALADIES

Autoimmune diseases are more prevalent in women than in men.[38] As many as 5 percent of US women have an autoimmune disorder, including life-threatening diseases such as multiple sclerosis, systemic lupus erythematosis, rheumatoid arthritis, and type 1 diabetes mellitus. The current medical model holds that autoimmunity occurs when the immune system misrecognizes the body's own tissues as invaders, leading the system, finely tuned to eliminate foreign agents, to chronically attack the body's tissues. The accumulated damage over time can cause significant disability and even be life threatening. Although palliative therapies are available, autoimmune diseases are typically incurable, and the etiology of autoimmune disease and the origins of its greater prevalence in women remain largely unknown.[39]

Some researchers, seeing a parallel between the self-on-self attacks of autoimmunity and mosaic female tissues made up of cells expressing the maternal or paternal X chromosome, have sought a mechanism for autoimmunity in X mosaicism (see fig. 6.2). The most basic X mosaicism hypothesis of female autoimmunity is that simple mosaicism of the X chromosome, in cases in which the X thereby produces two conflicting immune products, leads to autoimmunity. There is also a more sophisticated version of the X mosaicism hypothesis, which holds that if mosaicism is skewed so that an immunologically relevant organ such as the thymus gland contains a majority of one X, the immune system may

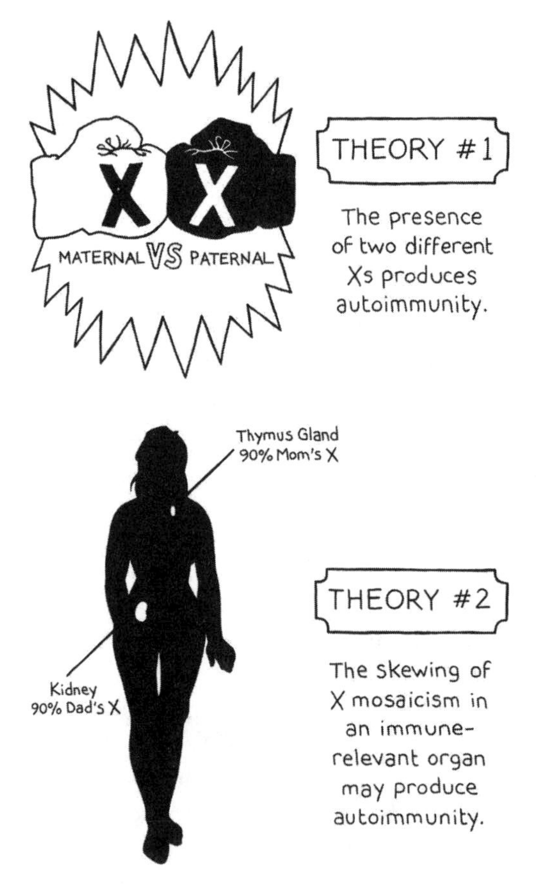

Figure 6.2. The X mosaicism hypothesis of female autoimmunity. Illustration by Kendal Tull-Esterbrook; copyright Sarah S. Richardson.

misrecognize tissues that carry the other X, leading to an autoimmune reaction.[40]

There are hormonal, environmental, and genetic theories of the prevalence of autoimmunity in women.[41] The X mosaicism theory stands as the leading genetic explanation of female autoimmunity.[42] Researchers promote the hypothesis as an exciting front in autoimmunity research. A 2005 editorial in the *Journal of Clinical Endocrinology and Metabolism*, for example, focuses a discussion of new autoimmunity research on the X chromosome: "Women are more complex in that they have two similar X chromosomes. . . . These chromosomes appear to battle it out within the mammalian cell to see which will predominate and actually function." "The phenomenon of X chromosome inactivation is attractive because of the unexplained female predominance of autoimmune disease in females."[43] Similarly, a 2000 *Nature Genetics* editorial related with great excitement new research on molecular genetic causes of autoimmunity in women discussed at a "Sex and Gene Expression" conference: "As a consequence of X inactivation, females are composed of two epigenetically different cell populations. [Scientists] speculated whether this could be a reason for the higher occurrence of autoimmune disorders in females." The editorial claims that geneticists have made "dramatic" findings related to the connection between the "mix of two different cell populations" and the higher incidence of autoimmune disorders in women. The article situates X mosaicism studies as part of a new wave of research that will provide "important clues to the molecular basis of autoimmunity."[44]

The notion that X mosaicism underlies female autoimmunity also regularly appears as authoritative medical knowledge in the health media. An Everydayhealth.com primer on "Women and Autoimmune Disorders" advises that "women, who have two X chromosomes in contrast to men's X and Y chromosome, are genetically predisposed to developing certain autoimmune diseases."[45] A "science mini-lesson" on X mosaicism produced by the Institute for Women's Health Research at Northwestern University portrays X mosaicism as a scientifically validated cause of female autoimmunity:

> One heart cell may inactivate the X chromosome from the mother, while another right beside it inactivates the X chromosome inherited from the father. This makes all genetic females a type of genetic mosaic. . . . Getting to the point, think about immune cells: they're these destroyer cells whose sole mission is to hunt down and kill any cells that don't match their DNA code. If an immune cell that has

> inactivated the maternal X chromosome meets a nerve cell that has inactivated the paternal X chromosome, that immune cell could be triggered to destroy the "invader." That, in the end, is how having two X chromosomes can lead to increased risk of autoimmune disorders in genetic females.[46]

Similarly, a 2004 Australian television feature titled "Hybrid Auto-Immune Women" reports that "women are a genetic mosaic, a patchwork of their mother and father" and explains that immune cells with the father's X chromosome "will attack some of those cells with the mother's X chromosome—which of course immediately means an autoimmune disease of some sort. This might be why women can have up to 50 times more autoimmune diseases than men."[47]

These theories, both in the scientific literature and in their more popularized forms, draw on gender-inscribed notions of the female body to link female autoimmunity to X mosaicism. As feminist science studies scholars have long noted, the concept of autoimmunity is already discursively gendered. Donna Haraway, Emily Martin, and Lisa Weasel have explored the relationship between immunity discourse and gendered metaphors and imagery, unpacking the parallels between "horror autotoxis" (the medical researcher Paul Ehrlich's 1957 term for autoimmunity) and traditional conceptions of femininity.[48] Haraway argues that immunological concepts and metaphors serve as "principal systems of symbolic and material 'difference'" in modern Western culture. She notes in particular the conception of immunity as a defense of the self against the other and the "awful significance" of autoimmunity that derives from the notion of a self divided against itself.[49] Building on Haraway's work, Martin has documented how gender inflects the concept of selfhood in biomedical writing on immunity. She found that popular and scientific texts employ images of male bodies that emphasize purity, individuality, selfhood, and defense to represent the immune system. Writes Martin, "this is a body whose boundaries are defined extremely clearly. Inside is only self; outside is only nonself. Should any foreign matter enter, it will be swiftly dispatched by the roving armies of the man's immune system."[50] As Martin notes, the greater susceptibility of females to autoimmune disease, leading to suggestions that females are biologically "hybrid" and "mixed-up," aligns with ideological notions of females as double, divided against themselves, contradictory, unstable, and lacking in unitary selfhood. She writes that "the greater sensitivity of women's immune systems becomes negatively tinged: Women's immune systems are so sensitive that they are apt to mix up self and nonself, painfully attacking self as if it were nonself."[51] Weasel also suggests that the construct of autoimmunity as a "threat to individuality" arises from and supports a set of pejorative conceptions of femininity. Weasel speculates that "the assumption that autoimmune disease arises from a poorly developed sense of self [may] reflect stereotypical assumptions about women's psychological development."[52] The biological mechanism of X mosaicism seems to confirm, via a shared field of metaphors and associations, these preexisting intuitions that female autoimmunity is rooted in women's essentially chimeric nature.

The principal evidence for the X mosaicism hypothesis of female autoimmunity comes from studies of skewed X mosaicism in women with autoimmune disorders—something it has only recently become possible to assess. In these studies, researchers look at the percentage of cells carrying the maternal or paternal X chromosome in tissues of women with autoimmune disorders compared to normal women. When one X predominates above a threshold of either 80 or 90 percent, the woman is deemed to have skewed X mosaicism.[53]

Scientists have invested substantial resources in empirically testing the theory that skewed X mosaicism is associated with autoimmunity in women. From these efforts, a somewhat higher rate of skewed X mosaicism among women with autoimmunity has been demonstrated in just two cases: scleroderma and autoimmune thyroid disorders.[54] It has not, thus far, been found in the cases of lupus, multiple sclerosis, type 1 diabetes, or juvenile rheumatoid arthritis; nor has it been found in the female-predominant and potentially autoimmune disorders of simple goiter and recurrent pregnancy loss.[55] There is conflicting, weak, or otherwise ambiguous evidence of an association with skewed X mosaicism in the case of primary biliary cirrhosis and adult onset rheumatoid arthritis.[56] Thus, there is not, as yet, strong evidence that skewed X mosaicism is implicated in the overall female predominance in autoimmunity.

Even if studies were to document high rates of X skewing in women with certain autoimmune disorders, this would not, in any case, constitute sufficient evidence that skewed X mosaicism predisposes women to these disorders or that women are more inclined, in general, to autoimmunity than men. Here are some of the problems:

- *Limited tissue sampling*: Almost all X mosaicism studies use blood samples, looking at peripheral lymphocytes rather than cell types within the immune reaction pathways or organ systems of interest. This limits their significance. For example, women with the skin disease scleroderma show skewed mosaicism in their peripheral lymphocytes, but the skewing the researchers observed in blood cells was not also found in the skin cells—the tissue of interest for the disorder in question.

- *Lack of controls for confounding factors such as age*: Rates of both autoimmunity and X skewing increase with age in women. To date, studies of X mosaicism pattern variation do not persuasively disambiguate aging and autoimmunity. Amos-Landgraf et al., in the most credible study of its kind, looked at patterns in 1,005 unaffected females, finding that skewing was relatively common among all women. The study reported that fully 25 percent of females had patterns skewed at least to 70:30 and concluded that "with advancing age, there is greater variation in X inactivation-ratio distribution."[57]
- *Lack of a mechanism*: While a few of these studies suggest that women with some autoimmune disorders more frequently have skewed X mosaicism, they cannot show that skewed X mosaicism causes autoimmunity. Degree of X skewing has not been found to be a predictive biomarker of autoimmunity nor of response to therapy, and it has not been experimentally linked to autoimmunity in animal models or in humans.

More fundamentally, the X mosaicism hypothesis does not explain enough specific features of female predominance in autoimmunity to stand, even hypothetically, as a leading candidate for a biological mechanism explaining the greater prevalence of female autoimmunity. Female X mosaicism cannot explain the following: why the incidence of autoimmunity, but not the severity of the disease, differs between males and females; why female predominance is much more pronounced among the cohort diagnosed with autoimmune disorders under age forty, with rates becoming more equal between the sexes as they age; why some autoimmune disorders are female-predominant, some are male-predominant, and others are sex-neutral; how X mosaicism interacts with the significant and well-documented role of environmental factors involved in sex differences in autoimmunity (such as chemicals in cosmetics or in the workplace); and finally, why there is wide variability in sex ratios of autoimmune diseases between different ethnicities, nations, and in developed versus less-developed regions of the world.[58]

Some of these problems are acknowledged in the scientific literature.[59] However, they seem to not at all affect researchers' convictions that X mosaicism mediates female autoimmunity. The X mosaicism theory of female autoimmunity regularly appears as authoritative medical knowledge in a variety of research, clinical, and health media contexts. Skewed X mosaicism remains the leading genetic theory of higher rates of autoimmunity in females. "Autoimmune diseases revolve around the sex chromosomes," writes Carlo Selmi. Zoltan Spolarics claims that "X-chromosome mosaicism represents an adaptive cellular system," bestowing females with "potentially two distinct regulatory and response arsenals" and predisposing them to autoimmunity.[60] But such assertions by biomedical researchers that the XX chromosome complement inclines women to autoimmunity are clearly unwarranted. Studies of associations between X mosaicism patterns and autoimmunity do not substantiate a causal link (or even an unambiguous association) between the two phenomena. It appears that X mosaicism is far from a general theory of, or a major factor in, higher rates of autoimmune disorders in females.

When a theory is widely held to be true despite a continued lack of empirical evidence, social and intellectual context may be relevant. In this case, a conviction that the X is the "female chromosome," combined with the X mosaicism theory's intuitive alignment with conventional cultural understandings of femaleness, helps to fill the gap between the droplets of evidence in favor of the X mosaicism model of female autoimmunity and the confident assertions of its validity. Sourcing notions of the X as female, and chimerism as feminine, X mosaicism theories of female autoimmunity present a poignant case of gendered conceptions of biological objects of analysis influencing scientific reasoning.

AN "ESSENTIAL" FEMALE TRAIT?

With the rise of genomics, the theory of X mosaicism is finding new interest as a key to women's health and biology. Barbara Migeon's recent book, *Females Are Mosaics*, is exemplary. Migeon is a leading American medical geneticist who played a critical role in establishing the field of medical genetics during the 1970s and 1980s at Johns Hopkins University. A self-identified feminist and a graduate of the progressive women's school Smith College, Migeon is active in the Society for Women's Health Research and a contributor to its journal, *Gender Medicine*. Migeon's book, along with her published articles and interviews, advances a series of striking claims. Migeon positions X-chromosomal genes and the phenomenon of X mosaicism as the basis of female biology and behavior. She argues that "the physical and intellectual attributes of females are determined to a great extent by their X-linked genes"[61]; X mosaicism "is likely to contribute to some of the gender differences in behavior"[62]; X mosaicism is a profound mediator of the course of disease in women[63]; and X mosaicism and X inactivation "is an essential part of female sex differentiation."[64]

At the heart of Migeon's theory of X mosaicism as a fundamental mechanism of sex differences and a hallmark of female biology and behavior is what I will call the cellular diversity hypothesis. Due to X mosaicism, Migeon writes, "a female has a greater variety of gene products in all her tissues than a male does."[65] Migeon estimates that between 60 and 200 X-chromosomal genes may be heterozygous in XX individuals, which she argues gives females "a little extra determinant . . . not possible in XY males."[66] According to Migeon, this cellular diversity underlies many stereotypic sex differences. "This extensive heterozygosity," she theorizes, "contributes additional variability and individuality to human females."[67]

Because of "cellular diversity," Migeon claims, females have greater "cognitive diversity," leading to behavioral differences between the sexes. Migeon proposes that "cellular mosaicism . . . is likely to contribute to some of the gender differences in behavior," including females' response to humor and differences in aggression, emotionality, and educational performance between males and females.[68] She writes that the X could explain why "from the first days of school, girls outperform boys, are more attentive, and are more persistent at tasks."[69] Molecular research on X chromosome mosaicism, Migeon argues, offers a promising platform for uncovering sex differences in the brain that studies of brain anatomy have not, thus far, revealed: "Despite dramatically different behavior between the sexes, surprisingly few anatomical differences have been identified," she writes. "[Perhaps] mosaicism for X-linked genes . . . may contribute to some of these sex differences in behavior."[70]

All of the claims described above are, at this point, purely conjectural. There are no studies demonstrating that cellular mosaicism generates cognitive and behavioral sex differences. Rather, Migeon builds her case for the potential role of cellular X mosaicism in brain and behavioral sex differences by referencing scientific research on the role of X mosaicism more broadly in female development, disease, and physiology. In particular, Migeon argues that X mosaicism explains variations in severity of disease among women and the predominance of certain diseases in women as compared with men.

For example, Migeon points to the case of Rett syndrome, an X-linked dominant disease causing severe mental and physical retardation.[71] Usually lethal in males, Rett is found in females carrying one functional and one disease allele. Rett has a wide variety of phenotypic expression. For instance, a hallmark of the disease is inability to speak, but some individuals are high functioning and able to speak some basic words. Researchers once believed that the severity of Rett in females depends largely on the proportion of cells with the mutated X. Takagi asserted in 2001, for instance, that "the XCI [X chromosome inactivation] pattern is the major determinant of the phenotype in females."[72] Rett was once regarded as an unassailable textbook case of female-specific disease mediated by X mosaicism pattern and regularly cited as part of the theoretical justification for X mosaicism studies of female-specific diseases. The hypothesis of X mosaicism as the cause of Rett in females and as a mediator of the disease's severity, however, is today in retreat after the arrival of genetic techniques for testing the hypothesis. Studies now show that skewed X mosaicism does not correlate with degree of severity in Rett. Instead, Rett is caused by mutations in the *MECP2* gene on the X chromosome. More than one hundred different Rett-causing mutations in the *MECP2* gene have been identified, and the evidence shows that the kind of mutation, not X mosaicism pattern, predicts disease severity. The retreat of the X mosaicism hypothesis of Rett syndrome has helped researchers recognize that there exist some men with Rett and even some men carrying *MECP2* mutations that are unaffected.[73]

Rett syndrome is not the only textbook case for the plausibility of a critical role played by X mosaicism patterns in female health and disease that has been overturned with further genetic analysis. Fabry disease, another commonly referenced case of the influence of X mosaicism on female disease phenotype, tells a similar story to that of Rett syndrome. Fabry is an X-linked metabolic disease seen primarily in men but evident in some women carriers as well. Researchers assumed that the occasional manifestation of Fabry disease in females was due to skewed X mosaicism. But recent studies persuasively show that, as in the case of Rett, X mosaicism in Fabry patients is random and that female Fabry patients with skewed mosaicism do not have more severe disease phenotypes.[74]

Migeon argues that "somatic cellular mosaicism . . . has a profound influence on the phenotype of mammalian females."[75] According to Migeon, X mosaicism "creates biological differences between the sexes that affect every aspect of their lives, not just the sexual ones."[76] Yet, cases such as Rett and Fabry do not provide strong evidence that X mosaicism produces "cellular diversity" with predictable, observable, or profound physiological effects in women, even in the case of X-linked disease. Even if such evidence were available, it would not be sufficient to demonstrate the claim that X mosaicism is a fundamental mediator of female brain, behavior, and physiological development. A genetic mutation that explains the cause of a disease—say, cystic fibrosis or Huntington's chorea—is not identical to a genetic explanation of normal function—here, mucus production or muscle control. Genes are part of complex developmental pathways, and their proximate influence on a particular trait is limited to only the step they control in a particular pathway. Thus, even if there were medically relevant sex-differential effects of the pattern of X-chromosomal cellular diversity, this does not imply, as Migeon argues, that the double-X determines or mediates normal sex differences in physiology and behavior.

Migeon even goes so far as to suggest that X inactivation is a necessary and "essential" female trait. "A single active X is absolutely required for female viability," writes Migeon. "X inactivation is one essential step in the female developmental program"; "X inactivation . . . is an essential part of female sex differentiation."[77] Migeon reasons that because the achievement of X inactivation is critical in early female development, it can be considered a developmental and phenotypic requirement for femaleness. The unique biology of the double-X is here elevated to a position similar to the uterus or the ovaries: an essential female trait requisite to function as a (fertile) biological female.

This line of reasoning, however, is clearly unsound. The result of failure to inactivate the second X in humans is not improper female development. It is nonviability—death. Moreover, a single active X is also absolutely required for *male* viability. Migeon's chain of reasoning equates *sex-limited* traits with sex-determining traits and elevates X mosaicism from a character or attribute of femaleness to an essential or defining quality of femaleness. That X mosaicism is a sex-limited trait exclusive to XXs does not imply that it is a central biological mediator of the normal female phenotype. For example, uterine cancer can occur only in women and testicular defects can occur only in men, but these genetically mediated sex-specific potentialities are neither necessary nor essential to femaleness and maleness. As Migeon points out, X mosaicism is present in XXY males, irrelevant in XO females, and, most notably, absent in 5 to 10 percent of phenotypically normal XX females. X mosaicism is best conceived as an *attribute* limited to double-X individuals, not an essential quality of biological femaleness.

A consequence of the ardent association of X mosaicism with femaleness is a striking neglect of biological context. Oriented as it is toward demonstrating how X mosaicism underlies sex differences, the theory of greater cellular diversity in females elides the fact that the main consequence of random mosaic X inactivation in females is to make them more like males. X inactivation serves to equalize X dosage between males and females, so that the cells of *both* sexes are functionally monosomic for X-linked genes. In cases of skewed mosaicism, women do not become more "female"—instead they develop diseases usually seen only in males, like hemophilia. While X chromosome mosaicism could be interpreted as a mechanism of equalization of X gene product between males and females, and skewed mosaicism as demonstrating that even diseases thought sex exclusive are present in both sexes, instead many women's health researchers choose to characterize the female double-X as a source of sex differences.

Moreover, cell mosaicism is common throughout the genome in both sexes, not limited to the X chromosome. Regions of the genome are silenced or upregulated by an epigenetic process known as genetic imprinting. Differences in genetic imprinting patterns, whether hereditary or triggered by developmental events or environmental inputs, lead to variation in gene expression between individuals and over the course of an individual's life. Genetic imprinting creates the potential for mosaic effects in any chromosome, across the genome, in both males and females.[78] Recently, a study by Gimelbrant et al. found that 5 to 10 percent of autosomal genes are monoallelically expressed with a cell-by-cell random choice of maternal or paternal expression, just like genes on the X in females. "Conservative extrapolation from our data," the authors wrote, "suggests that at least 1000 autosomal human genes are subject to random monoallelic expression."[79] Compared to this genome-wide mosaicism in both males and females, females are mosaic only for those loci on the X for which they are *heterozygous*—60 to 200 genes, at most, according to Migeon. Of these, only a few, if any, are likely to be functionally significant or even detectable, and then generally only in the rare case of skewing.[80] Gimelbrant et al.'s results imply, as the accompanying news article in *Science* concluded, that "each cell population displays a vast heterogeneity in patterns of mono- and biallelic gene expression, providing numerous combinatorial patterns of gene expression."[81] This finding is just the tip of the iceberg. With new research on the environmental and epigenetic tuning of each individual's gene expression, mosaicism is likely to become the rule rather than a novel sex-exclusive trait in human genomics.

In short, the claim that X mosaicism is central to female biology neglects biological context. Situating X mosaicism within our larger understanding of genetic variation makes this even more clear. X mosaicism is not the only mechanism leading to mosaicism of expression in the human genome. We are all mosaics, with enormous variation in expression.

The implications of this depend on biological context. Genetic mosaicism can occur in both males and females as a result of a variety of effects, including maternal and paternal imprinting, inactivation processes on the male Y chromosome and in other areas of the genome, and the dynamics of gene expression over time as we age or experience illness. A better way of conceiving of X mosaicism is as one potential mechanism in epigenetic control and tuning of gene expression. X chromosome mosaicism is one of a host of processes that can produce genetic variability between and within individuals. It is just a highly visible, tractable, prominent example of mosaicism—a clue to a larger genomic process operating in both males and females. Within this biological context, there is no meaningful way in which females can be described as mosaics and not males, or to claim, sans extensive additional documentation, that females are "more" mosaic than are males.

◇

As this chapter has shown, there exists a long-standing association between the X and femaleness that is the accumulated product of contingent historical and material processes. It is also inflected by beliefs rooted in binary conceptions of human male and female natures. While the presence of a single X in males and a double X in females does have different implications for male and female biology, historical and contemporary speculations of the relation between the X and femaleness show that the X has been overburdened with explaining female biology and sex differences. There is woefully little empirical evidence for the theory that X mosaicism is a source of sex differences. Conceptions of the X chromosome as the mediator of the differences between males and females, as the carrier of female-specific traits, or otherwise as a substrate of femaleness, however, remain powerfully persistent. The rich web of connections between X mosaicism and cultural notions of male and female differences helps to fill in the gaps in X mosaicism theories of sex differences, veil their empirical deficiencies, and glue their premises together.

Just as in the case of XYY supermale theories, which conceptualized the Y as offering a "dose" of maleness, the assumption that the X is the "female chromosome" shapes researchers' expectations. X mosaicism theories of female maladies, biology, and behavior are only a few examples. The results of a recent study published in the journal *Hormones and Behavior* surprised researchers. Examining the effect of an extra X chromosome on copulatory behavior in mice, they found that both male XXY and female XXX mice exhibit more masculine behavior. As the authors wrote, "*counter intuitively*, males with two X-chromosomes were faster to ejaculate and display more ejaculations than males with a single X. Moreover, mice of both sexes with two X-chromosomes displayed increased frequencies of mounts and thrusts."[82] They concluded that "a *perplexing* question from the current findings is: why should a female-typical sex chromosome complement increase male sexual behavior?"[83] Similarly, a 2001 *Nature Genetics* article reporting a large collection of genes for male sperm production on the human X chromosome was greeted by researchers as "*unexpected*," "*counterintuitive*," and "*intellectually surprising*."[84]

In "The Egg and the Sperm: How Science Has Constructed a Romance Based on Stereotypical Male-Female Roles," medical anthropologist Emily Martin famously examined gendered stereotypes of the male and female gametes in mid-twentieth-century English-language medical textbooks.[85] Attending carefully to the linguistic and representational conventions used in the textbooks, Martin found that the egg and sperm were consistently gendered female and male. Textbooks portrayed the egg and related biological processes in negative, feminine terms: as destructive, awkward, unwieldy, fragile, passive, and dependent. The sperm, in contrast, was productive, streamlined, efficient, aggressive, active, and activating. As Martin showed, the result was a scientifically distorted picture of the essential reproductive process of fertilization.[86]

Evelyn Fox Keller has termed this circular process of gendering objects of biological research—in which stereotypical conceptions of femininity and masculinity are sourced to describe objects or phenomena from which theories of sex differences are then derived—the "synecdochic" error in the sciences of sex:

> A basic form common to many [feminist analyses of science] revolves around the identification of synecdochic (or part for whole) errors of the following sort: (a) the world of human bodies is divided into two kinds, male and female (i.e., by sex); (b) additional (extra-physical) properties are culturally attributed to these bodies (e.g., active/passive, independent/dependent, primary/secondary: read *gender*); and (c) the same properties that have been ascribed to the whole are then attributed to the subcategories of, or processes associated with, these bodies.[87]

The sex chromosomes present a captivating example of this synecdochic error in the sciences of sex. To the pantheon of notoriously gendered objects of scientific knowledge—egg and sperm, testosterone and estrogen—we may now add the X and Y chromosomes.

CHAPTER 7

The Search for the Sex-Determining Gene

Once, history was told as a chronology of great men,[1] clinical drug trials were conducted using only male subjects,[2] and human origins were seen only through the weapons, tools, and hunting activities of cavemen.[3] Over the past thirty years, feminist scholars challenged these assumptions, leading to new research to fill these gaps and redress these biases. Feminist critiques of androcentrism, sexism, and heterosexism are now part of the mainstream research methods of the humanities and the social sciences. They are also, increasingly, influential in the biological and medical sciences.[4]

The discovery of the *SRY* gene in 1990 appeared to confirm a longstanding model of genetic sex determination—that of a single "master gene" on the Y chromosome that directs the development of the male gonads and thereby determines sex. By the late 1990s, however, this model fell as a result of challenges from all sides. Today the *SRY* gene is understood as one among the many essential mammalian sex-determining factors that are involved in the genetic pathways of both testicular and ovarian determination. Mammals require cascades of gene product in proper dosages and at precise times to produce functioning male and female gonads, and researchers recognize a variety of healthy sexual phenotypes and sex determination pathways in humans. In this chapter, I argue that feminist gender-critical perspectives played a role in this remarkable transformation in the genetic model of sex determination.

In the twentieth century, the X and Y were positioned as the genetic pillars of masculinity and femininity. But in recent decades, feminists have challenged these assumptions. This chapter and the next show how gender-critical perspectives have begun to enter into sex chromosome science and are now leaving a pronounced mark on the questions, central debates, and research models of the field. By direct exposure, osmosis, and demographic transformation, gender criticism has begun to enter into the standard practices and discourse of sex chromosome research, leading to changes in the content of the science.

FEMINIST GENDER CRITICISM

"Feminism" is a broad term that refers to a diversity of social movements and intellectual traditions. The focus of my inquiry here is on Western academic feminism in the 1980s, 1990s, and 2000s, a period in which feminist scholars forged critiques of common assumptions in research on gender roles, gender identity, sexuality, and sex differences. These feminist critiques were rooted in an analysis of "gender" as a cultural system of beliefs, assumptions, and practices inscribing norms and expectations for masculinity and femininity. In what follows, I use the terms "gender criticism" and "gender-critical" to refer to the critical intellectual orientation that these feminist scholars developed.

After the feminist epistemologist and philosopher of science Helen Longino, I define gender criticism as any intellectual practice that works to "make gender visible," to "reveal gender," or to "prevent gender from being disappeared" in an area of intellectual inquiry.[5] Gender criticism implies a stronger intellectual practice than merely a sensitivity to the gender *implications* of research. A gender-critical approach understands gender not merely as a system of power relations in the social world that research findings might support or challenge, but also as a power vector in the practice of research itself.

The term "gender criticism" highlights the critical discursive practices involved in "making gender visible." Gender criticism is a *critical orientation* or *practice* that attends specifically to how gender beliefs may influence human knowledge. Whether or not an individual identifies as "feminist," his or her research methods might rightly be identified as "gender-critical." Conversely, an individual may identify as "feminist" but not evidence "gender-critical" approaches in his or her research.

Gender criticism is not a distinct methodology, nor is it a research field unto itself. Rather, it is an intellectual practice that one adds to a larger methodological toolkit. As I conceptualize it, gender criticism is located within the tradition of skeptical empiricism. It may be employed by researchers in any field where gender is relevant, supplementing and working alongside quantitative and qualitative methods and shared practices of evidence, explanation, and interpretation in that research field. The aim of gender criticism is to produce more accurate and more empirically adequate knowledge.[6]

SRY, THE SEX-DETERMINING GENE

In 1959, analysis of human intersex individuals demonstrated that the genetic switch for male sex determination is located on the Y chromosome. It was not until the mid-1980s, however, when technologies for cloning, sequencing, and analyzing the human genome became cheaper, faster, and more ubiquitous, that a serious gene-discovery program was undertaken for the "sex-determining gene." At that time, research groups at the Massachusetts Institute of Technology in the United States, the Medical Research Council and the Imperial Cancer Research Fund in the United Kingdom, and La Trobe University in Australia began competing to analyze the Y chromosome and clone the sex-determining gene.

The sex-determining gene became a high-priority target in the early days of human genetic sequencing for several reasons. First, sex determination appeared to present a model system in which a single "master gene" controlled the development of an entire organ system. As geneticist Edward Southern wrote in 1987, "sex determination, as a model for the developmental process in mammals, is undoubtedly the principal reason for the intense activity of research on the Y chromosome."[7]

The sex-determining gene was also a low-hanging fruit. The Y chromosome is many times smaller than the other twenty-three chromosomes and houses only a few genes; it is a comparatively tractable target for genetic analysis. Through recombinant technology and deletion analysis in the 1970s, researchers had already isolated the sex-determining gene to a small region of the Y chromosome. Rapid sequencing technologies and straightforward micro-level deletion analysis of the Y chromosome of intersexed mice and humans promised to reveal the location of the crucial switch. As leading geneticist Peter Goodfellow of the Imperial Cancer Research Fund in London wrote in 1987, "the stage is set for cloning the mammalian sex-determining gene."[8]

Finally, the male sex-determining gene represented a holy grail of sex difference research. The prevailing theory held that two factors control sex difference: a gene that triggers gonad differentiation and sex hormones that direct the development of the gonads and secondary sex characteristics. Sex hormones having been well characterized by the 1970s, the discovery of the sex-determining switch would complete the account of the biology of human sex differences. Thus, the male sex-determining gene was a prestigious prize, a long-sought theoretical breakthrough that promised to answer persistent questions about male and female sex difference and open a new field of inquiry.

Sykes portrays the search for the male sex-determining gene in the late 1980s as a "hunt" and a "race," a "spectator sport where the prize for winning was the glory of being first."[9] The genetic search for the male sex-determining factor began in earnest in 1986.[10] Researchers used mouse models to probe a sex-determining region of the Y chromosome that had been isolated from intersex patient karyotypes. In 1987, David Page, a researcher at MIT and the Whitehead Institute in Boston, announced that a gene called *ZFY* satisfied the criteria for the sex-determining gene. Page named his proposed sex-determining gene "DP1007"; writes Sykes, "I am sure I am not alone in noticing the initials and a certain masculine resonance in the last three digits."[11]

Within a year, Australian researchers Jennifer Graves and Andrew Sinclair overturned the finding. Sinclair went on to identify the *SRY* (sex-determining region of the Y chromosome) gene for male gonad formation in 1990. In 1991, accompanied by a *Nature* cover with the "star mouse, swinging on a stick and sporting enormous testicles to prove the point,"[12] Goodfellow, Australian researcher Peter Koopman, and Robin Lovell-Badge, head of developmental genetics at the British National Institute for Medical Research, confirmed the sex-determining role of *SRY* by showing that a transgenic XX mouse would develop as a male if *SRY* is appended to one of its X chromosomes. Following on the heels of this work, in 1992 Page published the first genetic map of the Y chromosome.[13]

The *SRY* model of sex determination confirmed the anticipated model in which sex determination is controlled by a single "master gene" on the Y chromosome. Media coverage of the discovery of *SRY* added to the hype: "scientists now think they know what makes a male masculine," trumpeted the *New York Times*; "scientists believe that they have at last unraveled the secret of what makes a man," announced the *Guardian*.[14] The scientific community celebrated *SRY* as an example of "the astonishing power of modern molecular techniques to resolve longstanding and difficult questions in genetics with consequences that extend far across biology."[15] Textbooks immediately incorporated the *SRY* gene into accounts of sex determination. In 1992, the International Olympic Committee added a test for the *SRY* gene to its "gender verification" program for female athletes.

AN ANDROCENTRIC "MASTER GENE" MODEL OF SEX DETERMINATION

In the 1980s, the prevailing theory held that humans are bipotential (meaning they could be either sex) until six weeks after conception, at which time two biological switches initiate sexual dimorphism.[16] First, a gene on the Y chromosome triggers the development of the testes. Second, the testes begin producing two hormones, MIS (Müllerian Inhibiting Substance) and testosterone, which "masculinize" the fetus and initiate hormonal control of sexual development. The theory was first articulated in 1953 by Alfred Jost, whose 1950s research showed that errors in the development of a genetic male, either at the hormonal or the genetic level, caused mice to "revert" to a female developmental pathway. On this evidence, Jost hypothesized that the development of female gonads and secondary sexual characteristics is the body's "default" plan. In the absence of the two switches, a fetus will develop ovaries and become a phenotypic female. In 1959, cytogenetic studies of intersex patients by Charles Ford corroborated and extended Jost's view of sexual development.[17] Ford's research established that no matter the number of X's, the presence of a single Y causes male gonads to develop, confirming that the sex-determining switch is located on the Y chromosome (see chapter 4).[18] From Jost and Ford, then, the field of sex determination genetics inherited an androcentric framework for sex determination research: a gene on the Y chromosome initiates testes formation; testes formation is the crucial sex-determining event; and female sexual development proceeds as a "default" in the absence of this gene (see fig. 7.1). This theory led researchers in the 1980s and early 1990s to focus on isolating the "male-determining gene" on the Y chromosome and to see the question of *sex determination* as identical to the question of the genetics of *male testes determination.*

"Master gene" theories in developmental genetics were the second principal source for 1980s models of sex determination. The search for the sex-determining gene in the 1980s was not, as one might suppose, motivated by potential medical or "gender verification" applications. Rather, what drove much of the early interest in the sex-determining gene were its prospects for validating an emerging approach to general questions in developmental genetics. As Peter Goodfellow wrote in the

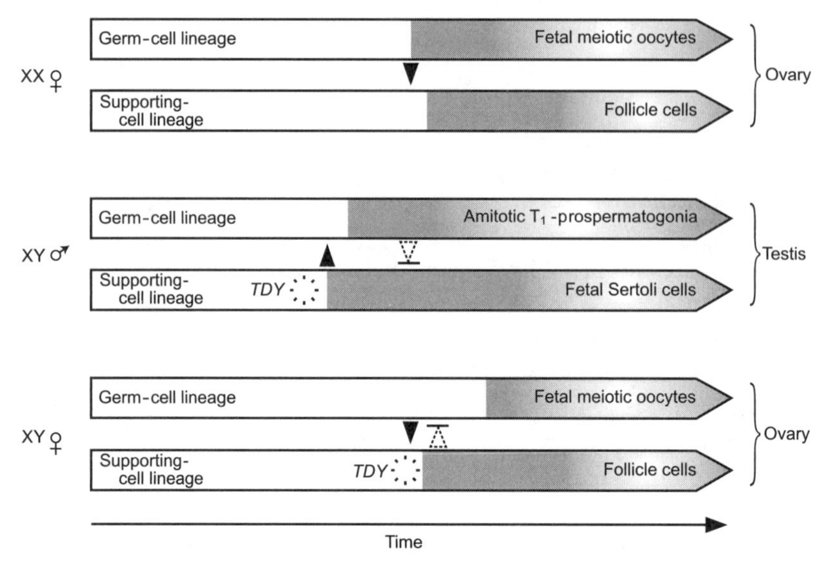

Figure 7.1. 1980s model of genetic sex determination. A single testes determining factor activated at the right time leads to male gonad formation. In its absence or failure, a female gonad is the default. Redrawn from Paul S. Burgoyne, "Thumbs Down for Zinc Finger?" *Nature* 342, no. 6252 (1989): 860–62, with permission from Nature Publishing Group.

introduction to a 1987 symposium volume on sex determination genetics, researchers' interest in *SRY* at this time was primarily "as a model for genetic control of development in mammals."[19]

Sex determination research in the 1980s found kinship with a school of developmental biology that modeled developmental processes as genetic hierarchies controlled by "master switches" in the genome. The then-prevailing paradigm for genetic control of development, as Goodfellow wrote, "assumes a hierarchy of regulatory genes," "an archetypical regulatory network." "In the simplest case, a master control gene directly regulates secondary genes which, in turn, regulate the expression of other genes."[20] Once the "master gene" that triggers this hierarchy is discovered, the identification of other genes involved in the hierarchy should be relatively simple. Under the assumption that testes determination must be an important and fundamental developmental process, such that all mammals would share a single, highly conserved genetic pathway, researchers saw the *SRY* as a perfect "archetype" of this hierarchical system. The aim of their research program was to clone *SRY* to build a simple model system for the elaboration of general theories in developmental genetics.

Master gene theories of genetic development, then, provided potent expectations that formed the backdrop for research in sex determination genetics. Researchers showed strong disciplinary allegiance to these theories and were invested in the *SRY* model as proof in principle for the value of their emergent field and its research program.

In what follows, I argue that feminist analysis of gender contributed to the downfall of this model and to the development of a significantly revised genetic theory of sex determination in the 1990s. Feminist biologists and science analysts anticipated the revised model earlier than others. Feminist theories of sex and gender lent intellectual resources to the model-reconstruction effort. And feminist gender criticism sharpened the tools of the field of sex determination genetics. It improved the level of critical discourse about the assumptions, language, and interpretive models of the field, and provided an analytical framework for articulating and making visible previously unattended gaps in knowledge.[21]

Developments both internal and external to the field facilitated the acceptance of gender criticism in the standard critical practices of sex determination research, a process I call "normalization of gender criticism." The term "mainstreaming" is often used to describe the integration of feminist ideas and methods into the dominant practices of a field, but in this instance the term "normalization" is more appropriate.[22] The field of sex determination genetics experienced a quiet and mostly unacknowledged shift in its epistemic practices. The term "normalization" highlights this feature of the entry of gender criticism into the field.

There were three stages in the progressive incorporation of gender criticism into sex determination genetics. First, cultural change in and around the field of sex determination genetics created the conditions for scientists' receptivity to gender criticism. This includes early feminist criticism from outside the field. Second, a respected female scientist working in sex determination genetics, Jennifer Graves, began to employ an explicitly feminist framework in her work. Graves introduced feminist criticism to the field and developed a formidable gender-critical alternative model of sex determination genetics. Third, over time, members of the larger sex determination research community came to see gender criticism as useful to their own thinking, incorporating feminist insights, even if they often did not explicitly articulate them as such. In this way, gender criticism became a part of the mainstream critical practices of the field.

A CHANGING PUBLIC DISCOURSE ABOUT GENDER AND SCIENCE

The 1980s initiated a period of intense public debate about gender and science. The National Institutes of Health's Office of Research on Women's Health opened its doors in 1990, raising the profile of women's and gender issues in American science. This period witnessed significant expansion of feminist science studies pedagogy and scholarship in the academy. Women also entered the profession of research biology, particularly genetics and developmental biology, in numbers that for the first time approached parity.

In large part as a result of these developments, the 1990s saw increasing challenges to dominant biological models of sex and gender. Biological claims about intersexuality and homosexuality came under particular scrutiny. Feminist science analysts retheorized these phenomena as part of the normal spectrum of human sex and gender. Over the course of the decade, this work found its way into sex determination genetics through a variety of channels. A mid-1990s controversy over sex chromosome testing for gender verification of female Olympians showed that the test could not definitively resolve cases of ambiguous gender. This ultimately led to the termination of the practice, providing sex chromosome researchers, whose expertise was sought, with a public crash course in the social impact of scientific definitions of sex and gender.[23] The intersex movement became increasingly visible during the 1990s, emblemized by the founding of the Intersex Society of North America in 1993. Sex chromosome researchers who studied and provided care for intersex and gender dysphoric patients were exposed to gender-critical perspectives and were drawn into political and medical advocacy in this community.[24] The "gay nineties" also changed the discursive context and conditions of sex and gender research in biology. The 1990s, for instance, saw the burgeoning of a genre of diversity-affirming science writing that highlighted the rich variety of sexual life in the natural world, in part as a response to dominant assumptions that homosexuality is "unnatural."[25]

Anne Fausto-Sterling's "Life in the XY Corral" offered the most detailed feminist critique of sex determination genetics in the 1980s. In this paper, Fausto-Sterling, a biologist, feminist science critic, and intersex patient advocate, analyzed gender beliefs in theories of sex determination. She argued that researchers had ignored explanatory gaps in their theories and failed to consider viable alternative models for sex determination. Her charge was threefold. First, by equating the genetics of testes determination with the genetics of sex determination, researchers had neglected to pursue parallel investigation into the genetics of ovarian development. Second, researchers had privileged male processes over female processes by accepting a highly resonant metaphor of "male as presence and female as absence." Male processes of sexual development were deemed a more interesting, complex, and dynamic object of investigation than female processes. Third, researchers had assumed that sex organizes into a "clear-cut" binary such that it can be unambiguously determined by genetic assay. Fausto-Sterling contrasted these conceptions of sex with feminist and social science concepts of sex and gender. An uncritical commitment to a binary concept of sex, she argued, "led researchers to ignore data which are better accounted for in approaches which accept the existence of intermediate states of sexuality."[26]

Fausto-Sterling concluded that these assumptions about gender had "prevented the articulation of a coherent theory" of sex determination. She urged an alternative model that includes both male and female developmental pathways and "permits the existence of intermediate states." Fausto-Sterling cited a neglected model of sex determination proposed by Eva Eicher and Linda Washburn that included ovarian development and posited that many genes must interact along complex and overlapping pathways to create male and female gonads. Fausto-Sterling's alternative model and that of Eicher and Washburn represent an early gender-critical model of sex determination.[27] Not a sex determination geneticist, Fausto-Sterling's critique registered little response from specialists, though it was widely cited among gender scholars. Judith Butler's celebrated 1990 work of feminist theory, *Gender Trouble*, featured a lengthy discussion of Fausto-Sterling's critique of Page's "master gene" model of the active genetic determination of maleness and the passive or default determination of femaleness.[28]

DIFFICULTIES WITH THE *SRY* MODEL OF SEX DETERMINATION

During the 1990s, the *SRY* model of sex determination encountered serious conceptual and empirical challenges. Jennifer Graves and Roger Short anticipated these challenges in a strong critique issued immediately after the announcement of the identification of *SRY*. "Will all mysteries of sex determination now be revealed? We think not," they predicted.[29]

Graves and Short raised several challenges to the *SRY* model of sex determination. First, *SRY* was insufficient to produce a fully sex-reversed, fertile transgenic mouse. Also, an X-linked gene was known to override the effect of *SRY* on testes determination. These and other empirical anomalies in the *SRY* paradigm affirmed that many more genes than just

SRY, perhaps even genes in distinct pathways, must interact to successfully decide sexual fate. In their critique, Graves and Short suggested that preference for a Y-chromosomal sex-determining mechanism had caused the role of the X chromosome in sex determination to be neglected, and they hypothesized that an X-dosage mechanism may interact with the *SRY* pathway to determine sex.

Second, there was no evidence of a target gene activated by *SRY* in the early stages of testes formation, suggesting a more circumscribed role for *SRY* in sex determination than the "master gene" and "gene hierarchy" theories presumed. *SRY* need not, as was widely assumed, be a direct, active inducer of testes formation. A more complex and interactive model of sex determination would better account for the lack of a gene target for *SRY*. Graves and Short held that, contrary to expectation, the evidence implicated *SRY* in a double-inhibition pathway. Rather than functioning as an activating switch, *SRY* stops other genes that would otherwise inhibit still other genes that cause testes development.

Third, Graves and Short challenged the developmental biologists' expectation that sex determination should be well conserved, universal, and nonredundant such that it could be explained by a single gene in mouse or man. They admonished sex determination geneticists to appreciate the diversity of sex determination processes, even among mammals.

In the early 1990s, scientists struggled to interpret research findings that were inconsistent with the *SRY* model of sex determination. The 1992 Boden Conference on Sex Chromosomes and Sex-Determining Genes, chaired by Graves, offers a window onto a field in transition as these questions came to a head. In the introduction to the conference volume and transcripts, Graves and coauthor Reed write:

> We are gradually getting *an uneasy feeling* that [the portrait of sexual determination given by Jost] is flawed. The history of studies of sexual differentiation exemplifies the truism to 'seek simplicity, then distrust it.' . . . *[W]e were not prepared* for the ambiguities and difficulties that would follow in trying to interpret the role of *SRY* in aberrant phenotypes and to ascribe downstream function to its gene product.[30]

Research on *SRY* confounded researchers' expectations about the biological phenomenon of sex dimorphism in several ways. One was the role of *SRY* in the direct induction of the testes. The conference transcript reveals researchers encountering a lack of fit between the *SRY* model of sex determination and the data, throwing their model-theoretical assumptions, and their description and interpretation of data, into turmoil:

> *Chairman:* But do the transgenic mice tell us that SRY is the only gene involved in testis determination?
>
> *Goodfellow:* The hoary old question of whether *SRY* can be the sex-determining gene because we know there must be other genes in the cascade, so it can't be the only gene! . . . I find it very compelling that all of the genetic information that you need to make a male is present in that 14kb [of the Y chromosome].
>
> *Monk:* It *sometimes* makes a male.
>
> *Burgoyne:* It only sometimes makes a male, even when it's expressed!
>
> *Goodfellow:* I give up![31]

Researchers also expected that the sex-determining gene would be well conserved in mammals, such that the sex determination process in mice could then be easily generalized to humans and other species. This expectation (a common assumption when working with closely related model organisms in molecular biology) proved unsustainable in this case.

> *Chairman:* One of the big surprises is how poorly conserved *SRY* is between humans and marsupials.
>
> *Foster:* Yes. We expected *SRY* to be well conserved. . . . We were expecting then—and right up until now—that [*SRY*], being a much more important gene and having a lot more selective pressure on it than any of the average house-keeping genes, would pop straight out and we'd find it on the marsupial Y chromosome.[32]

These and other inconsistencies between expectation and observation reveal a growing frustration in 1992 with the received model of sex determination. The conference's contributors, however, were unprepared at this early stage to formulate an alternative model of sex determination or to examine the broader assumptions that structured research in the field.

During the mid-1990s, researchers accumulated more anomalies in the *SRY* model and identified several other important genes in the sex determination pathway. In an early contribution, Ken McElreavey, Eric Vilain, and coworkers at the Pasteur Institute in France reviewed more than a hundred cases of human intersex subjects for whom *SRY* did not offer a sufficient explanation of their phenotypes. They hypothesized from these cases that there must be another major factor in sex determination, an "anti-testes" factor, which *SRY* acts to suppress. Opposing a "genetic hierarchy" concept of sex determination, they proposed a "regulatory gene cascade" hypothesis, in which many factors participate in pushing the balance of sex determination in favor of male or female, explaining the observed spectrum of intersex phenotypes.[33]

While articulating a nonbinary vision of the biology of sex and gender, McElreavey and Vilain's hypothesis also picked up on broader conceptual shifts in biology in the 1990s. Simple notions of genetic determinism and gene action increasingly fell short of providing adequate explanations of molecular-level phenomena. By the late 1990s, biologists began to move away from metaphors of "master genes" and "genetic programs" and toward nondeterministic, complex regulatory network approaches to biological explanation.[34]

In another significant mid-1990s finding, researchers identified two species of voles that lacked *SRY* but still reliably produced a fertile male phenotype. This confirmed that *SRY* was neither necessary nor sufficient to produce a male phenotype in all mammals. Comparative genomic evidence that *SRY* is poorly conserved, or highly variable in its sequence and target, even between organisms considered as genetically close as mice, chimpanzees, and humans, and that *SRY* is a relatively recently evolved gene corroborated this finding. Together, this research suggested that *SRY* may function differently from species to species and that it may also interact with other sex-determining mechanisms in the genome.[35]

The characterization of the genes *DAX1, SOX9, DMRT1,* and *WNT4,* all non-Y chromosomal genes that can override *SRY* to cause sex reversal, further contributed to pressure in the late 1990s for a revised model of sex determination. These and others in the expanding docket of genes known to be involved in sex determination increasingly challenged the "master gene" model of *SRY* gene action. A consensus began to emerge that *SRY* was far more "average" than expected, pointing toward a sex determination model of a "cascade"—or several cascades—of genes working in complex regulatory relation to one another.

JENNIFER GRAVES'S "FEMINIST VIEW" OF SEX DETERMINATION

Jennifer Graves is a leading scientist and a public figure in Australia. A member of the Australian Academy of Science, she has been described as a "National Treasure" and was tapped to direct Australia's high-profile effort to sequence the kangaroo genome. Graves is also a rare woman principal investigator in a male-dominated field, a marsupial researcher in a world of mouse models, and an Australian with comparatively little public funding in a research environment driven by lavishly endowed American and British labs. As a result, for much of her career Graves was somewhat of an outsider in the field of sex determination genetics.[36]

Graves's specialty is the comparative genomics of mammals and marsupials and the genetics of sex chromosomes and sex determination. She is well known for her lab's 1988 work debunking David Page's claim that his candidate sex determination gene was the mammalian sex-determining gene. Her critiques of Y chromosome-centric models of sex determination and her "Y chromosome degeneration theory" (see chapter 8) have also made her a polemical figure and attracted colorful media attention. As a result, as she said in an interview, "I unexpectedly became a ball-breaking feminist Y chromosome knocker."[37] It appears that Graves did not publicly self-identify as a feminist until 1999—the same year she was promoted to the Australian Academy of Science. In papers, talks, and interviews following her promotion, Graves began to place her ideas in a feminist framework. A 2001 profile described her as "concerned that a non-feminist view can [adversely] affect how science is done, particularly in her field that deals with what genes determine sex and sex-related characteristics."[38]

The 2000 paper "Human Y Chromosome, Sex Determination, and Spermatogenesis: A Feminist View" presents the clearest elaboration of Graves's feminist critique of the *SRY* model of sex determination. Graves argued that researchers' unreflective assignment of masculine qualities to *SRY* led them to ignore contradictory evidence and prefer an unsustainable Y-chromosomal model of sex determination over alternative models. Researchers clung to this model even when countervailing evidence should have led them to abandon it. Graves termed this the "Dominant Y" theory. In the paper, she outlined three ways in which this "macho" conception of *SRY* had misled sex determination research. She then proposed an alternative model of the role of *SRY* in sex determination.[39]

First, Graves argued that the Dominant Y model had led researchers to conceive of *SRY* as a transcendent "maleness" gene, a specialized master gene that reflects the ultimate refinement of male sex determination and is ubiquitous in nature. This caused researchers to expect that *SRY* would be well conserved and that it would act uniquely in the first stages of testes formation. Comparative genomic analysis, argued Graves, shows just the opposite. *SRY* is poorly conserved, shows a weak transcription signal, and appears to have different functions in different species. In addition, transgenic experiments demonstrate that the function of *SRY* can be replaced by other genes with a similar structure in the genome (such as *DAX1*). Instead, Graves argued, *SRY* acts as an important switch in sex determination only because of a contingency of molecular evolution and cannot be considered a "specialized" gene for testes determination. *SRY* may very well be a marginal autosomal allele that became integrated into the sex determination pathway by chance when the X and Y chromosome differentiated. Based on this evidence, *SRY* is better conceived, she suggested, as "a degraded relic of a normal gene that just got in the way of another gene."[40]

Second, Graves charged the Dominant Y model with uncritically attributing aggressive and agent-like qualities to the *SRY* gene. For instance, researchers presumed a model of Y chromosome evolution in which *SRY* "specialized" as a male-advantageous, and possibly female-antagonistic, gene—the result of a genetic sex war. A desire to see *SRY* in this light, she argued, led researchers to overlook the extent to which genes on the Y chromosome, including *SRY*, have homologues on the X chromosome, of which they are often "degenerated" versions. The model of the Dominant Y as an agent also led sex determination geneticists to assume that *SRY* acts as an "activator" at the top of a linear hierarchy. Wrote Graves, "this dominant action has traditionally been interpreted to mean that [*SRY*] codes for some kind of activator that turns on transcription of other genes in the male-determining pathway."[41] Graves argued that the attribution of the masculine quality of being "active" to *SRY* prevented researchers from imagining more complex models, or models in which the *SRY* gene serves instead as an inhibitor or "a spoiler that turns off genes."[42] Some models of *SRY* action went even further, attributing to the *SRY* gene the ability to "overrule" genes in the ovary-determination pathway. Once again, as Graves noted, this assumption was later contraindicated by empirical research documenting many examples of genes that can counteract the action of *SRY*, leading to sex reversal of normal XY individuals.

Third, just as Fausto-Sterling had some ten years earlier, Graves identified the Dominant Y model as androcentric, devaluing and neglecting female or female-identified biological processes and thereby leading to explanatory gaps in the theory of sex determination. For example, singular emphasis on the role of the Y chromosome in sex determination caused researchers to overlook or underrate potential contributions from the X chromosome, despite the prominence of X chromosome dosage mechanisms of sex determination in many other species and the discovery of a crucial sex-determining gene on the X chromosome. The genetic pathway of ovarian determination is another neglected female process. As Graves pointed out, no biological argument was offered for the assumption that ovarian development is a "default pathway"—and ovarian development is certainly just as interesting, contingent, and complex as testes development. "There are likely to be just as many genes required for ovarian differentiation and egg development, and so far we know rather little about these genes or how they are switched on in the absence of testis development," she wrote.[43]

A simpler and more explanatorily powerful model, Graves suggested, conceives of *SRY* as a degraded version of a gene on the X chromosome, occupying the role of a genetic switch in sex determination because it happens to be located on the male-exclusive chromosome. Graves emphasized that the genome may contain many genes redundant to *SRY* as well as alternative mechanisms of sex determination, which may involve the X chromosome and may interact with and overlay the *SRY* pathway. For Graves, sex determination is a highly contingent, error-prone, and always-evolving mechanism.

In this paper, Graves reiterated and built on arguments incipient in her work since her earliest critique of the *SRY* model of sex determination in 1990. It represents the first instance, however, of Graves's identification of her critique as a feminist one. The precise nature and source of Graves's feminist identification is not clear. Nonetheless, the label "feminism" enabled Graves to place her multifaceted critique within a systematic critical perspective. This systematic critical approach makes conspicuous the persistent gendering of biological phenomena and the ways in which male processes are valued over female ones in the *SRY* model of sex determination, revealing gender as a factor in both the construction of the model and its widespread appeal despite its inadequacies. A "feminist view," as Graves described it, placed a diverse set of critical insights that had motivated Graves's approach to sex chromosome and sex determination research for at least a decade into an easy-to-grasp organizing framework. Graves's "feminist view," then, is effectively presented as a relevant, well-motivated, and insightful critical perspective from which sex determination researchers might evaluate scientific models, identify potential sources of bias, and generate alternative hypotheses. Among several channels that carried feminist gender-critical sensibilities into sex determination genetics in the late 1990s, Graves's became one of the most forceful, direct, and prominent.

THE NORMALIZATION OF GENDER CRITICISM

Beginning around 2000, a marked shift of tone occurred in the sex determination genetics literature. As the *SRY* master gene model fell out of favor, questions and ideas once at the periphery flooded in from all sides, including gender-critical approaches. The shift was informal and not self-consciously feminist. Rather, a general awareness matured—not evident previously—of the pitfalls of androcentric and gender-dualistic thinking. Researchers took up and absorbed valuable feminist insights, often without realizing that they had done so. This gender-critical consciousness began to be adopted as a matter of course in the intellectual work of the field. I call this the normalization of gender criticism—one model of how feminist and critical perspectives might find reception and take root in a scientific field.

The growth and effects of gender criticism are abundantly evident in the set of research questions that have come to occupy the field, changes to the model of sex determination itself, and the framework used by contemporary sex determination geneticists to explain their research and describe the contribution of their work to biology and to society at large. Whereas "gender-critical" approaches are absent from the sex determination literature of the 1990s, Fausto-Sterling's 1989 critique of sex determination models is echoed by prominent researchers in the mainstream literature of the field today (though Fausto-Sterling is never cited). Researchers acknowledge the lack of research on the biology of female sex determination as a weakness in scientific theories of sexual development and sexual difference. In addition, in their current work, researchers seek to avoid language implying that male biological processes are active and dominant while female processes are passive and the default. When using the ideas of "sex" and "gender," researchers take pains to resist the implication that biological sex maps plainly or directly onto social conceptions of sex and gender. Sex determination literature emphasizes a plurality of sexual phenotypes and multiple pathways to normal sexual development. In a variety of ways, in their scholarship, public commentary, and pedagogy, genetic sex determination researchers signal their awareness of feminist critiques of the *SRY* model of sex determination and their sensitivity to the social consequences of scientific theories of sex and gender difference.

Two sources—transcripts of the 2001 Novartis Foundation symposium, "The Genetics and Biology of Sex Determination," and a set of interviews of prominent sex determination geneticists commissioned by the Annenberg Foundation for an online biology education project in 2004—provide a remarkable record of the normalization of gender criticism in this field.

Three noteworthy themes appear in the 2001 discussions at the Novartis conference: a new, broad consensus on the importance of research on ovarian determination in any sound model of sex determination; the replacement of the "master gene" conception of *SRY* by a multifactorial model of sex determination; and a call for a human-specific model of sex determination, acknowledging the distinctiveness and complexity of sex-gender systems from species to species and the special sensitivity required for research on the biology of human sex and gender.[44]

Whereas the research gap on ovarian determination is mentioned in scattered literature in the 1990s, in the 2001 conference it was repeatedly and urgently raised in papers and discussions. For example, Lovell-Badge and coauthors wrote:

> Considerable progress has been made over the last 11 years, such that it is now possible at least to formulate models of how sex determination may work in mammals. . . . However, we are no doubt still missing many relevant genes, in particular for the female pathway, both those that can be considered antitestis genes and those that are actively required for the specification of the cell types characteristic of the ovary.[45]

In a closing discussion about future priorities of the field, Koopman named ovarian development as a pressing problem for the field, acknowledging the gap in knowledge produced by the prior exclusive emphasis on the testes:

> In the coming decade, we are likely to see further progress in understanding one of the great black boxes in developmental biology, namely the molecular genetics and cell biology of ovarian development. Efforts to illuminate ovarian development have been overshadowed to some extent by progress in studying testis determination and differentiation.[46]

Male gonad formation was once the primary explanandum and "holy grail" of sex determination research. By 2001, a definitive shift had occurred. The goal of the research program was reconceived as the identification of the multitude of factors involved in gonad differentiation from a bipotential state. In the transcribed conference discussion, for example, Eric Vilain, now a clinical geneticist at UCLA, prompts researchers to keep in mind that "pro-male" factors are only one research target. "Proovary" and "anti-testes" factors (which, importantly, may be distinct) await characterization. Without these elements, Vilain argues, accounts of the genetics of sex determination remain incomplete. This perspective, reiterated throughout the conference proceedings by Koopman, Graves, Lovell-Badge, and Francis Poulat (of the Institute of Human Genetics, France), among others, reveals a widely shared conceptual transformation of the research problem of sex determination.[47]

Consistent with this, the Novartis transcripts also evidence a significantly revised estimate of the importance of the *SRY* gene in sex determination. The field's earlier attachment to an all-powerful master "maleness" gene now appears as an unmistakable blind spot in previous thinking. One (anonymous) discussant points out that the problem with the old model now appears obvious in light of empirical counter-evidence and basic principles of evolutionary theory, and wonders aloud why Graves's intervention was necessary to make researchers aware of the oversight:

> For model systems where there are genetic tests, we often isolate and identify particular genes, and assign them certain roles. We then tend to think, 'Ah, this gene must perform this function in a large number of organisms.' . . . We are terribly surprised when we get results such as Jenny Graves' demonstration that *Sry* is not the be-all and end-all of sex determination, when in fact this is probably a common theme in evolution.[48]

Goodfellow, who in 1992 claimed that the *SRY* gene contains "all of the genetic information that you need to make a male," in 2001 stated that it is likely that *SRY* must interact with another gene, and that this interaction itself requires the assistance of cofactors:

> I guess what I am saying is that we have ignored the cofactor molecules . . . for too long. This is why I was emphasizing the possibility that we may be looking at soaking up a cofactor that is needed for the expression of another gene.[49]

In 2001, the researchers assigned the *SRY* gene a far more modest role in sex determination. Poulat characterized *SRY* as a "box"—an interchangeable regulatory element: "We say that *SRY* is only a box. We can exchange this box with other boxes. . . . Basically we have a truncated SOX9 protein, which is also more-or-less only a box: nevertheless, in this case we have sex reversal." Similarly, Lovell-Badge et al. described *SRY* as "acting solely as an architectural factor." Reflecting both the shift to a nonbinary, multifactorial model of sex determination that includes both male and female gonad development, and the trend toward complex regulatory network models of gene action, the language of "master genes" is absent.[50]

Finally, the 2001 conference discussants were newly and keenly aware of the specificities of human sex determination genetics. Early enthusiasts championed *SRY* as a tool for Olympic gender verification and the determinant of "what makes a man a man." In 2001, researchers were far more cautious. For example, Vilain reminded colleagues that "a majority of patients with abnormal gonad development remain unexplained genetically."[51] The failure of the *SRY* model to fully explain human sex determination, researchers acknowledge, arose in part from a too-simple binary conception of sex difference and in part from inconsistencies between mouse models and humans. At first, glossing over discrepancies, researchers held to a theory of sex determination as a fundamental and therefore well-conserved mammalian developmental pathway, validating the generalizability of the mouse model system for sex determination research. Researchers in 2001, confronting the breakdown of this model, were more attuned both to distinctions between mouse and human systems of sex and gender determination and to the particular dangers of importing folk conceptions of sex difference as a simple binary into biological theories of sex determination. In 2001, we see human sex determination genetics developing into a distinct field of expertise, and specialists urging colleagues to be mindful of the specificities of the human sex phenotype. Vilain, for example, calls for a model of human sex determination that accommodates an "understanding [of] the tremendous phenotypic variability. . . . We often underestimate all manner of influences, from environment to genetic background." Short adds, "we mustn't be sucked into thinking that [human] sex determination begins and ends with the gonads."[52]

The transcript demonstrates that this new gender criticality is directly linked to increased awareness on the part of researchers of social and political issues raised by the intersex community. Responding to patient advocates, the researchers work to challenge their own assumptions about "normal" sex phenotypes and the naturalness and necessity of a male-female sex binary. They appreciate the need for care and precision in research design and language use in sex determination research. For example, in a transcribed discussion about recent research in human sex determination genetics, Goodfellow says:

> The dialogue that occurs between the medical profession and patient groups is something that the medical profession has to listen to. Not just with respect to this very difficult area, but generally.

> Treatment can reflect the social prejudices of the treaters. When a particular treatment is chosen because of the prejudices of the people who are performing that treatment, there has to be a social dialogue. The responsibility for the treatment of patients in the UK has changed in my lifetime. . . . Clearly, there is no easy solution to this problem, because unless social attitudes change dramatically we are dealing with individuals who fall outside societal norms. . . . [W]e would be wrong not to engage in dialogue with those to be treated.[53]

Goodfellow's alarm about the potential for "prejudices" to influence scientific practice, his sense of responsibility to the intersex community, his awareness of the power and contingency of social norms about gender, and the easy interjection of these issues into a theoretical discussion of sex determination models demonstrates the cross-talk that was occurring between the conception of gender as a spectrum advanced by the intersex community and the cognitive work of the field of sex determination research.

Interviews conducted by the Annenberg Foundation in 2004 with leading sex determination geneticists Holly Ingraham, David Page, and Eric Vilain offer a second source documenting the normalization of gender-critical approaches in the field of sex determination research. The interviews echo and elaborate the themes of the 2001 Novartis conference, while also presenting a more fine-grained picture of the integration of gender criticism into the models and epistemic practices of the field. These sustained first-person narratives reveal researchers' own evolving conceptions of sex determination and provide evidence of the broader intellectual framework in which these changes are understood by specialists.

The interviews demonstrate that today's model of sex determination, which involves both a spectrum of sexuality and an emphasis on gene dosage and regulatory mechanisms, is broadly undergirded by a gender-critical conception of human sex and gender (see fig. 7.2). Researchers explicitly link the new model to the development of a changed, more complex understanding of gender in the field, and the old one to a set of biased assumptions about the biology of sex. Vilain, for instance, describes the 1980s and 1990s conception of sex determination as "a simplistic mechanism by which you have pro-male genes going all the way to make a male." The model assumed that the male-determining gene contained all that was necessary, Page says, to "impose" masculinization on a bipotential gonad. Page describes this model as "extraordinarily male-biased"; "extremely biased in favor of the male"; and "the

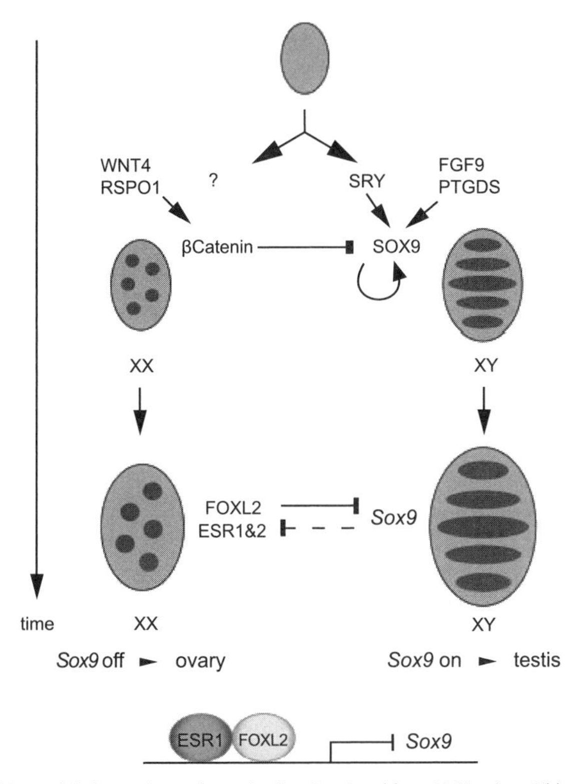

Figure 7.2. 2010s model of genetic sex determination. Reprinted from N. Henriette Uhlenhaut et al., "Somatic Sex Reprogramming of Adult Ovaries to Testes by Fox12 Ablation," *Cell* 139, no. 6 (2009): 1130–42, with permission from Elsevier.

most obvious hole in our understanding of the development of anatomic differences."[54]

Ingraham, Vilain, and Page all narrate the history of the master gene *SRY* model of sex determination as a lesson in the dangers of building untested, unreflective assumptions about gender into scientific theories. As Page relates, "biologists have been saying for half a century that female development is a default outcome that somehow all human or mammalian embryos are initially female and then have masculinity imposed on them. I don't think that the available data supports this idea." Vilain explains that "we used to think that females were the result of a default passive sex-determining pathway and we now know that is not true." Ingraham further suggests that the old model reflects biased interests of male researchers, who were less invested in characterizing "the active processes in females." She discloses, "I wish I could understand it because

I am female and I would like to know why I'm female and what are the active components to my gender assignment."[55]

When describing today's model, the researchers emphasize the inclusion of female developmental processes and a dynamic and nonbinary understanding of sex. For example, Vilain says:

> We [are] entering this new era in molecular biology of sex determination where it's a more subtle dosage of genes, some pro-males, some pro-females, some anti-males, some anti-females that all interplay with each other rather than a simply linear pathway of genes going one after the other.[56]

Similarly, Page says:

> Both the male pathway and female pathway are very active and require highly orchestrated, highly integrated sets of events, extremely complicated biochemical cascades that we're only beginning to understand.[57]

In these descriptions of the genetic model of sex determination, the researchers' emphasis on the complexity of sex determination, the active nature of both male and female processes, the parity of male and female genetic contributions to sex determination, and the interaction of male and female factors all reflect a deeper shift toward a gender-critical understanding of sex and gender. Vilain says, "there [are] many ways to define sex and each one of them [is] just as equally important as the other." Page says that "often we fall all over ourselves because of the limitations of the definitions we try to impose" on sex. He adds, "there is no such thing as a simple definition [of gender] and even within a scientific context, sex or gender has been defined at many different levels." Ingraham highlights the diversity of human gender identities, arguing that mouse studies imposed an idealized conception of gender on the research problem and that human sex determination must be contextualized in the phenotypic variability of sexual identity. "How are you going to find a transsexual mouse? Are you going to ask him?"[58]

◇

Today, gender criticism is part of specialist discourse in the field of sex determination research in a way that it was not in the 1980s and 1990s—and before. As Sinclair, who discovered *SRY* in 1990, said in an interview:

"I think humans like things to be ordered, and they get bothered about gray areas and when things become less clear-cut. But these days I don't think so much in black and white about male and female. Now I think of it all as being on a spectrum."[59]

It is possible to link this gender-critical perspective explicitly to the cognitive content of sex determination research. Here we have observed gender criticism come to play a part in the larger organizing conception that researchers use to think about sex determination, the descriptive language of sex determination, and the day-to-day work of evaluating hypotheses and interpreting data. In their own words (even if often painfully unaware of the contributions of feminism to their work), we see that researchers have found gender criticism valuable to their thinking, and we can observe gender analysis entering into the standard epistemic strategies for criticism and analysis in this field.

Sex determination research during the 1990s offers an excellent case study of the social and epistemological advancement of gender criticism in a scientific field. My focus on the gender dimension of sex determination research, of course, is not meant to imply that beliefs about gender were the *sole* factor shaping the *SRY* model of sex determination, nor that gender criticism was the sole motive force in the development of a new model. As I have made clear, gender criticism interacted with other factors, including advances in technology, new gene discoveries, and a broader rethinking of "master gene" theories in developmental biology over the past twenty-five years. Nonetheless, the contribution of gender criticism was clearly significant.

Modeling the entry of gender criticality into genetic research on sex requires a sensitive ear and a flexible framework for understanding the interaction of gender-critical perspectives and scientific research. The form and influence of gender criticality in genetic research on sex is heterogeneous and uneven. While models and descriptive language in sex determination genetics have been substantially reformed, many areas of genetic research remain untouched. Moreover, as I have indicated, gender-critical perspectives alone are not sufficient to account for changes in models of the genetic determination of sex. We must disentangle their interaction with technical and empirical factors in accounting for this shift. Finally, in contrast to social scientists and humanities scholars, those in the natural sciences are more resistant to understanding their practices as feminist or acknowledging the influence of feminist perspectives on their thought, a legacy of a long-standing ideal of good science as value-free and independent of politics. Thus, we must be willing, when warranted, to source certain ideas to gender-critical perspectives even when some of the principal actors involved would not understand it that way.

This chapter has pursued a focused line of inquiry: have feminist perspectives contributed to advancements in scientific knowledge in genetic sex research, and if so, how? Theories of genetic sex determination provide a stunning and nuanced example of the constructive contribution of the stance of gender criticality to scientific conceptions of sex. There is a second kind of question that we might ask, however, about the genetic science of sex after feminism—a question less explored. How have the cultural conceptions of gender sourced in genetic theories of sex changed since the advent of Second Wave feminism and the entry of feminism into mainstream popular culture? This question expands our inquiry beyond the question of feminist contributions to advancements in the science to explore, more broadly, how the specific content of contemporary postfeminist gender politics shapes and informs genetic research on sex today.

CHAPTER 8

Save the Males!

"Wouldn't the males in the room like to think that the Y has some more enduring contribution to maleness?" It is 2001, in Bethesda, Maryland. Y chromosome geneticist David Page looks out at the audience's young men—high school honors students. In the beat following Page's question, they visibly twinge with anxiety and anticipation. With a beaming smile, Page breaks the tension, reassuring the boys that new research in his lab has, fortunately, "intellectually rescued" the Y from "years and years of misunderstanding." The faces relax and nervous giggles titter around the room.[1]

Five years later, in a 2006 keynote at the International Congress of Human Genetics in Brisbane, Australia, sex chromosome geneticist Jennifer Graves has the women smiling and the men squirming. Says Graves, "There are two models for the Y chromosome. The model we were all brought up with was the Y as a macho little thing because if you have a Y you're male and that's it. . . . But our work on comparative mapping says that the Y is merely a wimp, a relic of the X chromosome. . . . This, of course, makes men very anxious." "Next thing you know," reports the *Australian Biotechnology News*, "all the women in the audience have broad smiles on their faces and the blokes are shifting uncomfortably, unnerved by the prospect of their fundamental redundancy. It certainly gets their attention."[2]

During the 1990s and early 2000s, two models of human Y chromosome structure, function, and evolution vied for acceptance. One promoted a view of the Y as the carrier of a sturdy crop of genes honed by selective pressure to specify maleness. The rival model conceived of the genes on the human Y as a snapshot of a rapidly degenerating chromosome, its sparse genes retained only because of their contingent enrollment in male-specific processes, not their functional specialization for maleness. Focusing on sex chromosome geneticists David Page and Jennifer Graves, this chapter tracks the development of these dueling models in the late 1990s and early 2000s, analyzing the gender dimensions of the debates over them.

Debates over Y degeneration illustrate how gender criticism—defined in chapter 7 as the practice of making visible the operation of gender conceptions in knowledge—and what I shall refer to as "postfeminist gender politics" are beginning to transform how gender operates in the contemporary life sciences. Gender politics are front and center in debates over Y chromosome degeneration, openly informing the language, research models, and empirical debates in this field. Each model of the Y chromosome invokes a substantial set of starting assumptions about sex and gender, and implicitly, about contemporary sexual politics. At the same time, both present well-motivated, testable scientific hypotheses that raise novel questions about the structure, function, and evolution of the Y chromosome.

As I shall argue, "gender bias" is inadequate for understanding how Graves's and Page's models differ, how each scientist uses political metaphors and cultural politics in his or her research, and what is at stake in the debate between them. Instead, I advance the notion of "gender valence" as a more perceptive framework than gender bias for taking account of the interaction between scientific knowledge and gender politics in this case. Whereas "gender bias" applies to cases in which gender conceptions operate invisibly and unreflectively in scientific practice in ways that introduce unproductive partiality, "gender valence" is appropriate to cases in which gender conceptions operate visibly and reflexively in ways that introduce productive partialities.[3]

"MEN ARE DOOMED"

The prospect that the human Y chromosome might be "degenerating" erupted into public debate in the first years of the twenty-first century. In a 2002 *Nature* concept paper, Jenny Graves and a colleague, Ross Aitken, predicted the extinction of the human Y chromosome in 10 million years:

> The original Y chromosome contained around 1,500 genes, but during the ensuing 300 million years all but about 50 were inactivated or lost. Overall, this gives an inactivation rate of five genes per million years. The presence of many genes that have lost their function (pseudogenes) on the Y chromosome indicates that this process of attrition is continuing, so that even these key genes will be lost. At the present rate of decay, the Y chromosome will self-destruct in around 10 million years. This has already occurred in the mole vole, in which the Y chromosome (together with all of its genes) has been completely lost from the genome.[4]

This was a back-of-the-envelope calculation, yet Graves's Y-extinction prediction raised several testable questions in genomics and evolutionary theory. If the Y chromosome disappears, what genes, if any, might replace the male sex-determining pathway in the mammalian genome? What role might a radical change in the sex-determining system play in speciation? What population-genetic dynamics might speed up or slow down degeneration on the Y?

Notably, Graves did not predict the disappearance of *males*. "A maleless world is not a necessary consequence of losing the Y chromosome," Graves has insisted.[5] Male-specific processes may also be performed by genes on the X or the autosomes. After all, XX/XO species that lack a Y, such as mole voles, still have males. Loss of the Y, if it did not lead to the extinction of the human species (both males and females), would more likely cause a new sex-determining system to evolve. Graves predicted that a new sex-determining pathway would probably begin to assert itself well before the disappearance of the Y, in response to increasingly low male fertility.[6]

Nonetheless, headlines around the world sounded alarms: "Is the Gene Pool Shrinking Men Out of Existence?" "The Male Malaise: Is the Y Chromosome Set to Self-Destruct?" "Men Are Doomed."[7] At the 2004 International Chromosome Conference in London, Graves faced off with another Y researcher, Dmitry Filatov of the University of Oxford, in the keynote event. "Leading Geneticists Debate the Fate of the Mammalian Y Chromosome," the conference press release touted. The idea surfaced in popular culture. Gwyneth Jones's novel *Life* and the comic book series *Y: The Last Man* called up the scenario of a threat to the Y chromosome to paint post-apocalyptic futures with dwindling populations of men. In short order, the Y chromosome degeneration hypothesis became a flashpoint for cultural anxieties around feminism and male social status. A highly mobile and strategic postfeminist identity of male victimhood, masculinist essentialism, and universal brotherhood has found a symbol in the Y chromosome. In the introduction to his 2010 book, *Manthropology: The Science of Why the Modern Male Is Not the Man He Used to Be*, the Australian science writer and archaeologist Peter McCallister asserts, "I love my brother males—every single one of those who, like me, carry the mark of our stunted, mutant Y chromosome on their brows."[8] In this discourse, the symbol of the Y as a "shriveled," "pathetic," inherently unstable chromosome, stands in for the "crisis of masculinity" in a postfeminist era. In *Y: The Descent of Men*, Steve Jones suggests that, as a result of women's advances, a "great loss of self-confidence . . . has swept across half the world" and "manhood itself is in full retreat":

> The figures are stark. At the end of the Second World War, husbands were in effective control of all family finances. As recently as the 1970s, British wives could not obtain credit without their approval. . . . Now three quarters of all married women have a job. In the 1960s they earned half what their husbands did but now the gap is far less.[9]

In *Adam's Curse*, Bryan Sykes proffers an even more alarmist hypothesis, arguing that humankind, having eclipsed men's social power, is now on the verge of losing males altogether. Throughout history, he argues, "patriarchal social structures where men seize and retain control" have helped along the crippled Y chromosome. Because of the decline of male power, coupled with environmental insults to sperm production and the near prospect of technologies for reproducing without the Y, Sykes foresees "the decay of the Y chromosome . . . inside every testis in the land." Sykes predicts that "men will become extinct" within 5,000 generations, or about 125,000 years.[10] Then, citing Valerie Solanas's 1967 *SCUM Manifesto*, which infamously, if ironically, called for the elimination of men, Sykes suggests that radical feminists seek to facilitate, and would celebrate, the extinction of the Y chromosome. It is a call to arms: "Men are now on notice," writes Sykes. In the closing pages of his book, Sykes outlines a program for genetic engineering to restore the Y before it is too late.[11]

Symbolically linking the degeneration of the Y chromosome to the decline of male social status after feminism, these authors make clear that anxiety over male decline in light of changes in the gender system is one context for the widespread media interest in Y degeneration theories. Fears of emasculation, matriarchal dominance, and male redundancy have, of course, been a recurrent cultural meme in Western societies, heightened since the rise of postindustrial urban economies and the advent of women's liberation movements.[12] While most of us, both men and women, celebrate advances in gender equality, changes in gender roles have shaken the foundations of many men's felt sense of power in the home, workplace, and political and economic spheres. Men's preferences and interests are no longer as culturally powerful as they once were, and some men experience this as a loss.[13] As such, the rapid and dramatic successes of Second Wave feminism have also been accompanied by "postfeminist" aftershocks, countermovements, appropriations, and backlash.[14] There is today a prominent discourse in which "feminism is constituted as an unwelcome, implicitly censorious presence" for men and in which "loss of power for men" is raised as a "possible consequence of female independence."[15] The notion that males, and masculinity, are endangered in a postfeminist age appears in diverse cultural arenas. As a much-discussed 2010 *Atlantic Monthly* cover article titled "The End of Men" (see fig. 8.1) reported, from the rise of Judd Apatow–style lad flicks featuring "perpetual adolescent(s)" who "cannot figure out how to be a man," to the new fashion of the slender-shouldered, innocent hipster, who prefers cuddling to sex, the notion that traditional masculinity is endangered is in the air.[16]

The supposed decline of males is now a serious subject of study in the social sciences. A 2011 report by the US Families and Work Institute titled "The New Male Mystique" reports that men are experiencing new pressures and confusion as they struggle to accommodate still pervasive "traditional views about men's role as breadwinners in combination with emerging gender role values that encourage men to participate in family life." The report concludes that "the new male mystique is harming men much in the ways that the feminine mystique harmed women."[17] In response to these concerns, activists have launched new organizations promoting men's rights, men have formed support groups to explore and deepen their sense of masculinity, and scholars have undertaken studies examining the poor educational and economic prospects for boys and men among more competitive, now ascendant females.

In biomedical science, too, the present-day vulnerability or weakening of men is a newly hot topic. Recent studies, hyped by the media, claim that human male sperm counts are dramatically declining worldwide;[18] that rising temperatures due to global warming will reduce the population of males (climate change is said to be harder on vulnerable male fetuses, causing the mother to spontaneously abort males at higher rates);[19] and that endocrine-disrupting environmental toxins are "feminizing" males, illustrated by male frogs found in suburban ponds growing female

Figure 8.1. The Atlantic, July/August 2010.

organs.[20] In another realm of biology, evolutionary psychologists affirm that men's biology—their size, strength, and aggression—is ill suited for the "postindustrial economy," which requires "social intelligence, open communication, the ability to sit still and focus."[21] Females, they claim, will prosper in this new economy, while in evolutionary terms, males will prove "remarkably unable to adapt."[22] As Martha McCaughey notes in her recent book, *The Caveman Mystique*, whereas once Western white men found any suggestion of affinities between their human behavior and that of apes and cavemen repugnant, curiously "today . . . many men find solace" in such comparisons common in popularized evolutionary psychology narratives.[23] Like the theory of Y chromosome degeneration, these many claims of male biological decline make explicit or implicit links to men's falling social status relative to women. McCaughey argues that these narratives of maleness in biological decline, or as biologically mismatched to a postfeminist age, are becoming increasingly popular as men lose their traditional hold on social, political, and economic power and status.

In characterizing present-day anxieties over the "decline of men," we must not paint too broad a stroke. Narratives about male decline are of wide interest today, but that does not mean that all individual men (or women) subscribe to them. Moreover, we must be historically and sociologically precise about the present cultural content of these narratives. Worries about the decline of males are not necessarily sexist, misogynist, or antifeminist. In his study of recent masculinist responses to feminist movements, the sociologist Michael Kimmel usefully distinguishes between antifeminism, which "require[s] the subordination of females," and pro-male backlash, which reasserts "the importance and visibility of masculinity" in the face of cultural changes brought about by women's increased power, countering feminization without specifically maligning women.[24] Kimmel's characterization of pro-male but not necessarily antifeminist fears of a "crisis of masculinity" best approximates the strain of postfeminist gender politics that is animated in contemporary scientific debates over Y chromosome degeneration.

In what follows, I argue that cultural anxieties about the status of masculinity after feminism are not just an entertaining media sideshow to the real scientific debate over Y chromosome degeneration. Rather, these anxieties create a receptive context that valences and informs the scientific debates over Y chromosome degeneration, providing a window into how postfeminist gender politics shape the genetic science of sex today.

DEFENDER OF THE ROTTING Y

David Page is a medically trained geneticist at the Massachusetts Institute of Technology (MIT) whose research has been at the leading edge of

American molecular biology and genetics since the 1980s. As described in chapter 7, Page is a Y chromosome researcher who was a key player in the search for the male testes-determining gene. In the late 1980s, he isolated the sex-determining region of the Y chromosome with elegant deletion studies demonstrating that a piece of the nonrecombining portion of the Y is attached to the tip of one of the Xs of an XX male and that the same region is deleted on the Y of XY females. In similar studies on infertile males, Page identified several Y chromosome mutations associated with male infertility. In 1992, he published the first map of the human Y chromosome. Over the next decade, Page's lab completed the painstaking work of sequencing the human Y.

Despite its diminutive size, the Y proved uniquely challenging to sequence. The Y's large amounts of heterochromatin (highly variable noncoding DNA), extensive palindromes (adjacent sequences that read as mirror reflections of each other, e.g., ATCGCT-TCGCTA), and repetitive elements challenged standard genome-sequencing methodologies.[25] Unlike the human X chromosome and the twenty-two autosomes, the human Y was excluded from the major collaborative Human Genome Project sequencing efforts. The sequencing of the Y chromosome reflected the initiative, determination, investment, and passion of a single scientist-entrepreneur. Published in 2003, the complete sequence of the human Y was recognized by *Science* as one of the top ten scientific breakthroughs of the year.[26] For this work, Page received the Curt Stern award for outstanding achievement in human genetics, and in 2005 he was inducted into the National Academy of Sciences and named director of the prestigious Whitehead Institute at MIT.

In the course of sequencing the Y, Page and colleagues developed a new model of the structure, function, and evolution of the human Y chromosome. The model posits that structurally, the Y is gene-rich; functionally, it is specialized for male-advantageous genes; and evolutionarily, it maintains stability and integrity through novel mechanisms of gene acquisition and recombination specific to the Y chromosome. This model suggests that genes on the Y chromosome are subject to positive selection for male fitness, and that in addition to the gene for testes determination, the Y should carry many other genes that specifically account for differences between males and females.

Page's model challenged the reigning view of the Y. Sex chromosome researcher Susumo Ohno, a geneticist at the Los Angeles City of Hope hospital, influentially argued in 1967 that the X and Y evolved from a pair of identical autosomes. Over time, they diverged so that they could no longer recombine, leading to loss and disrepair of the genes on the Y

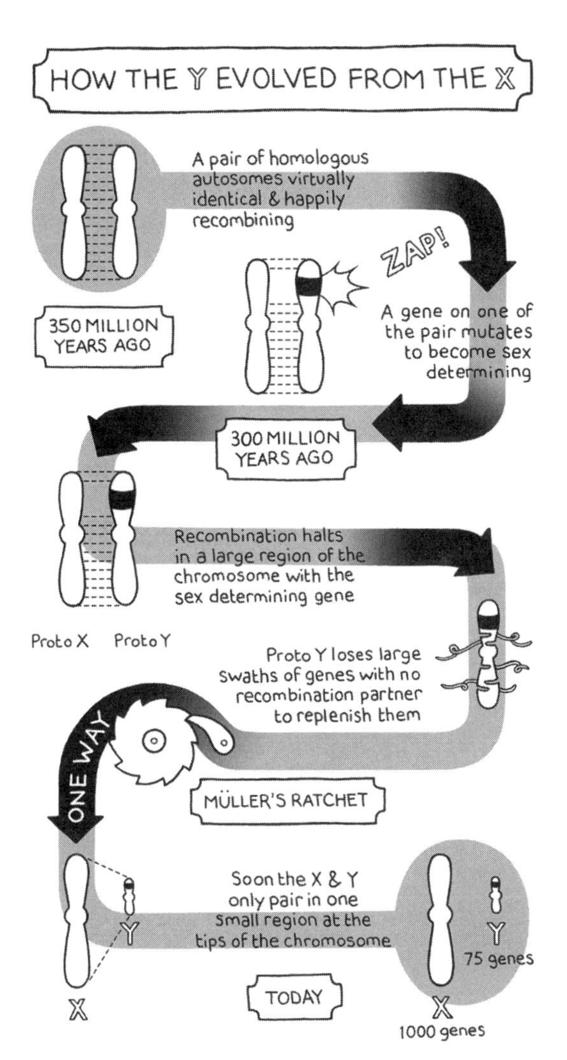

Figure 8.2. Illustration by Kendal Tull-Esterbrook; copyright Sarah S. Richardson.

(see fig. 8.2). Ohno's hypothesis predicted that because of suppression of recombination (a mechanism for chromosome repair and regeneration), high mutagenesis in the hostile environment of the testes, and populational sweeps caused by the irreversible progressive shedding of genes (a process known as Müller's ratchet), the Y's genes will degenerate over time, eventually lose function, and then disappear altogether. For decades, this theory of Y degeneration had framed research on Y chromosome structure, function, and evolution. Indeed, genetic and comparative genomic analysis of the X and Y chromosomes seemed to confirm Ohno's picture of sex chromosome evolution. The human Y, at perhaps one-twentieth the size of the X, has today lost most of its content through the processes of degeneration identified by Ohno. What remains controversial is the question of whether we can predict the eventual disappearance of the Y chromosome, or whether, as Page has argued, the Y has some tricks—such as special repair mechanisms or adaptation of male-advantageous genes—that will hold it back from complete degeneration and extinction.[27]

Describing himself as "the defender of the rotting Y chromosome," in media profiles Page presents as a careful investigator who has cut through dogma to forge a new understanding of the Y, reinstating it as a biologically significant object of investigation in the face of prevailing abject and denigrating attitudes toward it.[28] Page has long maintained that gender politics has something to do with predominant views of the Y chromosome. As a 2004 *Scientific American* profile of Page testifies,

> the very idea of investigating the Y chromosome offends those feminists who believe that it serves as nothing more than a subterfuge to promulgate an inherent male bias in biology. . . . Page can point to a long list of scientific papers with his name on them that demonstrate that the Y is an infinitely richer and more complex segment of the genome than ever imagined and one that does not fit neatly into the prejudices of gender-based interpretations of science.[29]

Opening a talk at the Whitehead Institute in 2003, Page said, "What I want to do today is to defend the honor of the Y chromosome in the face of a century of insults to its character and to its future prospects." These insults, he said, include cartoons that portray it as the source of such wifely complaints as "selective hearing loss," images of the Y as "shriveled," and dogma that the Y chromosome is a "genetic wasteland." "No other chromosome has to suffer these sort of insults on a regular basis," said Page.[30]

In Page's view, the Y, cast aside as a poorly behaved, degenerating chromosome of no biological value, is a victim of negative male stereotypes generated by feminists. Page portrays the Y as a harassed male surrounded by strong women, unpopular and unwanted in a postfeminist world. In a 2003 Whitehead Institute profile, Page is quoted as stating:

> The common perception of boys and girls, of men and women, greatly impacts biologists' perceptions of the X and the Y. . . . The

> science and the sexual politics become blurred. The idea of the Y as a shiftless, no-good degenerate chromosome is entirely too appealing and attractive to resist for reasons that have little to do with science and lots to do with sexual politics.[31]

Page's paper and talk titles include, "Save the Males!," "On Low Expectations Exceeded; or the Genomic Salvation of the Y Chromosome," and "The Evolution of Sex: Rethinking the Rotting Y Chromosome"—all in scientific publications and venues—and a 2000 profile describes him sitting in a lab meeting drinking "from his teal-colored mug imprinted with 'Save the Males.'"[32]

In public venues, Page often identifies the condition of the Y with that of the modern male. In a *New York Times* piece on the sequencing of the Y, Page is quoted saying that "the Y married up, the X married down," and "the Y wants to maintain himself but doesn't know how. He's falling apart, like the guy who can't manage to get a doctor's appointment or can't clean up the house or apartment unless his wife does it."[33] In a biographical profile, Page portrays his home life with a wife and three daughters as a "lone Y in a house full of Xs." In writings and talks, Page consistently calls the Y the "Rodney Dangerfield of the genome." In his images of the X and Y, the X is pink and the Y is blue.[34] In talks, he encourages male audience members to identify with the Y. In a 2003 talk at the Whitehead Institute, Page called on "all the men in the audience for just a minute, to sort of pay homage to their Y chromosomes. . . . You got your Y chromosome from your father, and he from his father, and he from his father, and his father, and his father in unbroken, unrecombined form."[35]

As noted in earlier chapters, the idea of a Y chromosome enriched for "maleness" genes is not entirely new. In the 1960s and 1970s, the Y was believed to house "male" morphological traits such as large stature and behavioral qualities such as aggression. The science that emerged in the 1990s built on this intuition, but it also drew creatively on a particular model of the evolution of the human genome that places "sexually antagonistic" processes at the center of action. Originating in the 1931 work of R. A. Fisher on dominance effects in genetics, and sporadically revived during the twentieth century, the hypothesis of genetic sex antagonism pairs a gene's-eye view of evolution with the idea that the battle of the sexes shaped much of the mammalian genome's history.[36] "Pro-female" and "pro-male" genes are seen as "fighting it out" at the molecular or genomic level, leading to differential selective pressures and possible pro-male or pro-female functional specialization at different loci in the genome. British evolutionary theorist Lawrence Hurst reinvigorated this theory in the 1990s.[37] In a pair of 1994 papers, Hurst theorized that the Y chromosome might, hypothetically, attract and retain male-advantageous and female-disadvantageous reproductive fitness genes, such as genes involved in spermatogenesis and male gonad development, or genes that, against the interest of the mother, increase the rate of fetal growth.

Citing Hurst's hypothesis, Page embarked in the mid-1990s on a comprehensive survey of the genes on the nonrecombining region of the Y (NRY). The 1997 *Science* paper "Functional Coherence of the Human Y Chromosome" revealed Page's findings; it was the first articulation of what I shall call his "gene-rich and specialized for maleness" model of the Y chromosome. The paper, described as a "blockbuster" by a reviewer, announced the discovery of twelve new genes on the Y chromosome, increasing the number of known genes in the NRY to twenty.[38] Explicitly countering the so-called wasteland model of the Y as a genetically inert, degraded X, Page and graduate student Bruce Lahn advanced a model of the Y as "gene-rich" and specialized for male reproductive fitness: "Ironically, it had been widely assumed . . . that these domains consisted of 'junk' DNA. To the contrary, our results argue that these Y-specific repetitive regions are gene-rich."[39]

Page and Lahn argued that there exist two "discrete classes" of genes on the NRY (see fig. 8.3). "Class I" genes have homologues on the non-inactivated X and are ubiquitously expressed "housekeeping genes," while "Class II" genes have no X homologue, specialize in spermatogenesis, are expressed exclusively in the testes, and exist in multiple copies (possibly, they argued, a protection against degradation and an indicator of their importance to male fitness). Page and Lahn asserted that Class II genes are evidence of functional specialization of the Y for male fitness-enhancing traits. The Y "favored the acquisition of testis-specific [gene] families, perhaps through selectively retaining and amplifying genes that enhance male reproductive fitness," wrote Page and Lahn. Citing Fisher and Hurst, Page and Lahn further hypothesized that Class II genes were evidence for a unique Y chromosomal mechanism for acquiring new male-advantageous genes during evolution. "We suspect that most of the NRY's transcription units do not date from the Y chromosome's common ancestry with the X chromosome but instead are more recent acquisitions," they wrote.[40]

Evidence for this claim had first arrived with Page's 1996 discovery that *DAZ*, a family of NRY genes involved in spermatogenesis, was closely related to a gene on human chromosome 3. Page and coauthors

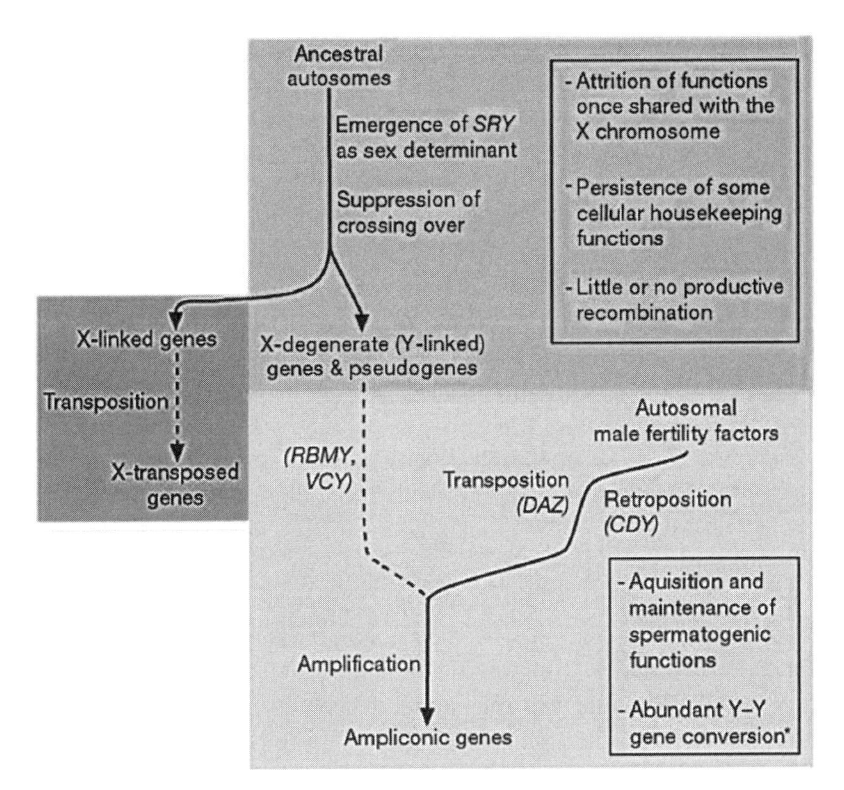

Figure 8.3. Page and Lahn's model of the evolution of the genes on the human Y chromosome. Reprinted with permission from the Macmillan Publishers Ltd.: *Nature* (Skaletsky et al., "The Male-Specific Region of the Human Y Chromosome Is a Mosaic of Discrete Sequence Classes," 423, no. 6942: 825–37), copyright 2003.

proposed that "a complete copy of [*DAZ*] was transposed from an autosome (what is now human chromosome 3) to the Y chromosome during primate evolution."[41] This finding, they argued, upsets the view of the Y as a shriveled X chromosome. It suggests that there is a secondary evolutionary process driving Y chromosome evolution: the acquisition of male-advantageous, and possibly female-disadvantageous, genes. As they wrote,

> the case of human *DAZ* challenges the prevailing view that most if not all Y-chromosomal genes were once shared with the X chromosome. . . . [O]ur results suggest that the Y chromosome's evolution and gene content may also be influenced by a process that is independent of the X chromosome. We speculate that the direct acquisition of autosomal genes that enhance male fertility is an important

component of Y chromosome evolution. Selective pressures would favor this process, particularly if the genes transposed to the Y were of little or no benefit to females, and most especially if they diminished female fitness.[42]

In direct challenge to the Y-degeneration hypothesis, Page theorized that the acquisition of "sexually antagonistic" and "male benefit" autosomal alleles for male fertility is an evolutionary mechanism for the protection of the Y chromosome from complete degeneration in the absence of recombination.[43] "Perhaps the rate of acquisition of male fertility genes approximates the rate of subsequent degeneration, resulting in an evolutionary steady state," he wrote.[44]

In 1999, Page and Lahn added another gene to Class II: *CDY*, a testes-specific Y chromosome gene family with no X-homologue, which was found to be "retroposed" from a gene on human chromosome 13. Retroposition is a process by which a DNA fragment is reverse-transcribed from an RNA molecule and inserted into a chromosome. Whereas in the Y degeneration model, retroposition had largely been considered one of the mechanisms of Y chromosome decay (through the accumulation of junk retroposed elements), Page and Lahn now suggested that it "has also provided a mechanism for gene building during the evolution of the human Y chromosome."[45] Page and Lahn added retroposition to the arsenal of molecular evolutionary processes that, they hypothesized, the Y chromosome had harnessed to maintain its evolutionary stability in the face of degenerative pressures. "Retroposition contributed to the gene content of the human Y chromosome, together with two other molecular evolutionary processes: persistence of a subset of genes shared with the X chromosome and transposition of genomic DNA harboring intact transcription units," they wrote. In contrast to *DAZ*, however, the relative of *CDY* on chromosome 13 did not have a role in testes development and spermatogenesis. Rather, *CDY* evolved testes-specific functions after retroposition. As they wrote, *CDY* "was fashioned from a fully processed, reverse transcribed cDNA which, we speculate, fortuitously integrated near an existing promoter or into an otherwise transcriptionally permissive locale on the Y chromosome." Thus, unlike *DAZ*, it was enrolled in male-specific processes only after random retroposition to the Y, and was not recruited to the Y because it was a "male benefit" or "female antagonist" gene.[46]

In 2003, Page and collaborators released the complete sequence of the nonrecombining region of the Y, again interpreting the results as corroboration of the model of the Y as gene-rich and specialized for maleness.[47]

The article announced that the NRY encodes twenty-seven distinct proteins, comprising three classes of gene:

1. X-transposed (2 genes)
2. X-degenerate (27 genes, 16 of which, including the *SRY*, are protein coding, and 11 of which are noncoding pseudogenes)
3. Ampliconic (9 protein-coding gene families, in multiple copies totaling 60 genes)

The "ampliconic" class is constituted by arrays of multiple copies of genes that are expressed exclusively in the testes. It is also partly made up of palindromes that, Page hypothesized, play a repair and regeneration role similar to that of recombination, as will be elaborated below. Figures depicted X-transposed genes as pink, X-degenerate genes as yellow (representing an ancient mix of male and female—presex, neutral, or neither-nor), and Y genes as blue. X genes were characterized as "housekeeping" and "ubiquitous" while Y genes "acquire" and "maintain" male-specific functions and experience "abundant" palindromic recombination.

Page and colleagues argued that this data refutes the degeneration model of the Y. Ohno's model had instigated an abject understanding of the Y as a "wasteland" and "a profoundly degenerate X chromosome." The application of "recombinant DNA and genomic technologies to the Y chromosome, culminating in molecularly based conclusions about its genes," Page and colleagues now argued, had overturned this "anecdotal" understanding of the Y.[48] Marking this shift, Page and coauthors announced that the nonrecombining region of the Y (NRY) should thereafter be known as the "male-specific region" of the Y (MSY):

> In recent years, we and other investigators have referred to the MSY as the NRY, or "non-recombining region of the Y chromosome." This usage reflected both awareness that productive X-Y crossing over did not occur in the MSY, and ignorance of the Y-Y gene conversion that is apparently commonplace there. We now refer to the NRY as the MSY, or "male-specific region of the Y chromosome," because it is recombinogenic and unique to males.[49]

Rather than a degenerate wasteland, Page's Y is a well-tuned product enriched with innovative mechanisms required for its survival in harsh evolutionary conditions. Not a passive victim of degeneration, this Y defends, recruits, and builds its genetic repertoire, and acts a central player in male reproductive fitness. The Y "acquired," "evolved," and "conserved" "genes that specifically enhanced male fertility."[50] The Y is "vigorous" in its response to evolutionary pressure and reveals "highly evolved and complex genetic structures."[51] As Page put it, the Y "contains gene-rich palindromes of unprecedented scale and degree."[52] In lectures and interviews, Page described the Y as "persevering and noble" and making an "enduring contribution to maleness."[53]

One implication of this model of the Y, Page further argued, is that there exist more "male-specific" genes than had been previously appreciated, genes that might contribute to differences between males and females. Page closed the 2003 *Nature* paper that announced the sequence of the Y with this very prediction:

> The common substitution of the Y chromosome for the second X chromosome dwarfs all other DNA polymorphism in the human genome. In decades past . . . biologists often judged this genomic dimorphism to be of limited functional consequence. . . . Now we must begin to reconsider this position, given the unanticipated number and variety of MSY genes, many of which are expressed throughout the body. . . . The present sequence of the MSY, and the emerging sequence of the X chromosome, offer the near prospect of a comprehensive catalogue of genetic and sequence differences between human males and females.[54]

Page's "gene-rich and specialized for maleness" model of the Y chromosome, then, supports a larger research program. This program seeks out male-specific genes, hypothesizes the contribution of these genes to maleness, and promises the genetic localization of male-female difference. This model of the Y speculates that MSY genes may control sexual dimorphism in more profound and wide-ranging ways than just testes development and spermatogenesis. Citing Page's findings, Paul Burgoyne of the National Institute for Medical Research in London, for example, recently suggested that MSY genes may hold the key to sex differences in "the brain and behavior" and illuminate what he described as "the accelerated development of male embryos."[55]

THE "WIMPY Y"

While Page was setting forth this robust model of the Y, quite a different picture of the human Y was emerging from comparative evolutionary genomics research. As genomes of model organisms across the animal kingdom became available in the 1990s, comparative genomics came of age as a powerful technique for genetic analysis. Using molecular, cytological, and bioinformatic techniques, comparative genomicists work to illuminate the phylogenetic distance between species and to trace the genomic rearrangements, deletions, and additions that produce or characterize speciation. Comparative evolutionary genomicists think on a different scale than molecular geneticists—in time spans of hundreds of millions of years and in terms of vast evolutionary design spaces. Comparative work in genomics is characterized by an appreciation for the happenstance and contingency out of which the genomes of different species evolve. The comparative genomicist tends to see genomes not as highly refined, efficient sets of genes required for the particular adaptive needs of a species, but as a palimpsest of fragments of ancestral genomes, some functional, some redundant—a colorful cut-and-paste job of rearranged chromosome blocks.

In the 1990s, Jennifer Graves developed just such a picture of the mammalian Y. As we learned in Chapter 7, Graves is a prominent Australian sex chromosome geneticist who was a key player in analysis of the sex-determining locus on the Y chromosome in the early 1990s, using comparative analysis of the Y across multiple taxa to confirm its ubiquity among mammals. Graves's work, however, also contradicted expectations that *SRY*, the male sex-determining gene, would be a specialized "master gene" for maleness. Graves revealed the sequence and expression of *SRY* to be poorly conserved across different sexual species, and she placed *SRY* within a large family of *SOX* genes that are distributed across the sex chromosomes and autosomes.[56] As such, Graves concluded, *SRY* may be seen as an X-degenerate gene that "arose well after the genesis of the mammalian sex chromosomes."[57] Looking at *SRY* through the long lens of comparative genomics led Graves to a particular view of the evolution of the sex-determining pathway in males. Rather than the "master gene" of testes development, Graves theorized that *SRY* is a descendant of "a developmentally important gene shared by partly differentiated ancestral X and Y chromosomes," a transcriptional fragment that subsequently became enrolled in testes determination.[58] *SRY*, she argued, "is more likely to reflect a by-product of a local inversion within a repetitive structure, rather than a brilliant evolutionary advance."[59]

In the mid-1990s, Graves began to extend her view of the sex-determining gene on the Y into a more general model of the structure, function, and evolution of the human Y chromosome. A 1995 finding that some species of rodents lack an *SRY* gene (and a Y chromosome), along with Page's emerging fine-grained research on the genetic content of the Y, prompted Graves to vigorously resurrect Ohno's original model of the Y as a degenerated X chromosome and to become its leading present-day defender.[60] Genetic research, she argued, now offered new support to Ohno's model. She emphasized two points: First, "of the dozen or so genes on the human Y chromosome, most have relatives on the X"; second, "genes on the differential region of the Y are remarkably variable in their presence, numbers, and activity."[61]

Graves argued that any remaining genes on the Y represent a tiny sample of the original X genes, in various stages of degeneration:

> Genes on the Y chromosome therefore seem to represent a small, non-identical but overlapping subset of genes on the X. Some, evidently expendable, genes seem to be dead or dying, while others appear to serve a male-specific function which ensures their survival over long periods of evolutionary time.[62]

SRY, the quintessential male-specialized Y chromosomal gene, served as exhibit A: "The *SRY* gene itself acquired its testis determining function in this haphazard manner," surviving among "a rather random subset of the genes on the X" because of a transcriptional plasticity that allowed it to be enrolled into the sex-determining pathway. *SRY*, in this framework, was a typical Y-borne gene that descended from an X-homologue and was maintained on the Y because of a fortuitously acquired role in male fertility.[63]

Rather than genes "specialized" for maleness, Graves regards the surviving genes on the Y as a mismatched set exhibiting a spectrum of gene degradation—nonexpressing pseudogenes and genes with high sequence and expression variation, present in multiple copies in different stages of disrepair and interspersed with swaths of junk DNA.[64] Page argued that the Y chromosome is gene-rich and specialized for maleness, a power-packed unit of genes recruited and functionally specialized for maleness. In contrast, Graves described the Y as comprised of X genes *retained* because of *relevance* to reproductive development. As Graves wrote, the Y's genes "are hard to understand in terms of efficient *function*, but they make excellent sense in terms of the *evolution* of the mammalian Y chromosome."[65]

In 1999, with Margaret Delbridge, Graves produced a table of Y chromosome genes identified to date. The table showed that nearly every human NRY gene active in spermatogenesis has a homologue on the X chromosome, and that all of these genes exhibit certain defining features of degeneration. On this basis, Graves argued that Page's distinction between Class I and Class II genes on the Y was unfounded. The present state of the human Y chromosome, she argued, confirms the Y-degeneration model's predictions. As she wrote,

> the human Y chromosome comprises more than 1% of the total human chromosome complement so it should contain several hundred genes. However, the Y chromosome seems to be something of a genetic wasteland. Only about 20 genes have been discovered on the differentiated region of the Y chromosome. Several of these genes are members of Y-specific gene families, some members of which are inactive. . . . The Y chromosome has a high content of repetitive DNA sequences which appear to have no functional role, and it is littered with pseudogenes.[66]

Page's finding that two male-specific Y genes had transplanted from autosomes, Graves argued, did not alter the overwhelming explanatory success of the degeneration model in accounting for the evolutionary history, present composition, and future trajectory of the human Y chromosome. These genes, too, are now subject to degeneration, she pointed out. Nor, she argued, does the determination that many of the remaining genes on the male-specific region of the Y are active in male reproductive development demonstrate the "functional coherence" of the Y chromosome claimed by Lahn and Page. In contrast to Page's view that the Y evolved under positive or sex-antagonistic selection as a male-enhancing genetic mechanism, she hypothesized that selective evolutionary processes had simply eliminated genes that could *not* be enrolled in male-specific processes.

In a 1999 lecture, Graves further sharpened her critique of Page's model of the Y as gene-rich and specialized for maleness.[67] Graves distinguished among three models of the Y chromosome: the old "dominant" Y, which foregrounds its decisive role in sex determination; Page's "selfish" Y, which emphasizes the Y's functional specialization in male reproductive fitness at the expense of female fitness; and Graves's "wimp" Y, the model of the Y as a degenerating relic of the X chromosome.[68] Tallying up the evidence for each model, Graves pointed out that Page had offered only three genes to support his "selfish Y" model. These genes, she argued, do not characterize the male-specific region of the Y: "It is evident that many or most of the genes on the Y chromosome, including three with suspected functions in spermatogenesis, have copies on the

X chromosome from which they were derived, as proposed by the wimp Y model."[69]

Graves argued that the "wimp" or X-degenerate model of the Y is broadly confirmed by multiple levels of genetic and genomic analysis and carries the most predictive and explanatory power:

> Of the three models of the human Y chromosome presented—the dominant Y chromosome, the selfish Y chromosome, and the wimp Y chromosome—the last model has the most explanatory power. . . . The genetic paucity and high content of repeated sequences and pseudogenes of the Y chromosome tells a tale of genetic degradation. . . . A mere handful of genes have clung to survival because of mutations that allowed them to adopt a male-specific function essential for sex determination or spermatogenesis.[70]

Overwhelming, consistent evidence, Graves concluded, demonstrates that "the Y chromosome is small and degenerate, and has few essential functions left."[71]

Like Page, Graves is a ham. She frequently uses the startling idea that the Y chromosome is going extinct as a hook to invite her audiences into the genetic science of sex chromosomes. The Y has a "use-by date," she is fond of saying. The mole vole, she writes, "provides a chilling reminder of the mortality of the Y chromosome."[72] The gendered language and imagery in her Y chromosome model reflects, at least in part, her interest in communicating to a wider public and stimulating interest in her research. A description of Graves's 2006 keynote at the International Congress of Human Genetics praised her "ability to convert difficult scientific concepts into language non-scientists can grasp" and described her as "delight[ing] the audience with [a] hilarious speech."[73] Graves portrays Page's "suggestion that all the interesting, male-specific genes on the Y were acquired, instead, from autosomes" as a transparently masculinist effort to elevate Y-chromosomal genes to "male-specific glory."[74] Painting a very different picture of the "poor little Y"[75] than Page, Graves describes its contents in the most abject terms, as comprised of "hard-core 'junk DNA,'" "pathetic relics," and "apparently disposable" genes.[76] Yet behind the cheeky language is a serious claim about gender politics in models of Y chromosome structure and evolution. Labeling her contrasting model the "Wimpy Y," Graves slyly suggests that there is something about the unmasculine image of the Y as a degenerate X that prevents Page and others from embracing it.

THE THEORY OF GENE CONVERSION

The spirited but low-level exchange between Page and Graves in the late 1990s became a full-blown scientific controversy and a topic of international headlines in 2002, after Graves's and Aitkins's hypothesis appeared in *Nature*. Page set out to marshal evidence that human Y chromosome degeneration has stabilized, slowed, and even stopped or reversed. He postulated a unique mechanism of genetic renewal on the Y chromosome called "gene conversion." With colleagues at Washington University in St. Louis, Page published findings in *Nature* in 2003 showing that one-quarter of the MSY is made up of sequence palindromes. Page hypothesized that the palindromes undergo gene conversion, in which the DNA folds into loops and the nucleotide sequence from one arm replaces that on the other, homogenizing the sequences (imagine the palindrome "a car, a man, a maraca," folded in half so that the matching letters are paired). He described gene conversion as a novel adaptive Y chromosome mechanism evolved to emulate recombination in the region of the Y that does not pair with another chromosome. Gene conversion, he speculated, has evolved to allow "genes in palindromes to resist, or at least retard, the evolutionary decay that is a hallmark of Y chromosome evolution."[77]

To Page, gene conversion was definitive proof that the Y is not going extinct. The Y "contains gene-rich palindromes of unprecedented scale and precision" and is "a place of abundant gene conversion," he wrote. The discovery of the phenomenon of gene conversion, he proclaimed, marks the demise of the "degenerate Y" model: "We no longer think of the Y chromosome as a land of no recombination and, hence, of inevitable gene decay."[78] Page's claim that the old model had at last been overturned generated much excitement. Science journals, as well as the *New York Times* and the *Boston Globe*, featured the news.[79] An item in *Science* quoted a Johns Hopkins geneticist who said: "In demonstrating the dynamic role of [the] Y chromosome, Page 'has brought a lot of honor to males. . . . I don't think people have appreciated [the nature of] the Y chromosome before.'"[80] Geneticist R. Scott Hawley similarly welcomed Page's findings in a mini-review in *Cell* entitled "The Human Y Chromosome: Rumors of Its Death Have Been Greatly Exaggerated." "Fortunately for those of us who value [the Y] chromosome," the "desolate picture of the Y chromosome" as a "genetic 'wrecking yard'" or "'wimp' among chromosomes" has been reversed by Page's findings, which portray the Y as containing "novel . . . chromosome elements that appear quite vigorous indeed." Reiterating Page's comparison of gene conversion to recombination, Hawley crowed, "whatever this region [of the Y] is doing, or not doing, it is very clearly recombining!" He concluded that "the obituary for the Y chromosome . . . may have been written rather prematurely."[81]

In 2005, Page extended his case with a study published in *Nature* showing that X-degenerate male-specific genes are more conserved in humans than in chimpanzees. Analysis of the primate genome, Page argued, showed that human Y chromosome degeneration had slowed, stopped, or reversed. As he wrote, the "greater conservation of coding" in the X-degenerate pseudogenes of humans indicates that "purifying selection has been a potent force in maintaining X-degenerate gene function during recent human evolution."[82] This showed, Page argued, that gene decay slows "as Y chromosomes evolve."[83] Rather than a model of Y evolution in which genes are lost randomly and in a linear, constant fashion over time, Page argued that gene loss is nonlinear and uneven. The media again clamored to spread the welcome news that males are safe from extinction. "I think we can with confidence dismiss . . . [t]he 'imminent demise' theory," said Page in a *CNN.com* article publicizing the findings.[84]

Graves first responded to Page in 2004. In an article titled "The Degenerate Y Chromosome—Can Conversion Save It?" Graves attacked a key supposition of Page's gene-conversion hypothesis of Y regeneration. As formulated by Page, gene conversion is a random transfer of genetic material from one arm of a palindrome to another. Graves pointed out that there is no evidence that gene conversion is biased toward replacing degenerate gene sequences with active ones. Thus, Graves argued, gene conversion is likely to be a cause of degeneration as often as it is a remedy to it:

> Gene conversion is not directional. The process is just as likely to substitute a mutated copy for an active copy than it is to resurrect inactive copies. Overall, it is hard to see how gene conversion within palindromes could increase the number of active copies of genes on the Y. In the absence of some mechanism that biases the direction of conversion, conversion would be expected to negate the cushioning effect of having large numbers of duplicates. . . . In fact, the numerous inactive pseudogenes, as well as the fifteen families of transcribed but untranslatable transcripts within palindromes, suggests that there are more casualties of the process than successes.[85]

Graves therefore disputed the comparison of gene conversion to sexual recombination. It's more like "masturbation," she quipped:

I would argue that gene conversion within palindromes is more like genetic masturbation than real sex. It does not offer interaction between different Y chromosomes, which is essential for the genetic health of a region of the genome.[86]

Graves concluded by reasserting and extending her prediction of human Y chromosome extinction. Citing Sykes, she pointed out that hers was "a conservative estimate."[87]

The hypothesis of palindromic gene conversion as a mechanism of recombinatory renewal on the Y chromosome remains a contested and unfolding story. Graves's argument that gene conversion will not regenerate genes on the Y chromosome unless it is shown to be directionally biased toward repairing genes, rather than also transferring damaging mutations, remains a persistent issue. In a 2003 MIT lecture, Page responded to an audience member who raised this question, saying, "I didn't take you through the data, but we have convinced ourselves that this gene conversion is occurring. If you ask me to tell you about the molecular detail of this process of gene conversion, we know nothing about it."[88]

Page has not backed away from his model of the Y chromosome, and nor has Graves. In a 2006 *Cell* paper, Graves restated her model of Y chromosome extinction, moving it from a back-of-the-envelope estimate to a more sophisticated model taking several variables into account (see fig. 8.4). First, she revised her predicted extinction date for the Y from 10 to 14 million years, in light of a recalculation of the number of genes on the X from 1,600 to 1,000. Since her initial predicted rate of Y chromosome degeneration was calculated based on the number of genes on the original X-homologue, this new, lower count entailed a slower rate of degeneration than originally estimated. Second, she sketched a revised model of Y chromosome degeneration that would allow for the likelihood that degeneration will not proceed uniformly and that the total size of the male-specific region of the Y will affect the rate of loss of function. Critics such as Filatov and Page had pointed out that as the Y gets smaller, "the rate might be expected to decay exponentially."[89] Similarly, additions to the X and Y chromosomes from the autosomes, which occurred in several instances during the evolution of the sex chromosomes, would speed the rate of decay for a period of time. Other factors affecting Y degeneration include generation time—in which mice outdo humans, for example—and variations in population size. Graves agreed: "Many factors feed into equations describing the rate of degradation of the Y chromosome, and these make it difficult to predict how near to extinction the human Y is." Nonetheless, "calculations of the rate of loss

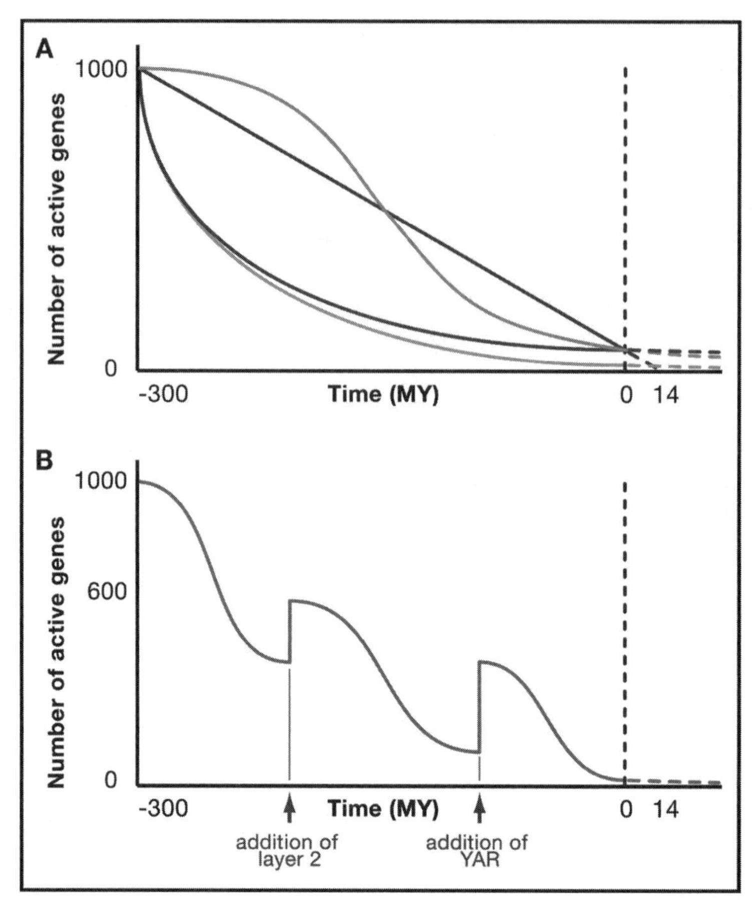

Figure 8.4. Models predicting the extinction of the Y chromosome. Reprinted from J. A. Graves, "Sex Chromosome Specialization and Degeneration in Mammals," *Cell* 124 (2006): 901–14, with permission from Elsevier.

of genes from the Y predict that, sooner or later, the Y will run out of genes altogether and disappear. This is not just a prophecy, but an observation in several different [heteromorphic sex chromosome] systems." "I challenge population and evolutionary geneticists to derive a meaningful model with predictive power," Graves concluded.[90]

A CASE OF GENDER BIAS?

Page sees the Y-degeneration hypothesis as a product of gender bias in science. He represents Graves's update of Ohno's theory as a politically motivated, unscientific hypothesis arising directly from feminist gender ideology. For example, in a 2005 *New York Times* piece on Page's "gene conversion" hypothesis of Y chromosome regeneration, Page accused Y chromosome degeneration adherents of improperly mixing gender politics and science: "The idea of the Y's extinction 'was so delicious from the perspective of gender politics,' Dr. Page said. 'But many of my colleagues became confused with this blending of gender politics with scientific predictions.'" A *Scientific American* profile similarly quoted Page asserting that "[the fascination with the 'rotting Y'] has to do with sexual politics." The profile continued, "biologist Jennifer A. Marshall Graves of the Australian National University in Canberra argues that gene conversion in the palindromes represents a form of 'genetic masturbation' that may not only fail to inhibit deleterious mutations but may even speed the process of the chromosome's decline. Page's terse response: 'Ah, rhetoric and theory unburdened by experimental data.'"[91]

At the same time, in his public lectures and media appearances, Page uses his status as "Mr. Y" to do gender politics, to express an anxiety about feminist political correctness, and to advance a particular postfeminist narrative about the status of men in our time. These discourses, characterized by Kimmel as "pro-male" but not necessarily "antifeminist," serve both as metaphorical resources for Page's model of the Y and, intentionally or not, as a way to increase the interest and reception of his research. In these ways, Page's Y chromosome model is valenced by the very particular pro-male masculinist politics of the postfeminist era.

Here emerges an asymmetry between Page and Graves. Page has, largely successfully, claimed the socially and epistemologically powerful standpoint of political neutrality. While Graves openly engages accusations of gender bias, as in a 2006 paper that opens with the line, "Is the human Y really disappearing, or is it simply the target of propaganda aimed at belittling men?," Page has never been compelled to respond to accusations that *he* is improperly "mixing politics and science." Graves has never claimed political neutrality—nor, perhaps, could she, for a variety of contextual reasons. As Graves memorably quipped, she is positioned as a "feminist Y chromosome ball breaker." The asymmetry between Page's positionality as a neutral, nonideological scientist and Graves's as a biased "feminist scientist" is maintained by persistent notions that feminist gender politics violate the parameters of good science.

There are many ways of conceptualizing scientific bias, but we tend to think of the term pejoratively, as suggesting a partiality of perspective that leads to errors in scientific reasoning. In chapters 5 and 6, I argued that presumptions that the X is the "female chromosome" and Y the "male chromosome" have contributed to just this kind of bias in genetic research on sex. Characterizing the debate between Page and Graves as a contemporary case of gender bias in science suggests that we are prepared to assert that the operation of gender conceptions here is, in some way, bad for the science. Such a conclusion would neglect contextual aspects of the interaction between social values, empirical evidence, and choice of scientific model in this case.[92]

Gender is not merely a subtext in the fray over Y chromosome models. It is a highly relevant contextual factor in the debate. Both Page and Graves are interested in producing an empirically adequate, explanatorily powerful model of the evolution of the Y. Reacting to what he sees as "political" attempts to denigrate males and the Y chromosome, Page is driven to validate a model of the Y as a sophisticated adaptive tool in the evolution of maleness. Page interprets data within the context of, and pursues hypotheses and research questions that would confirm, this model of the Y. Graves's feminist intuitions lead her to apply skepticism to Page's model of the Y as gene-rich and specialized for maleness. In her view, arguments that the Y chromosome is not degenerating, despite evidence to the contrary, betray an aversion to a model of the Y that is "wimpy" rather than one that is dominant and "selfish." She marshals evidence in favor of the Y chromosome degeneration theory and insists that it be taken seriously as a model of Y chromosome evolution.

Page's "pro-male" model of the Y led to meticulous scrutiny of the genes on the Y chromosome. It raised challenging and novel questions—such as whether some genes on the Y may be autosomal recruits rather than X-degenerates and whether the Y chromosome might have evolved mechanisms similar to recombination to secure its genetic integrity—that might not have been asked otherwise. Graves's feminist gender-critical perspective has generated novel hypotheses, countermodels, and critical approaches. It has inspired extensive comparative genomic investigation of Y-chromosomal genes and contributed to the validation of these methods for illuminating the structure, function, and evolution of the human genome. It has also raised fruitful questions—such as how to model and predict future Y degeneration—that may not have been asked otherwise.

As long as cross-charges of "gender bias" dominate the discourse between Page and Graves, important aspects of how gender is functioning in Y chromosome research remain problematically obscured. In the case of competing models of the evolution of the Y chromosome, one might ask: would the field move forward better—faster, or with fewer errors—without the gendered models? Graves's and Page's gender-inflected models have produced debate, clarified key issues, and opened new questions.

Perhaps Graves's and Page's research on the Y has been lively and productive at least in part *because* of the gendered models they have drawn on. We have here a case of competing biases, each productive in channeling particular programs of Y chromosome study. As these biases are the subject of active and open debate, they do not carry with them the same threat to scientific objectivity as do biases shared by an entire research community and thus invisible to its participants. For these reasons, "bias" emerges as a distinctly unhelpful framework for understanding how postfeminist gender politics interact with scientific debates over Y chromosome degeneration. As an alternative, I offer the concept of "gender valence" for modeling gender in science in this case.

The query of the sensitive gender analyst of science in such a case is not (or not merely) to diagnose bias. Rather, it is to precisely locate and reconstruct the content and context of the valencing gender conceptions and to assess the degree of criticality about the role of those gender conceptions in scientific research by the various actors involved. As the case of theories of Y chromosome evolution shows, changes in the gender system should not lead us to expect gender to one day disappear from the intellectual background of the science of sex, nor is it advisable to champion an unrealistic "gender-neutral" science. Gender conceptions—feminist, masculinist, antifeminist, and beyond—are an inevitable backdrop to the science of sex, and they can play a constructive role in science when they are subject to criticism. In light of this, our aim should be to construct a gender-critical practice of science in which debate about how gender conceptions valence scientific language, theories, and models is welcomed in the course of normal scientific practice.

◇

The role of gender in debates over the evolution of the Y chromosome presents a case study in the dynamics and historical specificity of the gender system viewed alongside rapid changes in scientific knowledge, institutions, and practices. In the 1990s, the success of feminist or gender criticism in reforming models of genetic sex determination led to greater awareness of how gendered assumptions can influence scientific knowledge. As we enter the postfeminist—and postgenomic—era of sex chromosome research, competing gender conceptions valence models of the Y chromosome in newly complex ways.

Not merely a source of humor and a means of interesting lay audiences in Y chromosome research, gender is also a factor in the choice of models, research questions, descriptive language, and global reasoning in studies of the evolution of the human Y chromosome. Gender works within Y chromosome science in ways that have much to tell us about present-day configurations of science, gender ideology, and feminist critiques of science. Rather than expecting gender conceptions to drop away from our scientific study of sex in the postfeminist age, we are better off acknowledging that all actors bring values to the table in scientific work on sex, and that these values can, when open to debate and scrutiny in an environment inviting a diversity of perspectives, play a constructive and clarifying role in evaluating the central claims and assumptions of our working scientific models. However, as long as asymmetries persist that lead us to regard critiques of feminist approaches in science as "gender neutral" while marking feminist gender-critical perspectives as "not properly scientific," this vision of transparent, reflective discussion of how gender conceptions can valence scientific knowledge will remain aspirational. With the knowledge that gender conceptions remain active, in a variety of ways, in our contemporary scientific research on the science of sex and gender, we are compelled to ask: how might we build a gender-critical genetics to suit the coming genomic age?

CHAPTER 9

Are Men and Women as Different as Humans and Chimpanzees?

In 2005, *Nature* announced the complete sequence of the human X chromosome. A headline-stealing paper in the same issue proclaimed that early genetic analysis of the X showed genetic differences between men and women to be far greater than previously thought.[1] "In essence, therefore, there is not one human genome, but two—male and female," stated coauthor Huntington Willard. *Newsweek* featured the finding, pronouncing that "The rift between the sexes just got a whole lot bigger. A new study has found that women and men differ genetically almost as much as humans differ from chimpanzees." Male and female genomes, the article concluded, have "an altogether different arrangement of gears."[2]

In the weeks following the publication of the article, authors Carrel and Willard promoted their findings as showing that males and females, like different species, have different genomes; that contrary to politically correct visions of a shared, universal human genome, males and females are more genetically different than had ever been conceived; and that genetics may hold the key to the "deep" differences between males and females. A *Los Angeles Times* piece quotes Willard saying "it is not just a little bit of variation. . . . This is 200 to 300 genes that are expressed up to twice as much as in a male. . . . This is a huge number." In a *New York Times* article in which Willard is quoted stating "men and women are farther apart than we ever knew," the *Times* writer is led to conclude that women are, indeed, "a different species."[3]

Sex chromosome geneticists and the media frequently present genetic analysis of the X and Y as offering, at last, the "real" story of—and the final word on—human sex difference. Mark Ross, head of the X chromosome sequencing effort, was quoted in *Science* as saying "now that we've got the sequence of both sex chromosomes, we can do a very detailed comparison [to] really ask the differences between male and female."[4] MIT Y chromosome researcher David Page concluded the 2003 paper detailing the complete sequence of the human Y by arguing that males and females differ genetically by approximately "2 percent" and predicting that the dogma of a single human genome would find its limit with sex:

> It is commonly stated that the genomes of two randomly selected members of our species exhibit 99.9% nucleotide identity. In reality, this statement holds only if one is comparing two males, or two females. If one compares a female with a male, the second X chromosome . . . is replaced by the largely dissimilar Y chromosome . . . . This common substitution of the Y chromosome for the second X chromosome dwarfs all other DNA polymorphism in the human genome.[5]

The sequence of the X and Y, he continued, offers "the near prospect of a comprehensive catalogue of genetic and sequence differences between human males and females."[6] Similarly, a 2003 *Boston Globe* article quotes Page saying "we all recite the mantra that we are 99 percent identical and take political comfort in it. But the reality is that the genetic difference between males and females absolutely dwarfs all other differences in the human genome."[7]

In this chapter, I critically analyze this vision of sex difference research in a genomic age. I do so by looking closely, through a gender-critical lens, at one recent effort to quantify genetic differences between the sexes: Carrel and Willard's dramatic claim that males and females are more different than humans and chimpanzees. Presenting an empirical and conceptual critique of these claims, I show how they systematically overstate differences between the sexes. I then examine the assumptions of what I call "thinking genomically" about sex differences. Recent genetic research on human sex differences attests to the emergence of a "genomic" concept of sex. I argue that genetic work on sex differences would do best to dispose of analogies between sexes and species, and free itself from the corresponding construct of distinct "male" and "female" genomes. In its place, I offer an alternative conceptualization of the sexes as a "dynamic dyadic kind," a concept with methodological implications for genomic research on sex.

"X-ESCAPEE" GENES

In their 2005 paper, Carrel and Willard advanced an "X-escapee" hypothesis of genetic differences between males and females. As discussed in chapter 6, one X chromosome in each female cell is permanently inactivated early in development, equalizing X chromosome dosage for males and females. It turns out, however, that some genes on the female's inactivated X may "escape" inactivation, continuing to express at some level in females (see fig. 9.1). X genes with Y homologues, for example, may escape inactivation in females in order to equalize levels of gene product in males and females. The escapee phenomenon was first demonstrated by Shapiro et al. in 1979 for the *STS* gene, which escapes inactivation on the X and for which there is a homologue on the Y. More recently in 1997, Carolyn Brown, with Willard and Carrel, identified

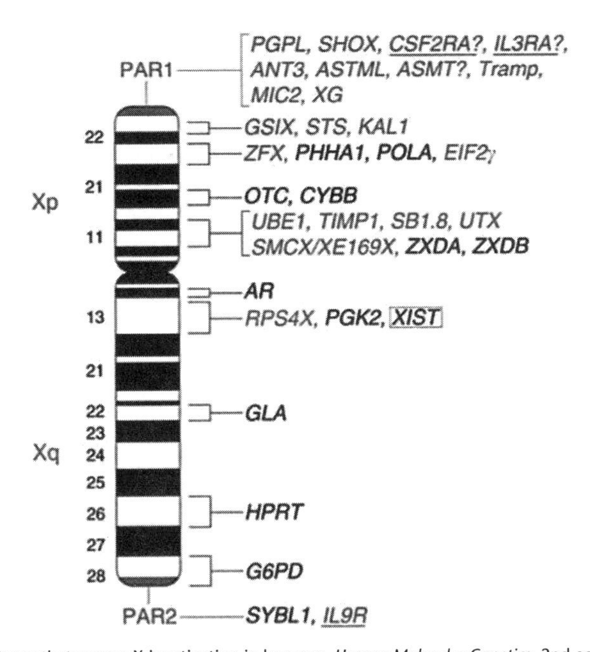

Figure 9.1. Genes that escape X inactivation in humans. *Human Molecular Genetics*, 2nd edition by Tom Strachan and Andrew P. Read. Copyright 2001. Reproduced with permission of John Wiley & Sons, Inc.

thirty-three genes that appear to escape inactivation and predicted that as many as one-quarter of X-inactivated genes may at least partially escape inactivation.[8]

In 2005, Carrel and Willard extended this result, using data from the newly completed human X sequence to produce the first comprehensive analysis of the extent of escape from inactivation on the human X chromosome. Using a clever experimental design—an in vitro assay of rodent/human fibroblast (connective tissue cell) hybrids, which allowed genes expressing from an inactivated X to be distinguished from those on the active X—Carrel and Willard were able to quantify and localize escape from X inactivation. They found a larger number of X-escapees than they expected. In fibroblasts, 15 percent of X chromosome genes permanently escape inactivation. In some women, the number is as high as 25 percent. This led Carrel and Willard to suggest that X-escapee genes represent a long-ignored piece in the causal picture of sexual dimorphism. As they wrote, "the female genome differs from the male genome in at least four ways":

> First, the Y chromosome endows the male with at least several dozen genes that are absent in the female. Second, the incomplete nature of X-inactivation means that at least 15 percent of X-linked genes are expressed at characteristically higher (but often variable) levels in females than in males. Third, a minimum of an additional 10 percent of genes show heterogeneous X-inactivation and thus differ in expression levels among females, whereas all males express a single copy of such genes. And fourth, the long-recognized random nature of X-inactivation indicates that females, but not males, are mosaics of two cell populations with respect to X-linked gene expression.[9]

X-escapees, Carrel and Willard concluded, "should be recognized as a factor for explaining sex-specific phenotypes both in complex disease as well as in normal, sexually dimorphic traits."[10]

The much-hyped estimate of 1 to 2 percent difference between males and females arose directly from this picture of genetic differences between males and females presented by Carrel and Willard. The reasoning goes as follows:

1. Fifteen to 25 percent of the genes on the inactivated female X chromosome escape inactivation, or as Willard stated, approximately 200 to 300 genes.

2. "Several dozen" genes (perhaps 50) are specific to the human Y chromosome.
3. These 250 to 350 genes may, in large part, specify sex differences.

If the going estimate of total genes in the human genome is 20,000 to 30,000, and 250 to 350 genes differ between males and females, then it might be said that males and females differ by approximately 1 to 2 percent of the total coding genome—more than the 1.06 percent difference between humans and chimpanzees.

HUMANS AND CHIMPANZEES

It is widely known that humans and chimpanzees share nearly identical genomes, a finding that stands as one of the most exciting revelations of the genomic age. It clinches the hypothesis of a close phylogenetic relationship between humans and chimpanzees, which in turn forms the foundation of our predominant theories of recent hominid evolution. That humans are "98 percent chimpanzee" is also among the most commonly known genetic factoids, featured in school textbooks, in genome-project literature for the general public, and as an object of humor in cartoons and on t-shirts and bumper stickers.[11]

Comparisons between humans and apes have a long history in biology and a prominent position in the history of scientific racism and sexism. "Chimp" is a racial and intellectual insult. From the eighteenth to the twentieth centuries, both white females and male and female blacks, as well as other minorities and marginalized groups, such as the Irish and the disabled, have frequently been claimed to be phylogenetically or morphologically closer to apes, or otherwise ape-like.[12] In the eighteenth century, women were typified as closer to nature—and thus to apes. Popular and scientific narratives and imagery depicted anthropomorphized female apes acting out gender-specific roles, and human women mingling with male apes.[13]

In the nineteenth century, physical anthropologists asserted that female brains are "closer in size to those of gorillas than to the most developed male brains," and that female skull structure was simian: "European women shared the apelike jutting jaw of the lower races." On the "Great Chain of Being," women sat below men, closer to the apes.[14] The nineteenth-century German physiologist Carl Vogt wrote that "we may be sure that, whenever we perceive an approach to the animal type, the female is nearer to it than the male."[15] Later, Italian criminal anthropologist Cesare Lombroso similarly embraced a hierarchy of closeness-to-animals in his atavistic theories of human behavioral "types." Women, animals, children, and the lower races were said to have a greater tolerance of pain. This of course was not seen as courage and endurance but "indifference" and "insensibility."[16] This distressing history underscores the importance of carefully interrogating contemporary reprisals of human-chimpanzee comparisons.

In the mid-twentieth century, cytogeneticists showed that humans and chimps carry an almost indistinguishable set of chromosomes. Later, molecular comparisons of humans and chimpanzees demonstrated that they share a nearly identical genome. Comparative analysis of human and chimpanzee protein structure in the 1970s found that human and chimp amino acid sequences differ by a mere 0.7 percent. Extensive analysis of aligned segments of coding DNA in the following decades, culminating in the first draft sequence of the chimpanzee genome in 2005, expanded estimates of overall human-chimpanzee divergence to 1 to 3 percent, with 1.2 percent becoming the generally agreed-on textbook statistic. It is not surprising that estimates of genetic differences between males and females were calibrated against the backdrop of the now widely accepted reference point of human-chimpanzee differences. And indeed, by one simple quantitative measure, human male and females could be said to be more genetically different than humans and chimpanzees. But this would, indeed, be too simple.[17]

There are a variety of levels at which the genetic similarity between humans and chimpanzees (or any two species) may be quantitatively and qualitatively analyzed. These include nucleotide sequence identity along aligned segments; large-scale structure and organization; microstructural elements and motifs; and gene expression profiles (see fig. 9.2). Visiting these various methods and the results that they produce reveals the complexities and limitations of comparative genomic methods for producing objective estimates of genetic differences.

The prevailing method for quantitative estimates of genetic difference is to compare, one to one, the identity of nucleotide or amino acid sequences along aligned segments of protein-coding DNA. In the 1980s, researchers compared 40,000 bases of coding genes on the human and chimpanzee chromosome 11. They found that humans and chimps differ by 1.9 percent. A variety of comparisons of this sort led researchers in the 1980s to estimate that human and chimpanzee genomes differ by anywhere from 1.8 to 3 percent. With the completion of a draft sequence of the chimpanzee genome in 2005, researchers were able to compare

FOUR METHODS FOR COMPARING GENOMES

Method #1 — Compare aligned sequences

HUMAN ...ATGCGCTAGTAGCTGCGAT...

CHIMP ...ATTCGCTAGTAGATGCGAT...

Method #2 — Compare large-scale gene organization and karyotype

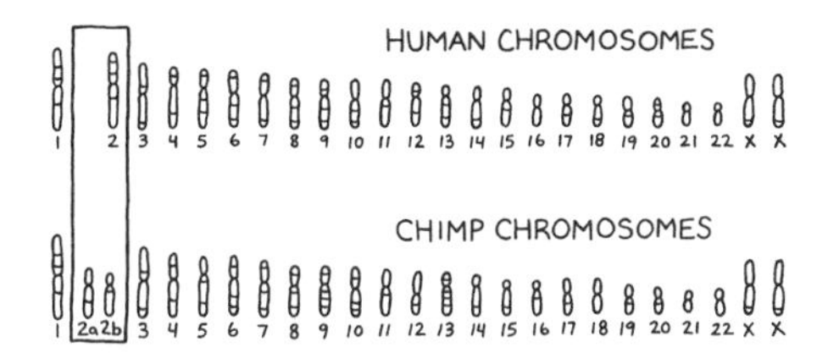

Method #3 — Compare microstructural motifs

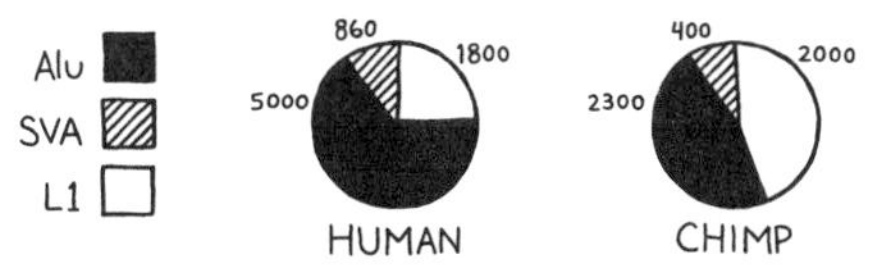

Source: Lock et al 2011

Method #4 — Compare levels of gene expression

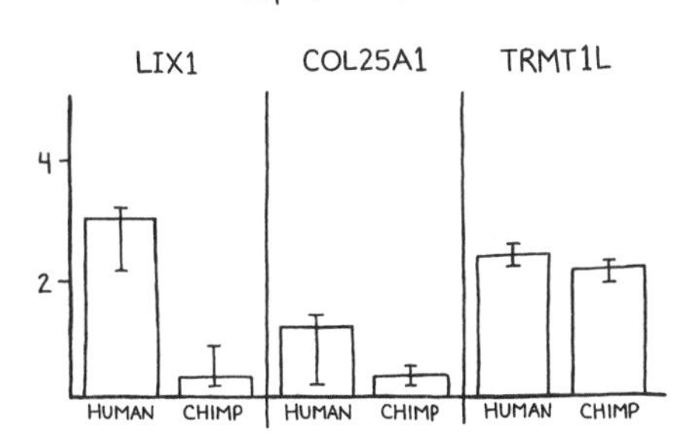

Differences in expression levels of LIX1, COL25A1, and TRMT1L genes in humans and chimpanzees.

Source: Bawand et al 2011

Figure 9.2. Ilustration by Kendal Tull-Esterbrook; copyright Sarah S. Richardson.

a far more extensive sample of 2,400 megabases (1 million nucleotides each) of aligned DNA sequence between humans and chimpanzees. They found a 1.23 percent rate of single-nucleotide substitutions. When variation within each species was factored out, they concluded that 1.06 percent or less of this represented fixed divergence between the two species. When synonymous nucleotide substitutions (DNA changes that do not lead to changes in amino acid coding) were pruned from the data, the rate of divergence between humans and chimpanzees along aligned segments of coding DNA was just 0.6 percent.[18]

Looking at single-nucleotide substitutions in aligned sequences alone, however, underestimates the amount of genetic difference between humans and chimpanzees. Consider the case of "indels," a kind of mutation that does not turn up in aligned sequence comparisons. Indels are insertions or deletions of one or more base pairs (but frequently large segments) that produce frameshift mutations in protein-coding regions. As such, they give rise to human- and chimp-specific sequences. With more sophisticated bioinformatics tools and the ability to examine longer segments of DNA, researchers are now able to estimate the impact of indel mutations on overall divergence between species. Indels, it turns out, result in three times more sequence divergence than single-nucleotide substitutions. Fifty-three known human genes are deleted entirely or partially in chimps, and all told, humans and chimps differ by an estimated 3,731 genes outside of aligned sequences. When taking indels into account, divergence estimates between humans and chimpanzees jump to 4 to 5 percent. In recent years (subsequent to the publication of Carrel and Willard's paper and the claim that males and females differ more than humans and chimpanzees, which was based on the earlier figure), this number has become the consensus estimate of overall genomic divergence between humans and chimpanzees.[19]

What we have discussed thus far represents analysis of coding DNA—one level of genomic comparison. Yet coding DNA represents less than 3 percent of the genome. Analysis of nucleotide sequence of coding DNA will fail to pick up macrostructural differences at the level of karyotype and large-scale gene organization—a second level of comparison. Unlike nucleotide sequence comparisons, macro-level differences are often difficult to represent quantitatively. Nonetheless, like nucleotide substitutions, large-scale events such as genomic rearrangements and expansions are functionally important. Scientists believe that structural rearrangements in the genome are a source of species-specific differences, a constraint on adaptive evolution, and a cause of speciation. While human and chimpanzee karyotypes are nearly identical, chimpanzees have forty-eight chromosomes and humans forty-six, with human chromosome 2 representing a fusion of two chimpanzee chromosomes (chromosomes 12 and 13). Chimpanzees also have approximately 1 percent more total DNA material than humans. The banding patterns on the tips of chimpanzee chromosomes differ from those of humans. Nine inversions—swaps of DNA from one arm of a chromosome to the other—also differentiate the chimpanzee and human genomes.[20]

A third type of comparative analysis examines the frequency of microstructural elements, such as repetitive motifs and virus DNA incorporated into the eukaryote genome, characterizing the composition of a species-typical genome. For example, in humans and chimpanzees, there are twice as many "Alu insertions" (sequences of DNA repeated at different locations on different chromosomes) in humans as there are in chimps, whereas "ERV class 1" elements ("endogenous retrovirus" elements, sequences in the genome that are derived from ancient viral infections) are more frequent in chimpanzees. In contrast, Line-1 (Long INterspered Elements—a type of retrotransposon copied from RNA to DNA and inserted, often in many copies, into the genome), SVA (composite retrotransposon elements unique to the primate family), and ERV class 2 elements are equally prevalent in the human and chimpanzee genomes. Some of these features, such as repeat expansions, Alu elements, retroviruses, and regions enriched for CpG dinucleotides, may have functional significance. Others are notable primarily as markers of genomic geography, indicating genomic changes since humans and chimpanzees diverged from their most recent common ancestor.[21]

Genetic sequence alone cannot predict gene expression—when genes are switched on, in which tissues, in coordination with which other genes, and at what levels of intensity. Gene expression profiles, which show levels of transcribed DNA, represent a fourth method of comparative genomic analysis. Genome-wide comparisons using DNA microarray technology—one of the latest technologies to transform the genome sciences—have become the focus of much recent work in comparative genomics.[22] Microarray studies produce full-color striated images, each its own spectacular sunset, showing changes in gene expression over time. Comparing which genes express in different tissues offers the promise of pinpointing precisely those genetic changes that result in functional differences between species. Since gene expression profiles capture the outcome of regulatory processes on gene activation and dosage, they also offer the hope of moving beyond the limits, noted above, of simply analyzing genomic sequence in search of genes specifying differences between species. Comparative gene expression profiling offers another layer of quantitative comparative genomic data that is not reducible to other levels of analysis. Because so-called "epigenetic" factors from an individual's life history and environment, such as nutrition and immunity, also affect gene expression, this measure has the potential to incorporate an ecological and developmental perspective into comparative genomics as well.

In sum, the genome is complex. The meaning of quantitative comparative genomic data is specific to the context, level, and method of analysis. Thus, ranking differences between human male and female genomes alongside differences between human and chimpanzee genomes is not a simple matter of counting genes that differ between them, as Carrel and Willard implied. Quantitative estimates of genomic divergence between humans and chimpanzees are based on differences detected along aligned segments of DNA. In contrast, genetic comparisons between males and females are based on a different kind of analysis. The claim that there are 50 genes unique to males, absent in females, is not a claim about divergence along aligned segments, but about unique genes. The claim that there are 250 to 350 genes expressing at higher levels in females is also not a claim about divergence along aligned DNA segments, but about relative levels of expression. Sex differences in the genome simply do not fit a comparative genomic model of analysis of aligned nucleotide sequence for quantifying species divergence. Valid comparative analysis requires comparing like kinds or categories; sex-specific gene expression is assessed along different dimensions than genomic differences between humans and chimpanzees. It is thus inaccurate and invalid to compare sex differences in the genome to genomic differences between humans and chimpanzees.

LOOKING FOR DIFFERENCES

Imagine that we set aside these concerns, however, and look strictly at Carrel and Willard's striking estimate of 250 to 350 genes that differ between males and females. A simple empirical critique may be made of the claim that males and females differ by "250 to 350 genes," or "1 to 2 percent." Indeed, as we shall see, it becomes apparent on examination that Carrel and Willard vastly overestimated the quantitative genetic differences between the sexes. They made assumptions at each level of reasoning, as well as in their model of sex difference, which systematically skewed their results in favor of the largest possible estimate of sex differences. Rather than 250 to 350 genes on the sex chromosomes that differ in functionally significant ways between males and females, the actual number is likely closer to a dozen.

Let's begin with the Y chromosome. Are there, as Carrel and Willard assert, fifty male-specific genes on the Y? When Carrel and Willard assert that there are "several dozen" male-specific genes, the assumption is that these are fully functioning true male-specifics coding for unique proteins. Recall that the Y chromosome evolved from an ancestral X (see chap. 8, fig. 8.2). Thus, the X and Y today share a large number of genes. They even exchange genetic material through recombination at the tips of the chromosomes. Most genes on the so-called male-specific region of the Y chromosome are descendants of genes that were originally on the X chromosome, or "homologues." A true male-specific gene must be a gene that (1) has sufficiently evolved since parting with the X to be, functionally, a different gene; or (2) has been recruited to the Y from another chromosome, or has otherwise independently arisen on the Y; or (3) has remained intact and functioning on the Y chromosome, but has disappeared from the present-day X.

Of the estimated sixty-nine genes on the Y, twenty-nine recombine with the X, and so are not male-specific.[23] Twenty-five outside of this recombining region have functional homologues (copies) on the X.[24] This leaves just fifteen that have no detectable X homologue.[25] Functional analysis of these fifteen male-specific genes shows that they contain a large number of duplicates and pseudogenes (ancestral genes that structurally resemble genes but are actually nonfunctioning). Additionally, roughly half of these genes are expressed specifically in spermatogenesis. They appear to be a family of very similar genes that are highly specific to the male gonad. (As such, they are likely to be of limited value for explaining global sex differences.)

What of the 200 to 300 escapee genes that are "doubled" in females? Carrel and Willard imply that escapees represent "extra" genes in females, leading to "double" the dosage of as many as 200 to 300 genes. But by and large, escapees express at *far* lower levels than their counterparts on the active copy of the X. In a follow-up study, Talebizadeh et al. found that "gene expression levels for a distinct gene that escapes from inactivation might be as low as 25% in the inactive X compared with the active X chromosome." Nguyen and Disteche similarly found that "only a few escape genes have a significant increase in expression in females, whereas most show a modest increase, no increase or even a decrease in expression"; moreover, "only one-fifth of human escape genes show expression from the inactive X chromosome that reaches 50% of that of the active X chromosome."[26]

Carrel and Willard also appear to have overestimated the number of genes that escape from inactivation. Just how many escapees there are has been a subject of much interest since the 2005 Carrel and Willard paper, and to a large extent, it remains an open question. Extensive subsequent investigation has failed, thus far, to replicate their striking result of 200 to 300 X-escapee genes. X-escapees that are candidates for explaining sex differences must be exclusive to the X. They must not be located on the PAR (the shared region of the X and Y). They also must not have an identical, fully functioning homologue on any other region of the Y. When these are ruled out, the numbers drop dramatically. Craig et al. found only *thirty-six* non-PAR escapee genes upregulated in lymphocytes (white blood cells). A more extensive in vivo (live tissue) study by Talebizadeh et al. found only *nine* non-PAR escapee genes that are expressed at a level of at least a 1.5 female-to-male ratio in at least three human tissues.[27]

Carrel and Willard's study used an in vitro (test tube) somatic cell hybrid system to assess escape from inactivation, while the Craig et al. and Talebizadeh et al. studies looked at in vivo expression on the inactivated X. The considerable difference in results indicates that while in vitro studies may establish the possibility of escape from inactivation, they will not sufficiently predict X expression in live bodies, in which other mediating factors and processes may work to suppress or otherwise equalize the effects of escape from X-inactivation.[28] "Factors other than the X-inactivation process (e.g., sex hormone influence, subject's age, tissue specificity and composition, and other factors that may alter the amount of mRNA production or degradation in a cell) may impact on expression outcome of X-linked genes," write Talebizadeh et al.[29] Variables such as age and nutrition will interact with sex difference as a factor in variation in X-chromosome expression. Carrel and Willard do not take this into account. Moreover, Nguyen and Disteche point out that in mouse studies, where there has been far more extensive profiling of escape from X-inactivation, researchers ultimately concluded that all known mouse X-escapees do have a Y paralogue (meaning that the X-escapee gene and a Y gene have nearly identical sequences and are derived from the same ancestor).[30] Future research may yet show more extensive equalization of gene dosage in human males and females, similar to mice.

A further limitation in studying the functional genetics of X-escapees is that the result varies depending on the tissue used. Which genes on the inactivated X are expressed in males and females will be highly specific to the tissue under consideration. Sexually differentiated reproductive organs show significant and stable gene expression differences between males and females, as one might expect. In contrast, the kidney, liver, and brain exhibit negligible differences, and these findings are often difficult to replicate.[31] This adds an important qualification to any *global* estimates of the quantitative, or percentage, genetic difference between males and females, a limitation that Carrel and Willard fail to address.

Furthermore, microarray technology, the most common system for measuring differences in gene expression, has well-known limitations for detecting the low-level differences in gene expression that characterize sex differences in most human tissues. Microarrays test the expression levels of thousands of genes at once. Such a system is poorly suited for detecting small differences in levels of expression, a level of resolution necessary to see the differences between the sexes. For these studies, the threshold for capturing expression differences must thus be brought very low. As a result, many false positives due to normal variation will show in the data, and the statistical power of predictions from microarray studies is greatly reduced. With large numbers of genes, the studies will tend to produce *some* evidence of difference; at the same time, they will have extremely low predictive power for identifying particular genes that differ between the sexes (see chapter 10).

A study of differences in gene expression in human male and female livers by Delongchamp et al., for instance, found that approximately 8 percent of genes showed different levels of expression between the sexes, but the estimated male-to-female gene expression ratio was small, the largest being 1.55. The authors concluded that "all estimates are less than 1.55 fold and they generally have narrow confidence intervals. Any gender [*sic*] differences that might be detectable through these arrays would be small." Unique genes and gene processes cannot be easily distinguished using this method: "There is evidence of a few small changes but the specific genes cannot be identified."[32]

Finally, Carrel and Willard reflexively assumed that genes that escape inactivation must do so because they play a role in sex differences. But, as Graves and Disteche have pointed out, this assumes that X escapee-ism is an adaptive mechanism "evolved to exploit differences in the expression of some genes between the sexes." It need not be the case that these genes are specially programmed to escape inactivation because of an important role in sex differences. Perhaps dosage equalization is not important for those genes, or perhaps the dosage difference is so small that the effect is null or negligible. Graves and Disteche note that the location of the escapees suggests that they may simply be the most recent genes added to the X, still in the process of being incorporated into inactivation:

> These escaping genes, expressed from both alleles (albeit at a reduced level from the inactive X) lie mostly on the part of the X . . . that was recently added to the X and Y. They are being slowly recruited into the X-chromosome inactivation system, but there appears to be no particular hurry, suggesting that their dosage inequality is not a huge problem that needs to be solved immediately.[33]

Graves and Disteche conclude that a role in sex differences is just one of several explanations compatible with the evidence about X-escapees.[34]

The most plausible picture of the role of genetics in sex differences is one in which genes play a role in hormone regulation, which in turn influences primary and secondary sex characteristics. As Rinn and Snyder point out, although there may exist some "hormone-independent sex differences," sex differentiation is principally caused by "sex hormone exposure at key developmental time-points," mostly during prenatal stages. Much subsequent sex-specific gene expression in the somatic organs is caused by dimorphic patterns in the release of growth hormone. The developmental and physiological context of gene action, rather than the number of genes, presence or absence of genes, or up- or down-regulation of genes, is therefore absolutely critical to understanding the role of genes in sex differences. If what matters is developmental and physiological context, the same gene expressed at the same level in males and females could have different effects; similarly, large male-female differences in gene expression could be wholly insignificant.[35]

In summary, there are not nearly as many genes that differ between males and females as originally estimated. Carrel and Willard's striking estimate of a 250 to 350 gene spread between males and females lies, at best, at the very extreme upper end of the range of likelihood. Carrel and Willard's high estimate of genetic differences between human males and females was presented as affirming a traditional gender-ideological view of the sexes as "Mars and Venus"—so unbridgeably different that they are like different species. In fact, the real finding is the strikingly tiny number of genetic differences between males and females. There are likely fewer than a dozen unique fully functioning genes on the sex chromosomes that express differently in males and females. Moreover, current research strongly suggests that both male-specific Y genes and female-upregulated X-escapee genes, in whatever numbers, are likely to be of limited value for explaining global sex differences.

"THINKING GENOMICALLY" ABOUT SEX DIFFERENCE

Having examined some of the empirical weaknesses of efforts to quantify sex differences in the genome, let us now turn to the conceptual model of biological sex differences advanced by Carrel and Willard. At issue is Willard's conclusion that "*in essence, therefore, there is not one human genome, but two—male and female.*" This arresting claim has gone unchallenged, inviting no formal critique from the genetics community, let alone a letter to the editor or a raised eyebrow. Do human males and females have different genomes? Can human males and females be compared as we might compare species or genetic populations?

The question of whether it is best to think of two human genomes rather than one is more complex than it first appears. It is not a simple factual question. Clearly it is possible to compare the genetic makeup of males and females using the high-throughput and bioinformatic tools of contemporary genomics, and to find some differences between them. Yet genetic and genomic differences between groups are not sufficient to establish the much stronger claim that these groups have "different genomes." In other cases in which there is variation between human populations (for instance, people of different continental ancestries), we do not conceive of group differences as different genomes, but as genotypic diversity within the human genome. The question is a model-theoretic one and our choice of how to answer it is based not solely on the empirical extent of genetic difference between males and females, but also on the explanatory aims at hand and the values and social aims of the researcher or research community.

Part of what gives Willard's italicized statement above its effect and significance is its startling reversal of the mantra of the 1990s Human Genome Project—that there is a single human genome and that humans are 99.9 percent identical. The idea of a single, universal human genome underpinned the Genome Project's central logic that sequencing a single human male would reveal the human genome, and that this would in turn illuminate the fundamentals of human biology and disease and unravel the natural history of the human species. From a genome sequencing perspective, the fact that female bodies have two Xs while male bodies carry X and Y matters not at all. The essential hereditary material of the human species is contained in a single haploid set of autosomes, plus one of each sex chromosome.[36] The power and importance of this idea of a single, shared *human* genome in late twentieth-century science and liberal social discourse cannot be overestimated.[37]

One worry, then, is that Willard's construction of sex differences in the genome as "two different genomes" inaccurately implies far greater genetic differences between the sexes than is the case. Sex differences in the genome are very, very small: of 20,000 to 30,000 human genes, marked sex differences are evident in perhaps a dozen genes on the X and Y chromosomes and, it is hypothesized, a smattering of differently expressed genes across the autosomes. Dogged searches for sex-based gene expression differences in dozens of tissues in the human body, including the brain, have yielded very small differences, often unreplicable.[38] In DNA sequence and structure, sex differences are localized to the X and Y chromosomes. Males and females share 99.9 percent sequence identity on the twenty-two autosome pairs and the X. The handful of genes on the Y are focused largely on male testes determination and spermatogenesis. Thinking of males and females as having different genomes exaggerates the amount of difference between them, giving the impression that there are systematic and even law-like differences distributed across the genomes of males and females.

But regardless of the number of genes that are found to differ between human males and females, I argue that we should resist the construction of genomic sex differences. "Genome" has become a concept used to fix research agendas and horizons and make salient certain ontological categories in the genetic landscape. At this time, "genome" is a powerful word with enormous authority and resonance and the ability to shape scientific research priorities. We have a choice. Genomic research on sex differences is just emerging. We have the opportunity to articulate how we would like to conceptualize these differences and, in doing so, to learn from history, heading off the inscription of a genomic conception of maleness and femaleness as distortive as the notion of the X and Y as the "female chromosome" and the "male chromosome."

SPECIES HAVE GENOMES

In common scientific parlance, a "genome" refers to the genetic code specifying a species. Textbooks frequently define a "genome" as a species' genetic "instructions." "Genome" refers to the complete gene complement—chromosomes, genes, and, increasingly, the relevant regulatory and epigenetic apparatus. This is the sense in which we can have a "Human Genome Project." The term "genome" is used, in this context, to refer to the entire genetic content of the species. This includes, in principle, the profile of genetic diversity within a species. Individuals in a species may have different variants of a gene—for example, eye color—but they still share the same genome. In lay terms, the human genome is what people mean when they refer to "the gene pool."[39]

Differences between species' genomes may be genetically quantified. The results can reveal phylogenetic relationships between species, which are often referred to as estimates of "genetic distance" or "genetic divergence." As we have seen, work of this sort has validated the hypothesis that chimpanzees are among the closest living human relatives. Comparisons between populations within a species can also be generated, but they use different methods. These comparisons make use of a set of highly formalized model-theoretic assumptions that permit making certain kinds of inferences about *geographic distance* and *population structure*. These kinds of studies have corroborated the "Out of Africa" hypothesis of human migration by showing that human genetic diversity maps onto human linguistic diversity and flows along the historical pathways of human migration and colonization. Since human populations have continuously interbred with one another and no population is homogeneous and genetically distinct, these comparisons are considerably more complex than species comparisons. Individuals of different racial and ethnic ancestries may be more or less likely to carry different common human gene variants depending on their ancestry, but they do not have different genomes.

In urging us to think genomically about sex differences, Willard situates genetic sex difference as a comparative genomics question, similar to the comparative genomics of species. Nowhere is this more evident than in the presentation of human-chimpanzee differences as a measuring stick for male-female differences. Pressing the idea that there is not one human genome, but two separate male and female genomes, Willard reveals the persistence of phylogenetic thinking in the background of biological models of sex difference. As such, the projection of phylogenetic and comparative genomic language and models into the study of sex differences in the genome makes salient the continuing influence of a long-standing ontological confusion about sexes, species, and populations as biological classes.

Strictly, of course, sexes are *not* accurately analogized to species. Species are the primary unit of classification in biological taxonomy. Commonly, species are defined as reproductively isolated interbreeding groups of organisms, as descendants of a single common ancestor arising from a speciation event, and/or by shared phenotype such as morphology.[40] Sexes are not lineages. Males do not produce males, females do not produce females. Males and females mate, and their male and female offspring carry a largely random combination of paternal and maternal genetic material. Because of sexual reproduction, the sexes do not meet any of the common criteria for a species, including interbreeding, shared common ancestry (monophyly), morphology, and spatial and temporal boundaries. Instead, human males and females are biological subclasses of a sexual species. As a class, males are not descended from a unique common ancestor, nor do they breed only with one another; the same goes for females. Male and female morphologies are differentiated but continuous. Human male and female species-specific activity is highly integrated and cohesive. Males and females occupy the same spaces and times. They exchange genetic material through reproduction and are interrelated to a high degree. The comparative genomics reference model of phylogenetic descent and divergence is, therefore, clearly not appropriate for describing sex difference.

Because species, populations, and genomes are the primary units of comparative work in biology, however, for many it seems natural, perhaps unavoidable, and possibly harmless for them to surface as analogues in comparative work on sexes. Certainly, when pressed, most biologists will readily agree that there are significant disanalogies between sexes and species. Yet many still see nothing wrong with comparing the global genetic differences between sexes with global genomic differences between species. I suspect that an intuition of ontological symmetry underlies this reasoning. On this view, sexes and species are similar kinds of kinds—ontologically, species and sex both hold status as low-level, core classificatory units in biology. An adherent of this view would argue, then, that while species and sexes are not the *same*, they are *comparable*. Thus, according to this view, while phylogenetic reasoning certainly does not apply to the sexes, a comparative genomics approach does not seem distortive or problematic as a background framework, model, or metaphor for comparing the sexes.

In the following, I show that phylogenetic, genomic models or metaphors of sex differences are not innocuous. "Thinking genomically" about sex difference misconceives the relational nature of sex difference, exaggerates the amount of differences between the sexes, and misconstrues the relationship between genotype and phenotype.

SEXES VERSUS SPECIES

Philosopher of biology John Dupré is one of the few to reflect deeply on the concept of sex in biological explanation. He offers a useful framework for approaching our question. Dupré identifies the ontological grouping of sex and species as a common conceptual confusion in biological explanation. The use of the category of "sex" provides, for him, a perfect example of the tendency of biologists to "extend the relevance of explanatory categories beyond their empirically warranted limits." Dupré argues for what he calls "categorial empiricism," "a plea for complete empiricism with regard to the explanatory potential of particular kinds."

Dupré would likely regard the genomic view of sex as problematic because it flattens biological ontology, obscuring and eliding the empirical constraints of different classifications, or kinds, in biology. As Dupré argues, as an explanatory concept in biology, "sex" both *crosscuts* and is *nested* in species. Sex may be understood, first, as a high-level generic kind in biology. Very few, if any, generalizations or laws may be made about sex at this level. For instance, among sexually reproducing species, many species with males and females have no X-Y sex-determining system. Sexual dimorphism, as well as mating, parenting, and sex-gender systems, varies so profoundly across species that sex ("maleness" and "femaleness") carries minimal explanatory value as a high-level kind. Second, sex may be understood as an empirical subdivision within a species. Here sex is a grouping that may have explanatory power with regard to many aspects of the behavior, organization, and natural history of a sexual species. Consider, for example, the diverse explanations of sex differences and reproductive roles specific to honeybees, mallard ducks, and humans.[41] Human and woman is "a nesting of natural kinds," Dupré argues, but if we are good categorial empiricists, "it would be absurd to suppose that man and woman, say, were 'better' kinds than humans," unless such a distinction is merited by the role it is playing in an explanation.

Dupré's "categorial empiricist" critique of analogizing sex and species provides one compelling reason why we cannot rank the differences between sexes alongside differences between species. Categorial empiricism, together with certain facts about the practical explanatory payoff of employing one category or another, provides good grounds for the view that human males and females can never be "more different" than humans and chimpanzees, no matter how many genes are found to differ between them. Because we are constrained by the empirical fact of the mutual existence of males and females in a species, even a finding that female humans and female chimps are more "genetically similar" to one another than are human males and females would not mean that sex is prior to species, nor would it require the radical revision in our system of biological taxonomy hinted at by Willard's claim.[42]

But nothing about categorial empiricism, however, rules out comparing the sexes globally in the *way* one might compare species globally. To see why this is inappropriate, we need to look to another type of ontological confusion at work in comparative genomic thinking about differences between the sexes. The historian and philosopher of science Evelyn Fox Keller has noted the persistent tendency of geneticists to think of the sexes as separable, individual classes—which she termed the "discourse of reproductive autonomy" in theoretical population genetics. Thinking of the sexes as autonomous, rather than reproductively interdependent, can introduce errors and distortions into genetic modeling. For instance, as Keller pointed out,

> even with random mating, the probability that a male will mate cannot in general be assumed to be one, but is contingent on the availability of females, and therefore depends on the proportion of males to females in the population; that is, on the relative viability of males and females.[43]

That is, the population dynamics of one sex cannot be modeled without the other. Keller urged biologists to innovate new models that would make visible the dynamics of sex in population biology. In the last decade, there have been important steps in this direction.[44] Much of biology, and even most population genetics research, however, continues to assume the reproductive autonomy of males and females, leaving sex fuzzy and undertheorized as an explanatory concept.[45]

Keller argued that the failure to theorize sex as an explanatory concept in biology was due to a prevailing model of heredity as residing in, and natural selection as acting on, the individual organism. In sexual species, individuals do not replicate themselves—a complicating factor for modeling population dynamics. Rather than dealing with the "complications of biparental inheritance," Keller showed, population geneticists have folded sexual reproduction into an individualist model of population genetics, represented most poignantly by the idealized assumptions, known as the Hardy-Weinberg equilibrium, that allele frequency is constant, population size is infinite, and mating is random. Keller suggested that sex has been undertheorized in biology because it poses an uncomfortable challenge to this predominant model of evolution. As she wrote, "for sexually reproducing organisms, fitness is *in general* not an individual property but a composite of the entire interbreeding population"; sexual reproduction "undermines the possibility of locating the causal efficacy of evolutionary change in individual properties."[46]

Sex is commonly formulated as a property of individual organisms. The sexes are constituted as two different classes of organisms marked by this property. But as Keller reminds us, from the perspective of population genetics and evolutionary theory, the sexes are properly conceived as units of two (or more, in some species). In calling for a theoretical intervention into population genetic models of sexual reproduction, Keller noted that a "*mating pair* . . . is a more appropriate unit of selection than the individual, but the fact is that mating pairs do not reproduce themselves any more than do individual genotypes." This insight that sexes, not as individuals, but in some substantial way, *as a class*, are paired and interdependent, forms the kernel of the concept of sex as a *dynamic dyadic kind* that I propose as a remedy and alternative to thinking of the sexes as individual, autonomous kinds in sexual species with two gametic sexes, such as humans.[47]

SEX AS A DYNAMIC DYADIC KIND

Classification of the particulars of nature into "kinds" is an integral part of scientific reasoning. However, what do we do with a kind that is at once one and two, like the genome of a sexual species? There are differences between properties of a population, properties of a pair, and properties of individuals. Thinking about "sex" requires paying attention to these differences. Sex is a relational property of individuals within a (sexual) population or species. From the perspective of genetic population modeling, sex is a highly relevant property of individuals in a population. "Males" and "females" are also biological subclasses of sexual species. While sexes are frequently explicitly or implicitly analogized to populations within a species, sex is not simply a property of individuals, nor is it simply a subclass. Because sexual reproduction is essential to the propagation of the species, the sexes are profoundly interdependent. The population dynamics and fitness of one sex cannot be modeled without that of the other. From the perspective of evolutionary and population genetic modeling, sex is an irreducible dyad. Moreover sex is relational. The sexes are not fixed and dichotomous subclasses within a population, but in dynamic interdependence and interaction with one another. Genetically, sex is a *dynamic dyadic kind*.

Understanding the male and female sexes as an irreducible dyad exposes a core difference between sexes and populations or species. Populations and species are usefully understood as "individual" kinds, rather than as dyads.[48] Like an individual organism, populations and species are continuous, cohesive, localized in space and time, integrated, causally connected, and related to a high degree.[49] The criteria of monophyly (or shared lineage) and reproductive interbreeding, which biologists require for a group of organisms to qualify as a species or population, capture these features of continuity and interrelatedness. For these reasons, species are, substantially, "individual" kinds.

The distinction between individual and dyadic kinds spells a critical ontological difference between sex as an explanatory category and population/species as explanatory categories. If sex is dyadic, one sex cannot be treated as an autonomous class, independent of the other. Sexes cannot be separated and compared globally as species and populations might. This would misrepresent the dynamics of sex as a biological kind. Another consequence is that studying the population dynamics or biology of one sex *without* an adequate understanding of the other clearly emerges as methodologically nonrigorous, underdocumented, and lacking in explanatory power.[50] Idealizing the sexes as different classes, or kinds, rather than as continuous, interdependent, interacting classes, contributes to lazy sex difference claims. The concept of sex as a dynamic dyadic kind presents a clear methodological constraint against these kinds of spurious comparisons. That sexes are not autonomous, individual classes but interdependent, permanently coupled, interacting, subclasses of species means that it is inappropriate to isolate and apply comparative genomics approaches globally to sexes as one might compare species or populations. Findings of sex differences in the genome must be conceptualized within the dynamic frame of contingent, reactive strategies of the human genome within differently sexed bodies.

But is it possible to embrace the dynamically dyadic nature of human genomes without simply reprising retrogressive binary thinking about the sexes? As gender theorists have well demonstrated, binaries invite dualistic, dichotomous thinking, so that it becomes difficult to think of a group of two without subsuming one into the other, ranking them, implying polarity or complementarity, or posing them as opposites.[51] Binaries tend to imply exhaustive categories and to drive reasoning toward the detection of difference as fixed polarity. In biology, binary thinking downplays the great similarities and continuities between the sexes. Binary thinking also effaces the existence of biologically intersex individuals, who may comprise as much as 1 percent of the human population. Finally, the notion of sex as dyadic may also seem at odds with the plurality of arrangements of sex in relation to gender observed in human social life. Many biologists and gender analysts of biology would like to see a working concept of sex in biological research that recognizes the nonbinary and diverse nature of biological expressions of sex.

I am deeply sensitive to these concerns, but I believe that they are misplaced in this case. Note that the concept of sex as a "dynamic dyadic kind" refers only to gametic sex—a gene's eye view of sex. While gender is, in many contexts, plastic and plural, gametic sex is, in two-sex species, dimorphic. In mammalian biology, the union of two different gametes—male and female—is required for reproduction, and the two sexes present reliably different morphology and behavior arising from their reproductive roles. That this two-ness of sex is perceived as a binary makes it unique and troubled in biological ontology. Species and populations, for instance, exist in a dynamic multivalent field, but in mammals there are two gametic sexes, and not more or fewer than two. Even intersex conditions do not change the assumptions of our genetic models of sexual reproduction and mating since they do not alter the two-gamete egg and sperm model of sexual reproduction. While it is essential to acknowledge the plurality and social contingency of sex and gender forms, the (present) necessity of male-female pairing for reproduction requires a different approach to the biological concept of sex. A strategic conceptual intervention into genetic research on sex differences must grapple with this two-ness of biological sex in a meaningful way while transforming it to avoid the pitfalls of binary thinking.

In contrast to the binary framework that underpins Willard's "genomic thinking" about sex difference in biology, the concept of sex as a "dynamic dyad" conceives of the sexes as knitted together and nonautonomous. The sexes are irrevocably, fundamentally paired; they are interdependent and interact; and the dynamics of one cannot be modeled without the other. The concept of sex as a dynamic dyad directs biological modeling away from sex binaries—the sexes as different kinds or populations—and toward a gene's eye view of collaboration, interaction, exchange, and interdependence between the sexes—the sexes as a dyadic unit. In these ways, the concept of sex as a dynamic dyadic kind advances a picture of sexual diversity quite different from the type of reductionist and ideological binary thinking about sex that often characterizes both popular and scientific discourse. A critical conceptual shift in thinking of this sort about the ontology of genomic sex will be required as we approach the genomic, and postgenomic, age of sexual science.

◇

Focusing on the claim that human males and females are more different than chimpanzees and humans, this chapter has engaged the new and challenging philosophical and biological questions posed by genomic sex difference research. The claim is not a straw man, but one put forward recently by prominent geneticists. It received significant uptake in scientific and popular media. Finally, it raises the history of problematic deployments of comparisons between apes and humans.

I have argued that males and females are not like different species or populations, and that comparative genomic and phylogenetic models are inappropriate for conceptualizing biological sex differences. Examining genetic divergence between the X and Y chromosomes and comparing gene expression levels in males and females across the genome leads to valid, testable functional hypotheses about genetic loci that may play a part in phenotypic differences between the sexes. However, while this kind of comparative work also permits estimates of *phylogenetic distance or divergence* when looking at species and populations, it does not permit these kinds of global, whole-genome estimates of the difference in the case of the sexes. Thinking genomically about a class of organisms implies that they are lineages or autonomous genetic populations. It invites creeping populational and phylogenetic thinking, in which difference becomes the more commanding "divergence" and "distance." Because sex is a dynamic dyadic kind, this model of sex differences is distortive. I find that there are not strong empirical, explanatory, social, or ethical reasons for genomicizing sex differences in this way. For these reasons, it is more advisable to refer to "sex differences in the human genome" than to a "male genome" and a "female genome."

Genomic thinking about sex is advancing at breakneck speed. "Dutch Scientists Sequence Female Genome," a *Biotechniques Weekly* newsletter recently announced, referring to a Leiden University initiative to sequence the DNA of a human female.[52] News coverage hyping this event raised no questions about the validity of the construct of a "female genome," nor did other scientists. That these recent claims have not attracted scrutiny from researchers represents a failure of criticality in this area of genetic research. As this book has shown, a lack of criticality about assumptions about sex and gender difference can contribute to bias in research and have harmful social consequences. With effort, however, it is possible for such a criticality to be nurtured and developed in a scientific community.

CHAPTER 10

Gender and the Human Genome

Sex Itself presents the history of a particular way of thinking about sex: as a binary that is ultimately biologically ordained in the human genome. This way of thinking crept its way into the casings of the vivid binary of the X and the Y chromosomes and over the course of the twentieth century became the foundation of the biological account of sex. The X and Y were not destined to become gendered objects of sexual science. The common use of the descriptor "sex chromosome" for the X and Y was neither obvious nor inevitable. It was a contingent product of interactions with major movements in genetics and the science of sex in the early twentieth century. The construct of the "sex chromosome" landed the X and Y in a class with famously gendered objects of scientific fascination such as the egg and the sperm and estrogen and androgen.

This sexing of the X and Y had important consequences for the subsequent development of genetic theories of sex and gender. Researchers ever since have sought—and seen—"sex itself" in the X and Y chromosomes. The tendency to see the sex binary as writ molecular in the X and Y chromosomes is most evident in the gendering of the X as female and Y as male. The cases of XYY supermale theories in the 1960s and 1970s and of X mosaicism theories of normal and pathological female biology show how this way of thinking about the sex chromosomes coheres and protects from criticism otherwise weak scientific hypotheses.

The notion of the X and Y as "sex itself" was updated in fascinating ways as late twentieth-century feminism, and postfeminism, redrew the terrain of gender politics. The X and Y, and theories of genetic sex more broadly, began to register changing conceptions of femaleness and maleness. The cases of the search for the male sex-determining gene and the debate over Y chromosome degeneration offer analytically challenging studies of the interaction of fluctuating gender conceptions with changes in science. These cases present an opening for a reformulation of "sex itself." Yet, the binary notion of sex also persists in robust ways. Examining the biology of sex differences at the dawn of the genomic age suggests that it may very well blossom again, elaborated in the "male" and "female" genomes.

LESSONS FROM THE SEX CHROMOSOMES

This book makes visible the operation of gender conceptions in human genetics. Gendered assumptions have led geneticists to bypass standard methodology, ignore alternative models, privilege certain research questions, and skew their interpretation of evidence. This history holds lessons for the ongoing practice of genetics and genomics today.

Two patterns of traditional gendered thinking persistently appear in genetic models of sex. First, genetic research on sex and gender privileges the question of sex differences, framing research questions and driving data interpretation toward documenting, modeling, and explaining differences between the sexes. As we have seen, throughout the twentieth century, and now in the twenty-first, geneticists have used sex chromosome studies to argue that there are new, additional, or "deeper" differences between males and females than once thought, to revive slumbering theories of sex differences, and to attempt to validate or quantify sex differences once and for all. As I showed in chapter 9, for example, interest in locating sex difference in the genome has led even to the suggestion that there is not "one human genome"—the humanist promise of the Human Genome Project—but two. The emphasis on *differences* between male and female genomes leads geneticists to exaggerate small differences as global differences, and can be a source of bias in genetic models of sex.

The second pattern of gendered thinking in sex chromosome research is the gendering of the X and Y. Researchers consistently animate the X and Y chromosomes with feminine and masculine qualities. The gendering of the X and the Y introduces a false symmetry and a rigid binarization into sex chromosome research. Thought of as "she and he," the X and Y are conceptualized as the sex binary writ molecular. The false symmetry accorded to the X and Y oversubscribes and overequalizes their importance to the determination of sexually dimorphic and reproduction-related traits.

History yields concrete insights about sex chromosome science that might be productively taken up by genetic researchers today. Three prescriptions follow from this book's analysis. The first is to reject the concept of different male and female genomes and formulate alternative frames for conceptualizing sex differences in the human genome. The second is to resist a sex chromosome-centric approach to the genetics of sex differences. The third is to consider sex-neutral alternatives to the terminology of "sex chromosomes" for the X and Y.

Sex Is a Dynamic Dyadic Kind

Humans are terrible with twos. Male and female, cat and dog, Macintosh and PC. We tend to order twos into binaries: complementary, opposite, or ranked with respect to one another. When imported into the science of sex, this kind of thinking can distort our reasoning, turning sex differences into commanding sexual dimorphisms. Conceptualizing sex differences in the genome as a "male genome" and a "female genome" walks right into this trap. History shows us the pitfalls of binary thinking. We can do better in a genomic age. We should seek a framework for understanding genomic sex dimorphism that appreciates sex as a distinctively dynamic, dyadic, relational kind of biological difference.

Language and terminology matter. A well-constructed and aptly named concept can transform thinking in a knowledge field. Presently, no ontological term adequately captures sex as an interactive and interdependent biological property of individuals and of populations. In the absence of a reorienting alternative, we fall back on notions of biological kinds such as "species" and "populations." As an alternative to thinking of the sexes as distinct, binary, species-like subpopulations, in chapter 9 I proposed sex as a "dynamic dyadic kind" as a creatively generative construct for orienting biological models of sex in a genomic age.

Values, both empirical and contextual, have a role in our choice of ontological frameworks, models, and descriptive language in science. Whether the concept of "genome" is apt, clarifying, and constructive for characterizing genetic sex differences is not settled by measuring quantitative genetic differences between the sexes. Rather, the conceptual framework we choose should reflect the roles that we want both sex difference and the concept of "the genome" to play in biological explanation and ontology. It should also reflect the importance that we place on countering harmful gender-ideological thinking in science and society.

At present, for instance, we choose to work with a concept of a "*human* genome" and we choose *not* to call haplotypes associated with racial and continental ancestry "genomes." Analysts of genetic research on race have argued strenuously that data should not be organized and marked by race, carving preconceived social ontologies into our genomic models and DNA sequence databases in the opening gestures of the human genomic era. I suggest we do the same with respect to sex difference.

Sex Chromosome-Centrism

There remains a persistent assumption that the sex chromosomes are, in fact, the locus of genes responsible for sex differences. Geneticists, clinicians, and science reporters commonly assert that genes on the sex chromosomes will elucidate the genetic substrate of sexual dimorphism. According to this view, genes or gene processes involved in primary and secondary sex characteristics should be located, or concentrated, on the sex chromosomes. This approach reflects the expectation, and key assumption, that the genotypic dimorphism of the sex chromosomes is the biological substrate of, or controls, the phenotypic dimorphism between male- and female-bodied individuals that we observe. This assumption, however, is not valid.

Sex chromosomes evolved as a sex-determining mechanism; they carry a critical switch in a larger pathway of genes that determine sex. That the X and Y carry the key genes involved in the characteristic sexual dimorphisms of a species, however, is dubious. We know that genes with sex-specific fitness effects need not be located on the sex chromosomes. There are many examples of autosomal genes critical to sex determination and sex differentiation. Genes essential to the maintenance of ovarian and testes differentiation, such as *DMRT1* on human chromosome 9 and *FOXL2* on human chromosome 3, are not located on the sex chromosomes.[1] Errors in these autosomal genes lead to sex reversal in chromosomally typical XX and XY individuals. We also know that extreme sexual dimorphism is observed in many species that lack sex chromosomes. Thus, sexual dimorphism may be reliably transmitted without a genotypic dimorphism such as sex chromosomes. Finally, the degree of sex dimorphism varies significantly among mammals with nearly identical sex chromosome complements. Degree of genotypic differentiation between sex chromosomes does not correlate with degree of phenotypic sexual dimorphism.

There is a theoretical case to be made, however, that where sex chromosomes are present, sexual dimorphism may be expected to be associated with X- or Y-linked effects. In a recent development of this hypothesis, Fairbairn and Roff argue that among organisms with the XX/XY sex-determining system, genes with differential fitness effects in males and females may be expected to accumulate on the X chromosome. The authors concede that "most genes contributing to [sexual dimorphism] for polygenic traits are autosomal."[2] But they point out that a particular kind of sex-specific gene, one that arises as a recessive trait on the X chromosome and benefits males, may be found preferentially on the X. Importantly, the theory predicts that X-linkage should contribute most to those sexually dimorphic traits limited to males. Indeed, studies have shown that genes for sex and reproduction in males are concentrated on the X chromosome (in contrast to the prevailing assumption that the X is the "female chromosome" discussed in chapter 6).

Fairbairn and Roff's analysis makes clear the qualified and hypothetical nature of the assumption that genes on the sex chromosomes underlie global sex differences. While genes that increase male fitness, such as those involved in spermatogenesis, are predicted to be concentrated on the X, there are no in-principle reasons to expect that most genes, or even the key genes, influencing sex differences in males and females will reside on the sex chromosomes. Focusing on the sex chromosome dichotomy in the genome excludes other sex-limited or sex-dichotomous expression patterns across the genome. These may include maternal and paternal imprinting and other sequence-independent, heritable epigenetic effects on gene action, regulation, and repair, as well as autonomous sex-specific regulation of gene expression by the sex hormones.

The assumption that "genes for sex" lie on the sex chromosomes is seductive—the simple, clear-cut genotypic dimorphism of the X and Y fits dominant views of the sexes as distinct, fixed dichotomous types. For this reason, sex chromosome-centrism pervades much recent work on the genetics of sex and gender, even though researchers often explicitly acknowledge that genes relevant to sex may be located anywhere in the genome. Attempts to estimate and localize genetic difference between human males and females using the X and Y chromosomes, as in the case discussed in chapter 9, represent an especially revealing recent example of this common error in genetic reasoning about sex difference.

Whither the Sex Chromosome?

Early twentieth-century geneticists argued that exceptionalizing the X and Y as "sex chromosomes" may distort reasoning about genomic structure, function, and evolution. These concerns are no less relevant today. As sex chromosomes, the X and Y are conceptualized not only as essential to or organized around sex determination and reproductive biology, but also as belonging to an exceptional set of "not-autosomes" sharing certain structural, functional, and evolutionary characteristics. But although they are both exceptional, the human X and Y are each so in a different way. Both males and females have an X chromosome, and the X chromosome is no more involved in sex-specific traits than many of the other autosomes. In contrast, the Y chromosome is male-exclusive, carries one-twentieth the coding genetic content of the X, and is involved in no biologically essential processes other than male testes determination (that is, a person can live a normal, healthy life without a Y chromosome). The Y chromosome is clonally transmitted in males—but that makes it more similar to mitochondrial DNA in females than to the X chromosome. The X has distinctive dosage-regulating mechanisms due to its doubled status in females, but as I argued in chapter 6, this is best seen as just a specialized version of imprinting and gene regulation mechanisms across the genome. None of these distinctive and interesting features of the X and Y are effectively captured by the term "sex chromosome."

We should examine whether the term "sex chromosome" remains apt. Consider parallels between the term "sex chromosome" and the contested term "sex hormone." Scientists and gender analysts alike have criticized the term "sex hormone" for distorting biological reasoning about sex and gender difference. The terms "sex hormone," "male hormone," and "female hormone" mislead by obscuring the role of this class of hormones in biochemical processes other than sex-specific and reproductive ones. As a result of these critiques, endocrinologists have moved away from this terminology and now instead use the terms "steroid hormones" or "gonadal hormones" for this class of molecules. The term "sex chromosome" leads to similar distortions in our understanding of the functional organization of the human genome. As discussed above, the symmetry ascribed to the X and Y because of their status as the sex chromosomes has oversubscribed their importance to the determination of sexually dimorphic and reproduction-related traits while undersubscribing other factors.

Calling the X and Y the "sex chromosomes" has consistently led to unclear thinking about sex differences. In chapter 4, I raised the question of what our models of the organization of the human genome, and the genetics of sex and gender, might look like had alternative, more sex-neutral terms such as "accessory chromosomes," "idiochromosomes," or "heterochromosomes" prevailed instead of "sex chromosome." The crucial features that distinguish the X and Y from the autosomes are their lack of homology and their unique inheritance pattern. Unlike autosomes, which pair with their homologues, the X and Y are a mismatched pair. Explanatory power and descriptive precision would certainly not be lost, and may even be enhanced, if we referred to sex not at the level of the genome but in terms of gametic sex: females as homogametic for sex and males as heterogametic, and the X and Y as the "accessory chromosomes" or "heterochromosomes" rather than the sex chromosomes. Geneticists should consider the merits of a more precise and sex-neutral terminology for the X and Y. "Accessory chromosomes" or "heterochromosomes" might be a good place to start.

THE POSTGENOMIC REVIVAL OF SEX DIFFERENCE RESEARCH

The "postgenomic age" may be defined as the period after the completion of the sequencing of the human genome and in which whole genome technologies are a shared platform for biological research across many fields and social arenas. The term specifies not just contemporary genome research but, more broadly, any biological research after the completion of the major genome projects that employs genomic technologies and draws on genomic knowledge.[3]

There are some central themes that characterize the postgenomic sciences. These include the transdisciplinarity, speed, and centrality of computational technology that mark the contemporary life sciences; an insistent abandonment of simple views of the gene and of gene action, coexistent with still abundant discourse about the power of genes; a turn toward the human as the central "model organism" of the life sciences; and a nearing of genomics to the clinic and to the consumer, provoking new unease about questions of access to health care and the ethics of genetic testing and enhancement technologies. The completion of the sequencing of the human genome also reenergizes long-standing social, political, and ethical questions about the implications of genomic technologies for human communities.

"Whole genome" technologies include human genome databases and biobanks; microarray chips for assessing the expression of hundreds of thousands of genes in human tissue at once and over time; rapid, inexpensive next-generation genome sequencing technologies; bioinformatic and computational advances in genome-wide association studies; and low-overhead mail-order mass sequencing and genome analysis facilities. These technologies are now ubiquitous in the basic life sciences. Genome sequencing costs will continue to plummet. The declining price and increasing speed of whole genome technologies are exceeding expectations, and a variety of private and public projects of global scale are now generating massive databases of genomic information.

Today, as biomedical scientists work in a transdisciplinary research environment framed in part by the interests and agendas of commercial pharmaceutical, biotechnology, and direct-to-consumer genetics enterprises, they are driven to apply human genomic data and technologies to locate variation in the human genome that may be marketable as a biomarker for disease, forensics, ancestry, or human enhancement.[4] Racial or biogeographical ancestry is among the principal tools for probing the human genome for differences.[5] As I shall argue here, sex, too, is becoming a prominent category for genomic analysis. Sex is a ubiquitous variable in biomedical research. This data is readily available, as genetic samples are routinely tagged with the sex of the donor. If not, genotyping will usually easily reveal the donor's chromosomal sex. This makes it easy to mine genetic data for sex differences even when it was not originally collected for that purpose. As such, human genomic data and technologies are poised to proliferate scientific claims of biological sex differences.

Genomics is increasingly becoming the descriptive mode for the science of sex differences. This new genomic approach to sex is not just the result of scientific discoveries. It is a byproduct of new descriptive and analytical genomic technologies sweeping through the biological sciences and producing reams of new information about human differences. It is also the result of funding and institutional shifts brought about by a new twenty-first century women's health agenda oriented toward the study of sex differences. Undergirded by an emerging sociotechnical landscape that includes commercial and pharmaceutical investment, these developments may contribute to the propagation of uncritical understandings of sex differences.

"SEX-BASED BIOLOGY"

In the late 1990s, some leaders of the US women's health movement began to rally around the cause of basic biological research on sex differences in health and disease. "Sex-based biology," as they dubbed the new agenda, would study the systematic ways in which women's biology differs from men's in order to identify male-female differences with implications for preventative care, diagnosis, and therapy.[6] Sex-based biology represents a shift in the traditional objectives of the women's health movement. Prior to the 1990s, the North American women's health movement focused principally on reproductive health issues such as routine gynecological care, access to the birth control pill, and legalized abortions.[7] Advocates of sex-based biology stress that "women's health" includes all aspects of a woman's health, not just reproductive matters, and have focused on promoting basic research on female physiology and health rather than on the clinical encounter.

The Society for Women's Health Research (SWHR), in Washington, DC, played a leading role in the development of sex-based biology. The SWHR is best known for its transformative work in the early 1990s leading to policies that required the inclusion of women in US National Institutes of Health (NIH) sponsored research and in clinical drug research submitted to the Food and Drug Administration (FDA).[8] In 1977, the FDA issued guidelines that limited the inclusion of women "of childbearing potential" in clinical trials. An expansive interpretation of the FDA guidelines led to the virtual exclusion of women from drug testing. The SWHR, leading a coalition of female physicians and scientists, women's health activists, and congressional policymakers, successfully reversed this policy. As a result of the SWHR's efforts, in 1993 the FDA rescinded its 1977 guidelines and required the inclusion of women in all drug efficacy trials. Encouraging basic research on sex differences became the central focus of the SWHR's efforts after the passing of the 1993 law. Current priorities of the SWHR include advancing studies of sex differences in autoimmune diseases, the brain, cancer, cardiovascular disease, diabetes, drugs/tobacco/alcohol, HIV/AIDS, mental health, musculoskeletal health, and obesity.

The SWHR's efforts to develop sex-based biology as a research field have been impressively savvy. With the aid of public and private funding partners, over the past decade the SWHR has accomplished, in rapid succession, several milestones in a long-term vision to institutionally establish the field of sex-based biology within the biomedical sciences. This includes the development of broad-consensus intellectual frameworks for the field; the designation of special public funding for sex difference research; the founding of conferences, organizations, awards, and fellowships to facilitate career advancement and exchange of ideas; the creation of new journals featuring research on sex differences; and support for the establishment of university-based research institutes committed to sex-based biology.

In 1996, the SWHR created a Committee on Understanding the Biology of Sex and Gender Differences and petitioned the Institute of Medicine (IOM), an influential research body known for its trendsetting and consensus-building reports on the state of medical knowledge, to produce a report detailing sex differences throughout the human body and supporting the need for more research. The result of the multiyear effort that followed was the widely publicized 2001 report "Exploring the Biological Contributions to Human Health: Does Sex Matter?," in which the IOM concluded that "sex does matter. It matters in ways that we did not expect. Undoubtedly, it also matters in ways that we have not begun to imagine."[9] The 288-page document, featuring a high-resolution image of the human sex chromosomes on its cover (see fig. 10.1), asserted that "every cell has a sex" and that sex differences should be studied in all parts of the human body, in all animal models, and in all areas of health and disease, "from womb to tomb."[10]

The IOM report provided the impetus and intellectual grounding for a series of conferences sponsored by the SWHR between 2000 and 2006. As their name indicates, the yearly Sex and Gene Expression (SAGE) conferences focused especially on sex differences at the genetic level. These conferences, with keynotes, poster sessions, and awards for junior and senior investigators, generated a community of self-identified sex difference researchers and helped propel the field forward. Titles of conference sessions included "Chromosome Disorders," "Gene Dosage in Sexual Development," "Genes in Testis and Ovary Development," and "Sex Chromosome Dosage Effects." Topics in basic genetic research on sex, which previously would not typically feature at a women's health research conference, were elevated by the SAGE conferences and connected to the social justice agenda of women's health.

Crowning this effort, in 2008, the SWHR founded the Organization for the Study of Sex Differences (OSSD), and in 2010, it launched a new interdisciplinary academic journal titled *Biology of Sex Differences*. The journal's mission statement designates sex-based biology as "a discipline in itself" and calls for a "sex-based conceptual framework" for the study of physiology and disease:

> Sex has profound effects on physiology and the susceptibility to disease. The function of cells and organs depends on their sex, determined by the interplay among the genome and biological and social environments. The study of sex differences is a discipline in itself,

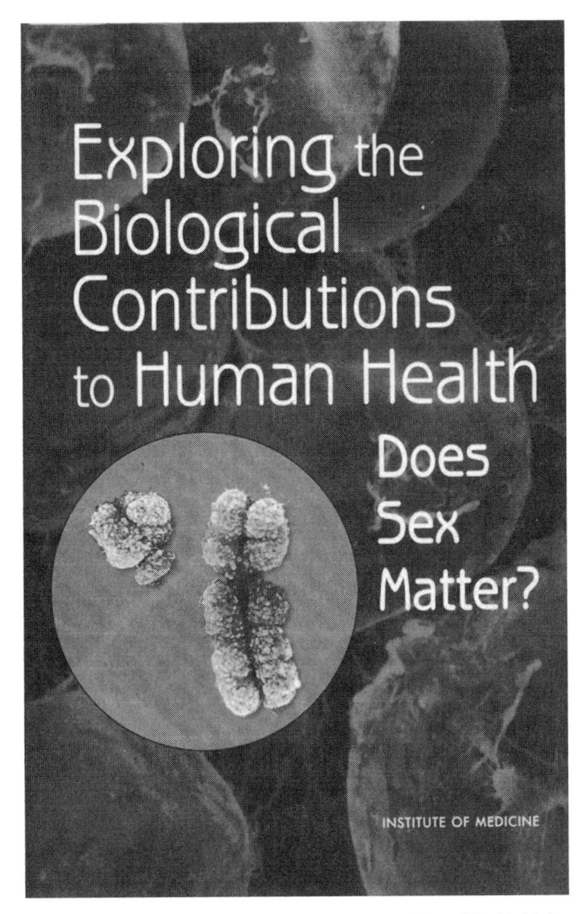

Figure 10.1. Cover of 2001 Institute of Medicine report, "Does Sex Matter?" with high-resolution image of the X and Y chromosomes. Reprinted with permission from the National Academies Press, Copyright 2001, National Academy of Sciences.

> with its own concepts and methods that apply across tissues. An understanding of the effect of sex on biology and disease requires an interdisciplinary approach in which the interaction of diverse sex-specific factors are studied at all levels, from molecular to the level of biological systems. The development of novel therapies for disease requires a sex-based conceptual framework.[11]

Genetics and genomics are at the core of the vision of sex-based biology. In its initial call for papers, *Biology of Sex Differences* named three priority topical areas, with "sex differences in the genome" at the top of the list, followed by "epigenetics."

These industrious efforts to establish the field of sex-based biology could not have been successful without major funding. As has been widely documented, the recent rise of interest in race-based medicine is in part a product of the financial innovations of the pharmaceutical industry, seeking to mine subgroup data for drugs that might be marketed to particular racial and ethnic communities.[12] The approval of a drug for specialized use in this way can extend the company's patent beyond the normal period, which can be worth millions, even billions, of dollars. The same, it appears, is true for sex-based biology. In addition to public funders such as the NIH and the National Science Foundation (NSF), the financial supporters of the OSSD and the SWHR include nearly every big name in pharma. This includes major international companies such as Amgen, BayerBoehringer Ingelheim, Boston Scientific, Eli Lilly, GlaxoSmithKline, Johnson and Johnson, Merck, NovartisPfizer, and Philips.[13] The promise that sex differences may point the way toward marketable sex-specific therapeutic regimes, which can allow the repackaging of drugs in new forms for new markets, is undoubtedly a major motivator for businesses to sign on to the agenda of sex-based biology.[14]

The determined efforts of these well-placed women's health advocates to promote sex difference research has promoted a new alliance between the women's health movement and basic research in the molecular life sciences, particularly genomics. Sex difference researchers from the brain, genomic, and endocrinological sciences have allied themselves with the women's health movement, finding community with one another and building funding bases and public venues for their research. From the glossy X and Y chromosomes on the cover of the IOM report, to the current logo of the SWHR, "Because seXX matters," to the central aims of the new *Biology of Sex Differences* journal, and the growth of the OSSD out of the Sex and Gene Expression conferences, the vision of a biology of sex differences rooted in the genome frames research in sex-based biology.

THE SEXOME

The movement for a sex-based biology has prompted a new industry of research on sex differences in the human genome—an object of study recently dubbed the "sexome" by one prominent geneticist.[15] Theories of the genetic etiology of diseases ranging from autism, schizophrenia, and major depression to cardiovascular disease and diabetes now invoke sex-biased gene expression or sex-dependent genetic pathways. Similarly, researchers are using genomic technologies to quantitatively characterize sex differences in gene expression in every tissue in the body, from the heart to the brain and the liver. These inquiries currently proceed along three principal lines:

1. *Gene expression studies* assess differences in levels of protein-producing activity of different genes in particular tissues.
2. *Genome-wide association studies* scan the whole genome to look for gene variants that are more frequently present in those with a particular phenotype.
3. *Genetic linkage and quantitative trait locus studies* analyze the association between known human genome variants and heritable traits.

Using these technologies, geneticists are able to ask a range of questions about sex differences in the genome. In the still-unsettled terminology of this emergent field, these questions include:

1. *Sex-dependent gene expression*: Are there sex differences in which genes express in human tissues?
2. *Sex-biased gene expression*: Are there sex differences in the levels of expression of genes?
3. *Sex-specific genetic pathways*: Are there differences in the network of genes involved in normal and pathological physiology in males as compared to females?
4. *Sex-specific linkage signals*: Are there sex differences in the underlying "architecture" of inherited genetic loci and human traits?
5. *Sex-specific heritability*: Does sex modify the penetrance and expressivity of an inherited trait?

In each of these areas, scientists are applying whole-genome technologies to redescribe known sex differences at the genetic level or to discover new sex differences not previously understood.

Slowly, geneticists are beginning to articulate an account of sex that brings genes to the fore. While speculations have long abounded about the X, Y, and sex, until recently, genes held a circumscribed role in typical sex differences, restricted mainly to sex determination in early fetal development. For much of the twentieth century, gonadal hormones—androgens and estrogens—stood as the linchpins of the grand theory of sex, gender, and sexuality. Scientists presented the so-called sex hormones as the molecular agents of everything from the development of the sex organs and secondary sexual characters to gender identity, sexual orientation, and gendered behaviors, interests, and career choices.[16]

Geneticists are now challenging this picture of sex differences. As one prominent group of sex difference researchers declared in 2012, it is time to overturn the "previous hegemony of hormone."[17] "Long ignored as an important factor, the importance of genotype is beginning to emerge," they wrote.[18] University of California, Los Angeles, geneticist Arthur Arnold and University of Maryland neuroendocrinologist Margaret McCarthy are among the leading intellectual architects of this new genomics-based framework for understanding the biology of sexual differentiation and sex differences. As they write,

> in the twentieth century, the dominant model of sexual differentiation stated that genetic sex (XX versus XY) causes differentiation of the gonads, which then secrete gonadal hormones that act directly on tissues to induce sex differences in function. . . . Recent evidence, however, indicates that the linear model is incorrect and that sex differences arise in response to diverse sex-specific signals originating from inherent differences in the genome.[19]

McCarthy and Arnold predict that sex chromosomal genes "act directly" and hormone-independently to influence "variability between males and females" far beyond mere reproductive differences.[20] As they write, "all sex differences must ultimately stem from the inherent imbalance of genes encoded by the sex chromosomes"; for example, "every cell in the brain of males may differ from those in females, by virtue of difference in their sex chromosome complement."[21] McCarthy and Arnold position the genome as a first-order line of study relevant to human sex difference research in any field. The linchpin and most prominent visual symbol of the research program to document the hormone-independent direct effects of the sex chromosome complement is the so-called four-core-genotypes model (see fig. 10.2). First developed by Arnold and colleagues in 2002, the experimental system employs mice genetically altered to have a male sex chromosome complement, but female gonads, and vice versa.[22] Analysis of the phenotype of these altered mice allows researchers to distinguish hormonal- and chromosomal-based differences between the sexes.

Whereas the genetics of sex was once largely limited to the sex chromosomes and gonad-determining genes, this new account of sex envisions the whole genome as imbued with sex-specific processes and sex

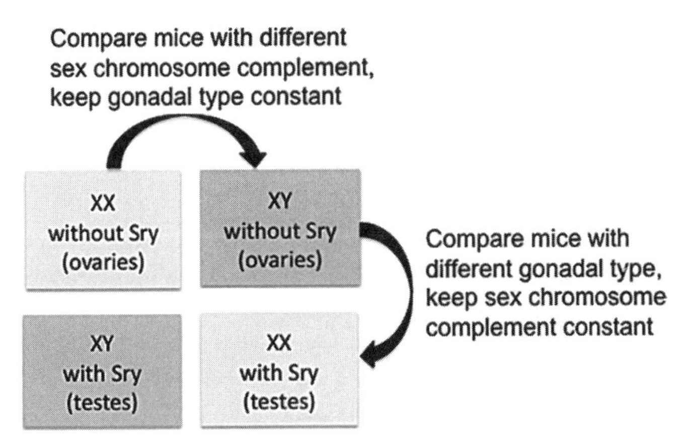

Figure 10.2. The four-core-genotypes model for assessing "direct sex chromosome effect" on sex differences. Reprinted from Margaret M. McCarthy, Arthur P. Arnold, Gregory F. Ball, Jeffrey D. Blaustein, and Geert J. De Vries, "Sex Differences in the Brain: The Not So Inconvenient Truth," *Journal of Neuroscience* 32, no. 7 (2012): 2241–47, with permission from The Society for Neuroscience.

differences in gene expression. Whereas hormones were once seen as the mediators of sex throughout the life course, geneticists, with new technologies to assess changes in gene expression over time in specific tissues, are reformulating the picture of hormone action. While hormones remain important, in this new account genes emerge as the coequal and even primary actors in sex differences, not second fiddle to the sex hormones.

THINKING CRITICALLY ABOUT GENOMIC SEX DIFFERENCE RESEARCH

It is curious that sex difference research has not been a more central focus of discussion among those studying the implications of genomics. Given the well-documented history of methodological problems with sex difference research, as well as of harmful abuses of sex difference claims by those who would limit women's opportunities, it is remarkable to find women's health activists promoting, with little qualification, sex-based biology's expansive picture of sex differences. As the sociologist Steven Epstein notes, "one of the most striking aspects of the movement for sex-based biology is its unabashed embrace of a thoroughgoing conception of difference between women and men."[23] Anne Fausto-Sterling suggests that the movement's focus on biological sex differences may result, in part, from its leadership by elite, politically connected, professionalized women's health activists not closely allied with feminist grassroots activism or critical theory. As she writes, "members of the feminist medical establishment, that is, those researchers and physicians for whom the activities and programs of the SWHR make eminent sense, see themselves perched on the forward edge of a nascent movement to bring gender equity to the healthcare system. These feminists work outside of an intellectual milieu that would permit the more revolutionary task . . . of contesting not only 'the domination of the body by biological terms but also [contesting] the terms of biology itself.'"[24]

It is a widely shared consensus among social scientists that genomics is transforming social relations. It is remaking categories of identity—concepts such as "blood relations," racial or geographic ancestry, citizenship, and subjectivity—as well as notions such as what it is to be normal and healthy, what rights we have to privacy and to health care, and what responsibilities we have to others to maintain and pursue information about our health.[25] Research on the emerging genomic concept of race, a topic with which I have also been deeply engaged, provides a helpful analogue for an investigation of gender in the age of the genome. This research shows the revival of old, debunked models of race in the language of genomics, and underscores the way in which trends such as pharmacogenomics, personalized medicine, and health movements for racial and ethnic minorities are supporting an often uncritical turn toward research on race-specific genetics. The same may be said of genetic research on sex and gender.

There are, of course, important differences between the sociohistorical and biological constructs of race and sex. However, as Epstein points out, there are also wide areas of congruence between the arguments that have been vociferously made against using race differences in genetic research and those that could be made about using sex.[26] That is, while efforts to build a sex-based biology are well intended, the results may be problematic. Using race or sex as a principal category for biological analysis may cause important differences within groups to be overlooked. This may lead to the use of therapies that do not provide generalized benefits to subgroups. Similarly, race- and sex-based studies can contribute to inaccurate and destructive notions of what race and sex are. In particular, a focus on the biological dimensions of race and sex differences may obscure the social and cultural factors at play. Because of the high epistemic status of science, biological race and sex difference claims may lend legitimacy to harmful racist and sexist ideologies, which often rest asser-

"Because my genetic programming prevents me from stopping to ask directions—that's why!"

Figure 10.3. Reprinted with permission from *The New* Yorker, September 23, 1991. Copyright Don Reilly/The New Yorker Collection/www.cartoonbank.com.

tions about racial or sexual inferiority on claims that races and sexes are biologically different (see fig. 10.3).[27] For all of these reasons, we should always think critically about the merits and risks of genetic sex difference studies.

COMMON PROBLEMS WITH GENETIC SEX DIFFERENCE STUDIES

Genomic sex difference research is already a burgeoning literature. There exist a number of excellent, well-designed genomic sex difference studies, particularly on basic biological systems in nonhuman animal models. However, as a body, this research raises alarming methodological concerns. Here I summarize findings by myself and others that show widespread problems in study design, description and interpretation of results, and frameworks conceptualizing the interaction between sex and gender in genomic sex difference studies.

The 2007 study by Patsopoulos, Tatsioni, and Ioannidis in the *Journal of the American Medical Association*, "Claims of Sex Differences: An Empirical Assessment in Genetic Associations," provides a useful starting point. The authors, well-known biostatisticians, analyzed 432 claims of genetic sex differences in 77 recent peer-reviewed research articles. Their findings are so remarkable that I summarize them in some detail here. They found that 55.9 percent of the published claims for which raw study data was available for reanalysis were not statistically significant. In addition, in all but 12.7 percent of the articles, the studies contained serious flaws in design. The authors established three uncontroversial criteria for an appropriately documented, nonspurious gene-subgroup interaction claim in genetic epidemiology: "First, the article had to address a genetic effect that was based on the same genetic contrast in both sexes. Second, it did not compare different subsets in the 2 sexes (eg, old men vs young women). Third, it needed to . . . report a nominally statistically significant . . . test that examined sex-gene interaction."[28] Using these criteria, they concluded that 74 of the claims were wholly spurious. Only 37 of the 432 claims "were stated to rely on a priori considerations, had raw data available that documented that they were indeed nominally statistically significant, and had analyses performed on the whole study sample."[29] In the end, Patsopoulos et al. found only *one claim* that met all of their criteria for an internally sound finding of sex differences and also had been replicated at least once by another study.

Patsopoulis et al. draw attention to problems of poor study design in recent genomic sex difference research. As the authors write, "in the analysis of sex-specific effects in genetic associations, investigators very often seem to fall into classic traps."[30] Among the traps that they identified: comparison of male cases directly with female cases, ignoring controls; comparison of male versus female cases with a given genotype, ignoring other genotypes; comparison of different genetic groups in male versus female cases; and comparison of one sex against a subgroup of the other sex.[31] In all of these cases, the findings are spurious because the study design did not allow comparison of similar subsets in males and females with respect to both genotype and phenotype.

As the authors note, sex-based analyses are likely to generate findings of sex differences. But those differences are likely to be merely chance, an artifact of the research technologies and study population used. Detecting a genuine gene-sex interaction requires a large study size to adequately power the subgroup analysis. None of the studies they surveyed were large enough to allow this. Because of this problem, genetic associations initially reported with great fanfare are routinely overturned.

As Patsopoulos et al. show, lack of replication is endemic in the genetic sex difference literature.[32] The upshot of their analysis is that gene-sex differences are likely to be rare, small, and difficult to corroborate. Addressing advocates of sex-based biology such as the SWHR, they conclude that while "some authors have argued for the need to perform and transparently report subgroup analyses, in particular those related to sex . . . the vast majority of claimed subgroup differences are likely to be chance findings."[33] The cautionary lesson that researchers should take from this study is that even in cases of well-documented, adequately powered studies of gene-sex interactions, researchers should evince an awareness that such findings require further study to confirm their validity and, furthermore, to determine whether they have meaning for biological function in living systems.

Supplementing the work of Patsopoulos et al., I conducted a more qualitative review of recent research on genetic sex differences in human tissues and diseases, examining basic empirical and methodological aspects of the studies as well as their conceptual framework for motivating, interpreting, and representing findings of sex differences. I analyzed the content of twenty highly cited research papers published between 2000 and 2010 that report genetic sex differences relevant to human physiology and disease.[34] My findings represent only a snapshot of a rapidly moving field. Nonetheless, they offer an enriching addition to the picture presented by Patsopoulos et al. Overall, I found a body of research that is pointedly focused on finding sex differences, while unreflective about the conceptual motivation for and larger context of these studies. Here are the three central findings:

1. MISUSE OF THE TERMS "SEX" AND "GENDER" AND LACK OF STUDY OF THE INTERACTION BETWEEN SEX-LINKED AND GENDER-LINKED FACTORS

The potential interaction of "sex" and "gender" is not analyzed in genetic sex difference studies. The terms "sex" and "gender," moreover, are used interchangeably and without definition. The titles of papers alone give ample evidence of this problem: "Genome-Wide Association Analysis of Gender Differences in Major Depressive Disorder"; "Gender-Specific Association of the Brain-Derived Neurotrophic Factor Gene with Attention-Deficit/Hyperactivity Disorder"; "Genome-wide Estimation of Gender Differences in the Gene Expression of Human Livers"; "Gender-Specific Gene Expression in Post-Mortem Human Brain." With titles like these, one would expect fascinating studies of the interaction of cultural gender roles and norms with gene expression, but instead these are all single-variable studies of sex differences in gene expression or genetic association.

Only one of the papers that I surveyed seriously considered gender as a relevant interacting factor with genes in producing observations of sex differences. In the review article "Contribution of Gender-Specific Genetic Factors to Osteoporosis Risk," Karasik and Ferrari note the "lack of consistent replication of chromosomal linkage and allelic associations in genetic epidemiological studies on osteoporosis." They observe that a potential reason for this is "the non-inclusion of a relevant effect of gene-environment (gene-gender) interaction in the assessment of the relationship between phenotype and genotype, which will introduce substantial noise into this relationship and may explain, in part, the conflicting results thus far reported in the literature."[35] Karasik and Ferrari cite extensive evidence that predisposition to fracture risk and osteoporosis is culturally mediated. They note that "expression and penetrance of the same genes are modulated by gender-specific environments, such as exposure to gonadal steroids, differences in physical activity, and muscle strength."[36] They conclude that "there is a need to consider gender-specific genetic and environmental factors in the planning of future association studies on the etiology of osteoporosis."[37]

That gender can produce biological findings of sex differences—that, as Fausto-Sterling puts it, "our bodies physically imbibe culture"—is well documented.[38] Consider the advent of Title IX in the United States, leading to newly athletic female bodies that stand, move, and experience injury in ways previously rare. Or the relationship between cultural acceptance of contraceptive use and the health status of women in a particular population. Or the once extremely low prevalence of lung cancer in women compared to men due to different gendered norms regarding smoking. Or even the mid-twentieth-century transition from a feminine beauty ideal of maternal curviness to one of slender preadolescent boyishness.

In her classic 2003 paper on the subject, social epidemiologist Nancy Krieger documents the various ways in which gender relations and sex-linked biology can interact in profound ways to determine differences in health outcome. In a series of examples, she shows how social gender relations such as "gender segregation of the workforce and gender discrimination in wages, gender norms about hygiene, gender expectations about sexual conduct and pregnancy, gendered presentation of and responses to symptoms of illness, and gender-based violence" can alter the picture of what we perceive to be sex-linked biological differences.[39] One simple example that Krieger offers is the "greater prevalence of contact lens microbial keratitis among male compared with female contact lens wearers." Although it is plausible that genetic differences make males and females more or less susceptible to keratitis, gender rather than sex is most likely the relevant variable in this case. As Krieger notes, males are "less likely to properly clean [their lenses] than women."[40]

In all of these cases, we see that gender dynamics can indeed be written on the body with consequences for health status as well as for what we understand to be typical biological development for males and females.

2. AN OVERWHELMING FOCUS ON DIFFERENCE IN THE DESCRIPTION, VISUAL REPRESENTATION, AND INTERPRETATION OF THE STUDY RESULTS

In the papers I surveyed, almost every study that starts out with a hypothesis predicting sex differences reports finding them. Few negative results are reported. The visual representation of findings of sex differences often lack error bars showing the range of variation within each sex and overlap between males and females. Most studies compare groups of men to groups of women but lack an intragroup analysis. Thus, from these studies, it is almost always impossible to know what the differences in gene expression are between women and between men, and what the range of overlap is between women and men. The result is that the differences between sexes may be exaggerated and the differences within a sex may be completely obscured. The danger is that sex differences may be exaggerated at the expense of important similarities between men and women and important differences within each group may be missed, with implications for both women's and men's health.

An example of this intensive focus on finding differences is a 2011 study by Zhang et al. that analyzed genetic sex differences in the human liver. The standard minimum for detecting differences in microarray gene expression studies is a 2-fold change, but this study used an extremely low threshold, a fold change of 1.15.[41] In the findings presented in the paper's tables, genes are labeled as either "female-biased" or "male-biased," with no indication of the variation among and between the sexes in levels of expression. Unsurprisingly, the authors report that they found an enormous number of sex differences in the liver: "more than 1,200 genes showing significant sex differences in expression."[42] The authors then draw far-reaching conclusions about the physiological and medical implications of these potential differences for women's and men's health. "These findings provide novel insights into human hepatic sex differences important for processes such as drug metabolism and pharmacokinetics, and could help explain sex differential risk of coronary artery disease," they wrote.[43]

The focus in this and other studies on quantifying and describing differences, rather than similarities, is remarkable. The body of sex difference research in genetics in fact shows an overwhelming picture of similarity, more than we might expect given the dramatic claims that are often made about differences between the sexes. As noted, only a handful of genes seem to express differently in males and females, and the differences in levels of expression are small. Yet similarities between men and women are not described in the text or represented in the accompanying illustrations. Janet Shibley Hyde, a psychologist known for her "gender similarities" hypothesis, argues that contrary to popular conceptions of gender difference, males and females are similar on most psychological variables. Hyde performed an extensive meta-analysis on psychological gender differences that showed that 78 percent were actually small or close to zero.[44] While Hyde's study treats claims of psychological sex differences, her novel focus on similarity rather than difference is relevant to genetic sex differences, as this is likely an area in which the large swaths of genetic similarity between the sexes is just as provocative a biological result as findings of differences.[45]

3. NEGLECT OF RELEVANT INTERACTING VARIABLES SUCH AS AGE, WEIGHT, AND HORMONES THAT ARE KNOWN TO AFFECT GENE EXPRESSION

On the whole, the studies surveyed tend to exclude relevant biological variables that interact with sex such as age, weight, nutrition, stress, hormone levels, and environmental exposures in their study design and data analysis. These variables are known to affect gene expression.[46] In the case of sex, hormonal profile is particularly important. Different hormonal profiles characterize women who are preadolescent, on a normal menstrual cycle, taking oral contraceptives, pregnant, lactating, menopausal, and postmenopausal. These hormonal differences may contribute to variation in gene expression. While a number of the genetic sex difference studies that I reviewed directly examined gene-hormone interactions, very few looked at these interactions in a dynamic in vivo system reflecting the functional significance of changes in gene expression within a complex biological environment. Beyond hormones, few other variables are regularly incorporated into sex different studies.

Only one study in my sample offered a comprehensive vision of what kinds of interacting variables would need to be considered to validate a genuine, functionally relevant genetic sex difference. Isensee et al.'s 2008 "Sexually dimorphic gene expression in the heart of mice and men" looked at the interaction of sex and age as modifiers of sex- and growth-steroid mediated cardiac gene expression. The study found that changes in gene expression correlated with both sex and age. Highlighting the complexity of analysis of sex-specific gene expression patterns in humans, the authors conclude that "in future studies, more time points will have to be considered to distinguish between individuals at childhood, puberty, adulthood and senescence. . . . In addition, the use of synchronized females may allow the identification of more estrogen-responsive genes to elucidate the complex actions of female sex hormones on the cardiovascular system. This represents a challenging task especially in humans, as it requires the analysis of a large number of individuals."[47] Isensee et al.'s study presents a striking exception to the main body of genetic sex difference studies, which make no effort to control for such interacting factors. Their approach might serve as a model for the rigorous study of genetic sex differences in human systems.

Of course, I agree with the SWHR: sex differences should be a standard consideration for medical researchers. Research on genetic sex differences may reveal findings critical to women's—and men's—health. At the same time, genetic studies consistently show only small differences between males and females and great variation among females, as well as males. Moreover, there are risks to women's health if sex differences are exaggerated. An excellent example of this, cited by Patsopoulos et al., is the long-held claim that aspirin is effective for stroke prevention in men but not women, which turned out to be wrong: women can benefit from aspirin, too.[48] Finally, a singular focus on sex differences can overshadow other medically relevant variables that cut across sex. For example, drug metabolism and action may be profoundly mediated by body size, a matter of interest to very large or very petite people, whether male or female.

Thus, while sex may be a relevant variable in some cases, finding differences between the sexes should not be an end in and of itself. Sex difference research should be grounded in valid medical research questions, motivated by sound biology, and rigorously designed. As we advance into the genomic age, we need shared standards for conducting sound sex difference research in genetics. The record shows that much recent genetic sex difference research is underpowered and poorly designed. If accepted without caution, genetic sex difference findings could have harmful medical consequences and reinforce problematic conceptions of biological sex differences.

Concrete methodological guidelines for a more gender-critical genetics follow from the above analysis of the methods and larger context of recent genetic sex difference research. Here I will focus on just two key points that merit emphasis. First, *rigorous study designs are required to substantiate claims of male-female sex differences in the human genome.* It is important to recognize that studies of genomic sex differences are a form of subgroup analysis. Subgroup analyses are subject to a range of standard methodological considerations, including how well matched and well powered, in size, the subgroups are, and whether the data set was designed with studying sex differences in mind or the analysis is being done retrospectively using data collected for other purposes. As Patsopoulos et al. write, "ideally, sex differences should be based on a priori, clearly defined, and adequately powered subgroups. . . . Even then, results should be explained with caution and should be replicated by several other studies before being accepted as likely modifications of genetic or other risks."[49] While these elements of study design are foundational in epidemiology, they seem not to be rigorously observed when studying sex differences. This may have something to do with the historical and ideological propensity to think of sex differences in an essentialized and highly binary manner—in terms of the mean, or average, of each sex, and not *variance.*

Second, *genetic sex difference studies must distinguish clearly between sex-linked and gender-linked factors in reporting findings of biological sex differences.* In genetic research on sex differences, the term "gender," which specifies social-contextual aspects of male and female behaviors, roles, and expectations, is almost ubiquitously used instead of, or interchangeably with, the term "sex," which specifies biologically based traits of males and females.[50] Researchers should recognize the conceptual distinction between sex and gender and define how they are using each term in a given study. They should also acknowledge the possibility that observed biological sex differences may not be inborn, but rather a result of interacting gender factors. This interaction is unexplored in almost all of the recent genetic sex difference studies that I surveyed. Any human sex difference study that does not rule out, or actively document, sex-gender interactions, should meet with rigorous critique, both at the funding and the publication stage.

◇

Biological conceptions of human sex difference have changed over time, shifting with the privileged theories, methods, and interests of the dominant biomedical research program. Reproductive organs, blood color and density, skeletal morphology, brain size and lateralization, and hormones have each been claimed as the "essence" of human sex difference. For much of the twentieth century, genes and sex hormones held a friendly détente, genes responsible for throwing the initial switch to determine sex, hormones for all the complexities and riches of sex difference. Today, genomics is increasingly the preferred language for explaining and describing sex and gender differences.

Stephen Jay Gould's *Mismeasure of Man* (1996) traces two centuries of attempts to establish a biological basis for the argument that nonwhites are less intelligent than whites. Gould shows how each attempt to do so was discredited, then replaced a few years later by another research program looking for racial differences in new physical substrates, with new technologies, and with a new set of terminology. The history of attempts to establish the biological essence of sex difference and the physical and mental inferiority of women follows a similar trajectory. Each time a new research program emerges, the claim is made that at last the difference between the sexes can be located, measured, and quantified, and that the differences between the sexes are greater than ever previously imagined.[51]

"Geneticization" has transformed many fields of scientific inquiry over the past decade. Research questions and explanatory hypotheses that can be translated into genetic terms have a high epistemic profile at this time and are reshaping many areas of biology. Once primarily a subfield of endocrinology, sex difference research is increasingly turning to the genome to test its theories. Human geneticists are now in hot pursuit of the population of genes and gene pathways that contribute to sex dimorphism, fertility, gender identity, and sexual preference. Traditional ideas about sex and gender difference are seen as newly rediscovered in the data of the genome. Old concepts and theories, such as the greater male variation theory of sex differences in intelligence discussed in chapter 4, are revisited and recast in genetic terms as genetic data and methods gain prestige. The geneticization of biological conceptions of sex difference, however, has gone largely unnoticed by science analysts. We must develop a level of awareness, rigorous debate, and criticality about claims of sex and gender difference in genetics similar to what has been achieved, with some success, in the area of genetic research on race and ethnicity. This requires thinking deeply about the empirical, model-theoretic, and ethical foundations of the concept of sex difference at work in contemporary genomics.

For scholars of gender and science, this book offers an approach to gender analysis of science that focuses on the rich and multivalent interactions between gender conceptions and science, including, but not restricted to, the question of gender bias in science. I have called this approach "modeling gender in science." Gender stands as an omnipresent and epistemically relevant source of background norms and beliefs in the theories, models, and descriptive language of sex chromosome research. Yet the workings of gender in each of the cases examined in this book are different, and all cases stretch beyond questions of bias, where bias is understood primarily in a pejorative sense as a distorting partiality of perspective. From the standpoint of modeling gender in science, one can also see gender conceptions as resources, doing constructive, cohering work to drive forward empirically testable theories—such as those of X chromosome mosaicism as a fundamental mediator of female biology, the male Y chromosome as carrying genes for aggression, or the *SRY* as the master gene of sex determination. The notion of the X and Y as "sex itself," similarly, may be seen as a focusing framework for investigations of the genetics of sex in the age of the hormones and prior to the advent of the genome sequencing projects. In short, sometimes it is most precise and useful to understand gender conceptions as biasing. Sometimes, merely valencing. Sometimes, it is useful to also see them as doing constructive, even clarifying, work in the evaluation of competing scientific models. The context—in particular, the degree to which such gender conceptions are themselves visible and subject to critical debate—matters when modeling gender in science.

The project of modeling gender in science, by decentering the question of bias, not only points to a broader frame for analyzing gender in the sciences, but also to a more constructive approach to addressing the uncritical use of simplistic conceptions of sex differences in the sciences. Scholars forging the historical and philosophical study of gender in science over the past thirty years have principally focused on documenting and analyzing cases of egregious scientific bias or error produced by sexism, androcentrism, and heterosexism in science. This literature has often been read as implying a particular remedy: find cases of gender bias, expose them, and surgically excise the bias from the science. The goal of analysis of gender in science, on this view, should be a science untarnished by gender ideology.

The story of the workings of gender conceptions in sex chromosome science, as I have told it, suggests that this ideal of unbiased, value-neutral science is misconceived. As the case studies in this book demonstrate, gender beliefs are a key source for scientific models of sex and gender in genetic science. Gender ideology is dynamic, persistent, and ever-present in genetic and genomic research on sex and gender; it cannot be surgically or permanently excised from the science. Rather than seeking to somehow eliminate gender in science, we are better advised to focus on modeling the many roles of gender assumptions in particular areas of the sciences to develop gender-critical methods for and approaches to science. The question is not "how can we get all of this gender politics out of genetics?" but rather "how can we enlarge and critically hone our ideas about gender, which are central to our scientific theories of sex?"

My insistent plea in this book for a more gender-critical practice of genetics follows from this insight. We should not seek a simple prescription for how scientists should conceptualize sex and gender differences. Instead, we should cultivate a mode of ongoing critical and open-minded engagement, reflection, and dialogue surrounding research on the biology of sex and gender. This is my hope for gender in a genomic age.

ACKNOWLEDGMENTS

I began writing this book in 2005, after the cover of *Nature* reporting the sequencing of the human X chromosome caught my eye. Skimming the scientific and journalistic articles describing the achievement, I suddenly realized that the history of the sex chromosomes had not yet been written—and also that X and Y chromosome science was intriguingly rich with gendered language and imagery.

During the writing of this book, I consulted with and interviewed many geneticists, visiting them at their laboratories or chatting over coffee at scientific conferences. These interviews were invaluable to my understanding of the history and science of the sex chromosomes, and I am grateful to these scientists for their interest, support, openness, and generosity with their time. Thanks, in particular, to Arthur Arnold and Eric Vilain of the University of California, Los Angeles; Nick Barton, Brian Charlesworth, and Deborah Charlesworth of the University of Edinburgh; Ian Craig of the UK MRC Social, Genetic and Developmental Research Center; Mark Feldman and Joan Roughgarden of Stanford University; Jenny Graves of LaTrobe University and the Australian National University; Vincent Harley of Monash University; Peter Koopman of the University of Queensland; David Page of MIT's Whitehead Institute; Barbara Migeon of the Johns Hopkins School of Medicine; Ursula Mittwoch of University College London; Mark Ross of the Sanger Institute in Cambridge, UK; and Huntington Willard of the Duke Institute for Genome Sciences and Policy. Special thanks to Art Arnold and Peter Koopman for reviewing the book's illustrations and to Jane Gitschier of the University of California, San Francisco, and Jenny Graves for permission to recreate their humorous cartoons of the sex chromosomes.

Many forms of support, for which I am deeply grateful, made this book possible. Helen Longino, Londa Shiebinger, Robert Proctor, and Michael Friedman served as important mentors in the early stages of research. Numerous colleagues gave helpful comments on portions of the written manuscript, including Rene Almeling, Louise Antony, Linnda Caporael, Soraya de Charaderevian, Shari Clough, Angela Creager, Michael Dietrich, Scott Gilbert, Nathan Ha, Jennifer Hamilton, Becky Holmes, Evelyn Fox Keller, Barbara Koenig, Linda Layne, Susan Lindee, Elisabeth Lloyd, Erika Milam, Afsaneh Najmabadi, Aaron Panofsky, Katharine Park, Joanna Radin, Svati Shah, Shauna Shames, Banu Subramaniam, and Eviatar Zerubavel. I also thank the many audiences who heard this work in talk and lecture form and offered valuable feedback.

Research, editorial, and administrative assistance was provided by Meredith Bircher, Noam Cohen, Lori Kelley, Will McGinnis, Monica Moore, Jenna Tonn, Kate Womersley, and students in the 2010 Gender and the Human Genome summer undergraduate research program at the University of Massachusetts, Amherst, and the Five Colleges. Librarians at Stanford University, the University of Massachusetts, Amherst, the California Institute of Technology, and Harvard University also provided critical assistance. Thank you also to Karen Merikangas Darling and Abby Collier of the University of Chicago Press and to the Press's anonymous peer reviewers. The Seattle-based graphic artist Kendal Tull-Esterbrook created the book's wonderful illustrations.

The research that led to this book was funded by Stanford University's Program in Modern Thought and Literature, the Stanford Humanities Center, an American Fellowship from the American Association of University Women, the Mary Anne Bours Nimmo Fellowship, the Clayman Institute for Gender Research, and a junior faculty publication grant from Harvard University. *Signs*, *Biology and Philosophy*, *Synthese*, and Stanford University Press graciously allowed use of previous writing that forms the basis for some of the material presented here.

Personal thanks to Ashley Burczak, Ju Yon Kim, Nora Niedzielski-Eichner, and Kaja Tretjak, and to my family, Douglas, Suzanne, and Alexa Richardson, for their support. This book is dedicated with love and gratitude to Richard Olson.

NOTES

Chapter 1

1. Thomas Hunt Morgan and Calvin B. Bridges, *Sex-Linked Inheritance in Drosophila* (Washington, DC: Carnegie Institution of Washington, 1916). The phrase "sex itself" bears striking resemblance to the concept of "life itself," which has a long and rich history in the twentieth-century life sciences and has recently been taken up by social scientists in reference to new regimes of knowledge production and consumption in twenty-first century biomedicine, e.g. Nikolas Rose, *The Politics of Life Itself: Biomedicine, Power, and Subjectivity in the Twenty-first Century* (Princeton, NJ: Princeton University Press, 2006). My use of the phrase here, however, is entirely in homage to Morgan's concise summary of the concerns of this book, with no intention of importing the larger theoretical construct of "life itself" that Rose and others have developed.

2. Elizabeth Pennisi, "Mutterings from the Silenced X Chromosome," *Science* 307, no. 5716 (2005).

3. Patrick Geddes and J. Arthur Thomson, *The Evolution of Sex* (London: Walter Scott, 1889); Patrick Geddes and J. Arthur Thomson, *Sex* (London: Williams & Norgate, 1914); Cynthia Eagle Russett, *Sexual Science: The Victorian Construction of Womanhood* (Cambridge, MA: Harvard University Press, 1989).

4. Nelly Oudshoorn, *Beyond the Natural Body: An Archaeology of Sex Hormones* (New York: Routledge, 1994); Chandak Sengoopta, *The Most Secret Quintessence of Life: Sex, Glands, and Hormones, 1850–1950* (Chicago: University of Chicago Press, 2006); Adele Clarke, *Disciplining Reproduction: Modernity, American Life Sciences, and the Problems of Sex* (Berkeley: University of California Press, 1998); Anne Fausto-Sterling, *Sexing the Body: Gender Politics and the Construction of Sexuality* (New York: Basic Books, 2000); Rebecca M. Jordan-Young, *Brain Storm: The Flaws in the Science of Sex Differences* (Cam-

bridge, MA: Harvard University Press, 2010); Alice Domurat Dreger, *Hermaphrodites and the Medical Invention of Sex* (Cambridge, MA: Harvard University Press, 1998).

5. E. J. Vallender, N. M. Pearson, and B. T. Lahn, "The X Chromosome: Not Just Her Brother's Keeper," *Nature Genetics* 37, no. 4 (2005): 343; J. A. Graves, J. Gecz, and H. Hameister, "Evolution of the Human X—A Smart and Sexy Chromosome that Controls Speciation and Development," *Cytogenetic & Genome Research* 99, no. 1–4 (2002).

6. C. Gunter, "She Moves in Mysterious Ways," *Nature* 434, no. 7031 (2005): 279.

7. David Bainbridge, *The X in Sex: How the X Chromosome Controls Our Lives* (Cambridge, MA: Harvard University Press, 2003); Natalie Angier, *Woman: An Intimate Geography* (Boston: Houghton Mifflin, 1999); Natalie Angier, "For Motherly X Chromosome, Gender Is Only the Beginning," *New York Times*, 1 May 2007.

8. See, e.g., P. S. Burgoyne, "The Mammalian Y Chromosome: A New Perspective," *Bioessays* 20, no. 5 (1998); Angier, *Woman: An Intimate Geography*; Angier, "Motherly X Chromosome"; J. A. Graves, "Human Y Chromosome, Sex Determination, and Spermatogenesis—A Feminist View," *Biology of Reproduction* 63, no. 3 (2000); Bainbridge, *The X in Sex*.

9. Page quoted in Maureen Dowd, "X-Celling over Men," *New York Times*, 20 March 2005.

10. Bainbridge, *The X in Sex*, 56, 58, 145.

11. Bryan Sykes, *Adam's Curse: A Future without Men* (New York: Bantam Press, 2003), 283–84, 242–43, 244.

12. L. Carrel, "'X'-Rated Chromosomal Rendezvous," *Science* 311, no. 5764 (2006).

13. Matt Ridley, *Genome: The Autobiography of a Species in 23 Chapters* (New York: HarperCollins, 1999), 107. See also John Gray, *Men Are from Mars, Women Are from Venus: A Practical Guide for Improving Communication and Getting What You Want in Your Relationships* (New York: HarperCollins, 1992).

14. Elisabeth Pain, "A Genetic Battle of the Sexes," *ScienceNOW Daily News*, 22 March 2007, http://news.sciencemag.org/sciencenow/2007/03/22-04.html; Bainbridge, *The X in Sex*, 83.

15. Angier, *Woman: An Intimate Geography*, 26.

16. *The XX Factor: What Women Really Think* (blog), *Slate*, http://www.slate.com/blogs/xx_factor.html.

17. *What a Difference an X Makes!* (Society for Women's Health Research, 2008), http://www.womenshealthresearch.org/.

18. See, e.g., Joan H. Fujimura, Troy Duster, and Ramya Rajagopalan, eds., "Race, Genomics, and Biomedicine," special issue, *Social Studies of Science* 38, no. 5 (2008): 643; Barbara A. Koenig, Sandra Soo-Jin Lee, and Sarah S. Richardson, *Revisiting Race in a Genomic Age* (New Brunswick, NJ: Rutgers University Press, 2008); Ian Whitmarsh and David S. Jones, *What's the Use of Race? Modern Governance and the Biology of Difference* (Cambridge, MA: MIT Press, 2010). Articles of particular interest include Duana Fullwiley, "The Molecularization of Race: U.S. Health Institutions, Pharmacogenetics Practice, and Public Science after the Genome," in *Revisiting Race in a Genomic Age*, ed. Barbara A. Koenig, Sandra Soo-Jin Lee, and Sarah S. Richardson (New Brunswick, NJ: Rutgers University Press, 2008); Joan H. Fujimura and Ramya Rajagopalan, "Different Differences: The Use of 'Genetic Ancestry' versus Race in Biomedical Human Genetic Research," *Social Studies of Science* 41, no. 1 (2011); Jonathan Kahn, "Exploiting

Race in Drug Development," *Social Studies of Science* 38, no. 5 (2008); Catherine Bliss, "Genome Sampling and the Biopolitics of Race," in *A Foucault for the 21st Century: Governmentality, Biopolitics and Discipline in the New Millennium*, ed. Sam Binkley and Jorge Capetillo (Cambridge, MA: Cambridge Scholars, 2009); Jonathan Kahn, "Patenting Race in a Genomic Age," in *Revisiting Race in a Genomic Age*, ed. Barbara A. Koenig, Sandra Soo-Jin Lee, and Sarah S. Richardson (New Brunswick, NJ: Rutgers University Press, 2008); Adele E. Clarke et al., "Biomedicalizing Genetic Health, Diseases and Identities," in *Handbook of Genetics and Society: Mapping the New Genomic Era*, ed. Paul Atkinson, Peter Glasner, and Margaret Lock (London: Routledge, 2009).

19. Russett, *Sexual Science*; Anne Fausto-Sterling, *Myths of Gender: Biological Theories about Women and Men* (New York: Basic Books, 1985); Barbara Ehrenreich and Deirdre English, *For Her Own Good: Two Centuries of the Experts' Advice to Women* (New York: Anchor Books, 2005); Londa L. Schiebinger, *Nature's Body: Gender in the Making of Modern Science* (Boston: Beacon Press, 1993).

20. Pascal Bernard and Vincent R. Harley, "Wnt4 Action in Gonadal Development and Sex Determination," *International Journal of Biochemistry & Cell Biology* 39, no. 1 (2007).

21. John Money and Anke A. Ehrhardt, *Man & Woman, Boy & Girl: The Differentiation and Dimorphism of Gender Identity from Conception to Maturity* (Baltimore: Johns Hopkins University Press, 1972). For discussions of Money and the Johns Hopkins clinic, see Dreger, *Hermaphrodites*; Katrina Alicia Karkazis, *Fixing Sex: Intersex, Medical Authority, and Lived Experience* (Durham, NC: Duke University Press, 2008); Jordan-Young, *Brain Storm*; Fausto-Sterling, *Sexing the Body*; Suzanne J. Kessler, *Lessons from the Intersexed* (New Brunswick, NJ: Rutgers University Press, 1998).

22. Money, in particular, emphasized that in cases of intersex or the desire of a transgender individual to transition to living as a different sex, the primary consideration should be the individual's gender identity, and that biology should be therapeutically guided to be as consistent with a person's gender presentation as possible. Taken to its extreme in the famous John/Joan case, this protocol has become controversial today (see John Colapinto, *As Nature Made Him: The Boy Who Was Raised as a Girl* [New York: HarperCollins, 2000]). Nonetheless, researchers continue to recognize the importance of aligning therapeutic interventions with a person's gender identity in cases of disorders of gender identity or sexual differentiation. See Bernice L. Hausman, "Demanding Subjectivity: Transsexualism, Medicine, and the Technologies of Gender," *Journal of the History of Sexuality* 3, no. 2 (1992): 289.

23. Gayle Rubin, "The Traffic in Women: Notes on the 'Political Economy' of Sex," in *Toward an Anthropology of Women*, ed. Rayna Reiter (New York: Monthly Review Press, 1975).

24. This point about the ironic role of the sex/gender distinction in hardening the separation of biology from culture has been made by Anne Fausto-Sterling, Elisabeth Grosz, and Elizabeth Wilson, among others. Fausto-Sterling, *Sexing the Body*; E. A. Grosz, *Volatile Bodies: Toward a Corporeal Feminism* (Bloomington: Indiana University Press, 1994); Elizabeth A. Wilson, *Neural Geographies: Feminism and the Microstructure of Cognition* (New York: Routledge, 1998).

25. Hausman, "Demanding Subjectivity"; Joanne J. Meyerowitz, *How Sex Changed: A History of Transsexuality in the United States* (Cambridge, MA: Harvard University

Press, 2002); Dreger, *Hermaphrodites*; Alice Domurat Dreger, *Intersex in the Age of Ethics* (Hagerstown, MD: University Publishing Group, 1999).

26. As an example, transgender peoples' personal narratives of gender transitioning often mention experiencing the painful rebuke that they cannot ultimately change their chromosomes. As Bernice Hausman writes, "while a transsexual can become an effective representative of that other sex, the opposition between sexual signifiers remains as a reminder of his/her crossing over. Genetic sex cannot be altered; other secondary sex characteristics remain" (Hausman, "Demanding Subjectivity," 301). I use the term "transgender" inclusively here to refer to transsexuals who transition to a different sex than that with which they were born, as well as individuals who are gender fluid in various ways that may or may not include medical or surgical transitioning. See Julia Serano, *Whipping Girl: A Transsexual Woman on Sexism and the Scapegoating of Femininity* (Emeryville, CA: Seal Press, 2007).

27. Cynthia Kraus, "Naked Sex in Exile: On the Paradox of the 'Sex Question' in Feminism and in Science," *NWSA Journal* 12, no. 3 (2000): 151, 157.

28. Key recent literature on gender in genetics and genomics includes Fiona Alice Miller, "'Your True and Proper Gender': The Barr Body as a Good Enough Science of Sex," *Studies in History and Philosophy of Biological and Biomedical Sciences* 37, no. 3 (2006); Joan Fujimura, "'Sex Genes': A Critical Sociomaterial Approach to the Politics and Molecular Genetics of Sex Determination," *Signs* 32, no. 1 (2006); Nathan Q. Ha, "The Riddle of Sex: Biological Theories of Sexual Difference in the Early Twentieth-Century," *Journal of the History of Biology* 44, no. 3 (2011); Catherine Nash, "Genetic Kinship," *Cultural Studies* 18, no. 1 (2004); Amade M'Charek, "The Mitochondrial Eve of Modern Genetics: Of Peoples and Genomes, or the Routinization of Race," *Science as Culture* 14, no. 2 (2005); Sarah S. Richardson, "When Gender Criticism Becomes Standard Scientific Practice: The Case of Sex Determination Genetics," in *Gendered Innovations in Science and Engineering*, ed. Londa Schiebinger (Palo Alto, CA: Stanford University Press, 2008); Sarah S. Richardson, "Sexes, Species, and Genomes: Why Males and Females Are Not Like Humans and Chimpanzees," *Biology and Philosophy* 25, no. 5 (2010); Sarah S. Richardson, "Sexing the X: How the X Became the 'Female Chromosome,'" *Signs* 37, no. 4 (2012); Sarah Richardson, "Gendering the Genome: Sex Chromosomes in Twentieth Century Genetics" (PhD diss., Stanford University, 2009). See also Anne Fausto-Sterling, "Life in the XY Corral," *Women's Studies International Forum* 12, no. 3 (1989); Christopher Koehler, "The Sex Problem: Thomas Hunt Morgan, Richard Goldschmidt, and the Question of Sex and Gender in the Twentieth Century" (PhD diss., University of Florida, 1998); Bonnie Spanier, *Im/partial Science: Gender Ideology in Molecular Biology*, Race, Gender, and Science (Bloomington: Indiana University Press, 1995); Ingrid Holme, "Beyond XX and XY: Living Genomic Sex," in *Governing the Female Body: Gender, Health, and Networks of Power*, ed. Lori Stephens Reed and Paula Saukko (Albany: State University of New York Press, 2010); Helga Satzinger, *Differenz und Vererbung: Geschlechterordnungen in der Genetik und Hormonforschung 1890–1950 [Heredity and Difference: Gender Orders in Genetics and Hormone Research, 1890–1950]* (Köln: Wien Böhlau Verlag, 2009); Kraus, "Naked Sex in Exile."

29. This presumed contrast between genetics as "fixed and static" and endocrinology as "dynamic and fluid" is also noted by Holme, "Beyond XX and XY," 276.

30. Eugene Thacker, *The Global Genome: Biotechnology, Politics, and Culture* (Cam-

bridge, MA: MIT Press, 2005); Barry Barnes and John Dupré, *Genomes and What to Make of Them* (Chicago: University of Chicago Press, 2008).

31. Helen Thompson Woolley, "A Review of the Recent Literature on the Psychology of Sex," *Psychological Bulletin* (1910); Helen Thompson Woolley, "The Psychology of Sex," *Psychological Bulletin* 11, no. 10 (1914); Charlotte Perkins Gilman and Mary Armfield Hill, *The Man-Made World* (1911; Amherst, NY: Humanity Books, 2001); Ruth Herschberger, *Adam's Rib* (New York: Pellegrini & Cudahy, 1948); Eliza Burt Gamble, *The Sexes in Science and History: An Inquiry into the Dogma of Woman's Inferiority to Man* (1916; Westport, CT: Hyperion Press, 1976); Eliza Burt Gamble, *The Evolution of Woman; An Inquiry into the Dogma of Her Inferiority to Man* (New York: G. P. Putnam's Sons, 1894); Antoinette Louisa Brown Blackwell, *The Sexes throughout Nature* (New York: G. P. Putnam, 1875).

32. Steven Goldberg, *The Inevitability of Patriarchy* (New York: Morrow, 1973). See also texts such as David P. Barash, *The Whisperings Within* (New York: Harper & Row, 1979).

33. Agenda for "Genes and Gender" symposium. Series I, Folder 1.1. Conference I, 1976–77. Archive of the Genes and Gender Collective, Schlesinger Library, Radcliffe Institute for Advanced Study, Cambridge, MA.

34. Registration forms for conference. Series I, Folder 1.1. Conference I, 1976–77. Archive of the Genes and Gender Collective.

35. Contributions to the series include Myra Fooden, Susan Gordon, and Betty Hughley, *The Second X and Women's Health* (New York: Gordian Press, 1983); Ruth Hubbard and Marian Lowe, *Pitfalls in Research on Sex and Gender* (New York: Gordian Press, 1979); Anne E. Hunter, Catherine M. Flamenbaum, and Suzanne R. Sunday, *On Peace, War, and Gender: A Challenge to Genetic Explanations* (New York: Feminist Press, 1991); Suzanne R. Sunday and Ethel Tobach, *Violence against Women: A Critique of the Sociobiology of Rape* (New York: Gordian Press, 1985); Ethel Tobach and Betty Rosoff, *Genetic Determinism and Children* (New York: Gordian Press, 1980); Ethel Tobach and Betty Rosoff, *Challenging Racism and Sexism: Alternatives to Genetic Explanations* (New York: Feminist Press at the City University of New York, 1994); Georgine M. Vroman, Dorothy Burnham, and Susan Gordon, *Women at Work: Socialization toward Inequality* (New York: Gordian Press, 1988).

36. Ethel Tobach, "Famous People Letter," 21 August 1978. Series II, Folder 2.4. Genes and Gender I, 1978–81. Archive of the Genes and Gender Collective.

37. E.g., Dorothy E. Roberts, *Killing the Black Body: Race, Reproduction, and the Meaning of Liberty* (New York: Pantheon Books, 1997); Rayna Rapp, *Testing Women, Testing the Fetus: The Social Impact of Amniocentesis in America* (New York: Routledge, 1999); Charis Thompson, *Making Parents: The Ontological Choreography of Reproductive Technologies*, Inside Technology (Cambridge, MA: MIT Press, 2005); Sarah Franklin and Celia Roberts, *Born and Made: An Ethnography of Preimplantation Genetic Diagnosis* (Princeton, NJ: Princeton University Press, 2006); Mary Briody Mahowald, *Genes, Women, Equality* (New York: Oxford University Press, 2000).

38. In this book, I use the terms "chromosomal sex" and "genotypic sex" to refer to the chromosomal phenotype (e.g., XX/XY). I use the term "genetic sex" when referring to explanations of sex differences based on specific genes and gene sequences and to research on the genetic basis of sex prior to the Human Genome

Project. I use the term "genomic sex" to refer to whole-genome conceptions of sex differences and to research on human genetic sex differences after the Human Genome Project.

39. Michael S. Kimmel, *The Gendered Society* (New York: Oxford University Press, 2000), 1.

40. See, e.g., Evelyn Fox Keller, *Secrets of Life, Secrets of Death: Essays on Language, Gender, and Science* (New York: Routledge, 1992); Evelyn Fox Keller, "The Origin, History, and Politics of the Subject Called 'Gender and Science,'" chap. 4 in *Handbook of Science and Technology Studies*, rev. ed., ed. Sheila Jasanoff, Gerald E. Markle, James C. Petersen, Trevor Pinch (Thousand Oaks, CA: Sage, 1995); Evelyn Fox Keller, *Reflections on Gender and Science* (New Haven, CT: Yale University Press, 1985).

41. Keller, *Secrets of Life*, 17.

42. Ibid., 18.

43. Feminist scholars who have explored this terrain include Lorraine Daston, Elisabeth Grosz, Donna Haraway, Sandra Harding, Ludmilla Jordanova, Genevieve Lloyd, Emily Martin, Carolyn Merchant, Londa Schiebinger, and Elisabeth Wilson.

44. Fausto-Sterling, *Sexing the Body*.

45. Elisabeth Anne Lloyd, *The Case of the Female Orgasm: Bias in the Science of Evolution* (Cambridge, MA: Harvard University Press, 2005).

46. Joan Roughgarden, *Evolution's Rainbow: Diversity, Gender, and Sexuality in Nature and People* (Berkeley: University of California Press, 2004); Joan Roughgarden, *The Genial Gene: Deconstructing Darwinian Selfishness* (Berkeley: University of California Press, 2009).

47. Jordan-Young, *Brain Storm*; Cordelia Fine, *Delusions of Gender: How Our Minds, Society, and Neurosexism Create Difference* (New York: W. W. Norton, 2010).

48. M. Marchetti and T. Raudma, eds., *Stocktaking: 10 Years of "Women in Science" Policy by the European Commission, 1999–2009* (Luxembourg: Publications Office of the European Union, 2010); "Recruit and Advance: Women Students and Faculty in Science and Engineering," in *National Academies Press* (Washington, DC: National Academy of Sciences, 2006).

49. See, e.g., "Doctorate Recipients from U.S. Universities: 2009," US National Science Foundation, http://www.nsf.gov/statistics/doctorates/.

50. Thomas S. Kuhn, *The Structure of Scientific Revolutions* (Chicago: University of Chicago Press, 1962); Ludwik Fleck, *Genesis and Development of a Scientific Fact* (1935; Chicago: University of Chicago Press, 1979).

51. The concept of "modeling gender in science" was first developed in Sarah S. Richardson, "Feminist Philosophy of Science: History, Contributions, and Challenges," *Synthese* 177, no. 3 (2010). Feminist science studies scholars have emphasized the importance of developing analyses of science that go beyond the question of bias. See, in particular, Deborah Findlay, "Discovering Sex: Medical Science, Feminism and Intersexuality," *Canadian Review of Sociology and Anthropology* 32, no. 1 (1995); Kraus, "Naked Sex in Exile"; Helen E. Longino and Ruth Doell, "Body, Bias and Behavior: A Comparative Analysis of Reasoning in Two Areas of Biological Science," *Signs* 9, no. 2 (1983); Louise M. Antony, "Quine as Feminist: The Radical Import of Naturalized Epistemology," in *A Mind of One's Own: Feminist Essays on Reason and Objectivity*, ed. Louise M. Antony and Charlotte Witt (Boulder, CO: Westview Press, 1993).

52. Historians who have made valuable contributions to an account of the role of sex chromosomes in the development of the chromosomal theory of heredity include Elof Axel Carlson, *Mendel's Legacy: The Origin of Classical Genetics* (Cold Spring Harbor, NY: Cold Spring Harbor Laboratory Press, 2004); Sharon E. Kingsland, "Maintaining Continuity through a Scientific Revolution: A Rereading of E. B. Wilson and T. H. Morgan on Sex Determination and Mendelism," *Isis* 98, no. 3 (2007); Garland E. Allen, "Thomas Hunt Morgan and the Problem of Sex Determination, 1903–1910," *Proceedings of the American Philosophical Society* 110, no. 1 (1966); Scott Gilbert, "The Embryological Origins of the Gene Theory," *Journal of the History of Biology* 11, no. 2 (1978).

53. Peter S. Harper, *First Years of Human Chromosomes: The Beginnings of Human Cytogenetics* (Bloxham: Scion, 2006); M. Susan Lindee, *Moments of Truth in Genetic Medicine* (Baltimore: Johns Hopkins University Press, 2005).

54. H. A. Witkin, D. R. Goodenough, and K. Hirschhorn, "XYY Men: Are They Criminally Aggressive?," *Sciences (New York)* 17, no. 6 (1977).

55. Z. Spolarics, "The X-Files of Inflammation: Cellular Mosaicism of X-Linked Polymorphic Genes and the Female Advantage in the Host Response to Injury and Infection," *Shock* 27, no. 6 (2007); Barbara R. Migeon, *Females Are Mosaics: X Inactivation and Sex Differences in Disease* (Oxford: Oxford University Press, 2007); C. Selmi, "The X in Sex: How Autoimmune Diseases Revolve around Sex Chromosomes," *Best Practice & Research: Clinical Rheumatology* 22, no. 5 (2008).

56. Haraway's 1970s articles on feminist primatology are collected in Donna Jeanne Haraway, *Primate Visions: Gender, Race, and Nature in the World of Modern Science* (New York: Routledge, 1989).

57. Londa L. Schiebinger, *Has Feminism Changed Science?* (Cambridge, MA: Harvard University Press, 1999).

58. A. Jost, P. Gonse-Danysz, and R. Jacquot, "Studies on Physiology of Fetal Hypophysis in Rabbits and Its Relation to Testicular Function," *Journal of Physiology (Paris)* 45, no. 1 (1953).

59. P. Berta et al., "Genetic Evidence Equating SRY and the Testis-Determining Factor," *Nature* 348, no. 6300 (1990); P. Koopman et al., "Male Development of Chromosomally Female Mice Transgenic for SRY," *Nature* 351, no. 6322 (1991).

60. K. McElreavey et al., "A Regulatory Cascade Hypothesis for Mammalian Sex Determination: SRY Represses a Negative Regulator of Male Development," *Proceedings of the National Academy of Sciences USA* 90, no. 8 (1993); J. A. Graves, "The Evolution of Mammalian Sex Chromosomes and the Origin of Sex Determining Genes," *Philosophical Transactions of the Royal Society of London B: Biological Sciences* 350, no. 1333 (1995); W. Just et al., "Absence of SRY in Species of the Vole Ellobius," *Nature Genetics* 11, no. 2 (1995).

61. D. C. Page, "Save the Males!," *Nature Genetics* 17, no. 1 (1997); D. C. Page, "2003 Curt Stern Award Address: On Low Expectations Exceeded; or, The Genomic Salvation of the Y Chromosome," *American Journal of Medical Genetics* 74, no. 3 (2004); D. C. Page et al., "Abundant Gene Conversion Between Arms of Palindromes in Human and Ape Y Chromosomes," *Nature* 423, no. 6942 (2003); J. A. Graves, "The Degenerate Y Chromosome—Can Conversion Save It?," *Reproduction, Fertility, and Development* 16, no. 5 (2004); J. A. Graves, "Recycling the Y Chromosome," *Science*

307 (2005); J. A. Graves, "Sex Chromosome Specialization and Degeneration in Mammals," *Cell* 124 (2006); J. A. Graves and R. John Aitken, "The Future of Sex," *Nature* 415, no. 6875 (2002).

62. Sykes, *Adam's Curse*; Steve Jones, *Y: The Descent of Men* (London: Little, Brown, 2002).

63. Steven Epstein, *Inclusion: The Politics of Difference in Medical Research* (Chicago: University of Chicago, 2007).

64. A. P. Arnold and A. J. Lusis, "Understanding the Sexome: Measuring and Reporting Sex Differences in Gene Systems," *Endocrinology* 153, no. 6 (2012).

65. Many scholars have commented on the turn to a search for human differences following the completion of the human genome sequencing projects. See, e.g., Adam Bostanci, "Two Drafts, One Genome? Human Diversity and Human Genome Research," *Science as Culture* 15, no. 3 (2006).

Chapter 2

1. Edmund B. Wilson, "Studies on Chromosomes III. The Sexual Differences of the Chromosome-Groups in Hemiptera, with some Considerations on the Determination and Inheritance of Sex," *Journal of Experimental Zoology* 3, no. 1 (1906): 28.

2. Robert E. Kohler, *Partners in Science: Foundations and Natural Scientists, 1900–1945* (Chicago: University of Chicago Press, 1991).

3. For a nice discussion of the ambiguous and unstable notion of the "sex problem" in early twentieth-century genetics, see Christopher Koehler, "The Sex Problem: Thomas Hunt Morgan, Richard Goldschmidt, and the Question of Sex and Gender in the Twentieth Century" (PhD diss., University of Florida, 1998).

4. Even today, research on "sex" tends to run together a range of biological questions and phenomena, including reproduction and fertility, sex determination, sex differentiation, gender identity, sexual identity, and sexuality. Confusion between "sex" and "gender" in scientific sex research continues to be a considerable problem in scientific research (see chapter 10).

5. Here I follow the terminology of Jane Maienschein, who characterizes three main views of sex determination active in the period between 1890 and 1910: (1) externalist, in which sex is determined by external conditions acting on the organism in the course of development; (2) internalist, in which sex is determined by environmental, maternal, and hereditary factors in the cytoplasm or nucleus; and (3) hereditarian, with sex caused by solely inherited determinants in the nucleus. See Jane Maienschein, "What Determines Sex? A Study of Converging Approaches, 1880–1916," *Isis* 75, no. 3 (1984).

6. John Farley, *Gametes and Spores: Ideas about Sexual Reproduction, 1750–1914* (Baltimore: Johns Hopkins University Press, 1982), 218.

7. See, e.g., H. E. Jordan, "Recent Literature Touching the Question of Sex-Determination," *American Naturalist* 44, no. 520 (1910).

8. Roy Porter and Lesley A. Hall, *The Facts of Life: The Creation of Sexual Knowledge in Britain, 1650–1950* (New Haven, CT: Yale University Press, 1995), 157.

9. Patrick Geddes and John Arthur Thomson, *The Evolution of Sex* (London: Walter Scott, 1889), 39.

10. Ibid., 47, 50–51.

11. Rudolf Virchow, *Cellular Pathology: As Based upon Physiological and Pathological Histology. Twenty lectures delivered in the Pathological institute of Berlin during the months of February, March and April, 1858* (London, 1860).

12. Charles Darwin, *On the Origin of Species, 1859* (Washington Square: New York University Press, 1988).

13. Walther Flemming, *Zellsubstanz, Kern und Zelltheilung* (Leipzig: Vogel, 1882).

14. Aryn Martin, "Can't Any Body Count? Counting as an Epistemic Theme in the History of Human Chromosomes," *Social Studies of Science* 34, no. 6 (2004): 925; Heinrich Waldeyer, "Über Karyokinese und ihre Beziehungen zu den Befruchtungsvorgängen," *Archiv für mikroskopische Anatomie und Entwicklungsmechanik* 32 (1888).

15. August Weismann, *Das Keimplasma; eine Theorie der Vererbung* (Jena: Fischer, 1892).

16. The provisional hypothesis of the pangene was first laid out in the 1868 work *Variation of Animals and Plants under Domestication*, and later elaborated in the 1871 *Descent of Man*. Charles Darwin, *Variation of Animals and Plants under Domestication*, The Works of Charles Darwin (1868; New York: New York University Press, 1988); Charles Darwin, *The Descent of Man and Selection in Relation to Sex*, 2nd ed. (1871; New York: D. Appleton, 1897).

17. Ernesto Capanna, "Chromosomes Yesterday: A Century of Chromosome Studies," *Chromosomes Today* 13 (2000): 6.

18. H. Henking, "Uber Spermatogenese und deren Beziehung zur Entwicklung bei Pyrrhocoris apterus L.," *Zeitschriftffur wissenschaftliche Zoologie* 51 (1891).

19. Thomas H. Montgomery, "The Spermatogenesis in Pentatoma Up to the Formation of the Spermatid," *Zoologische Jahrbucher* (1898). Montgomery is here quoted in C. E. McClung, "The Accessory Chromosome: Sex Determinant?," *Biological Bulletin* 3, no. 1/2 (1902).

20. Ibid.

21. C. E. McClung, "A Peculiar Nuclear Element in the Male Reproductive Cells of Insects," *Zoological Bulletin* 2, no. 4 (1899): 187.

22. McClung, "Accessory Chromosome," 63.

23. Ibid., 74–75.

24. Ibid.

25. Edmund B. Wilson, *The Cell in Development and Inheritance*, 2d ed. (1896; New York: Macmillan, 1906).

26. Marilyn Bailey Ogilvie and Clifford J. Choquette, "Nettie Maria Stevens (1861–1912): Her Life and Contributions to Cytogenetics," *Proceedings of the American Philosophical Society* 125, no. 4 (1981): 309.

27. Ibid., 300.

28. Ibid.

29. N. M. Stevens, "A Study of the Germ Cells of Certain Diptera, with Reference to the Heterochromosomes and the Phenomena of Synapsis," *Journal of Experimental Zoology* 5, no. 3 (1908): 370.

30. N. M. Stevens, *Studies in Spermatogenesis with Especial Reference to the Accessory Chromosome*, vol. 36(1) (Washington, DC: Carnegie Institution, 1905), 13.

31. Stephen G. Brush, "Nettie M. Stevens and the Discovery of Sex Determination by Chromosomes," *Isis* 69, no. 2 (1978): 167.

32. Edmund B. Wilson, "Studies on Chromosomes II. The Paired Microchromosomes, Idiochromosomes and Heterotropic Chromosomes in Hemiptera," *Journal of Experimental Zoology* 2, no. 4 (1905): 539.

33. Wilson, "Studies on Chromosomes III," 2.

34. See Sharon E. Kingsland, "Maintaining Continuity through a Scientific Revolution: A Rereading of E. B. Wilson and T. H. Morgan on Sex Determination and Mendelism," *Isis* 98, no. 3 (2007); Scott Gilbert, "The Embryological Origins of the Gene Theory," *Journal of the History of Biology* 11, no. 2 (1978).

35. N. M. Stevens, *Studies in Spermatogenesis: A Comparative Study of the Heterochromosomes in Certain Species of Coleoptera, Hemiptera and Lepidoptera, with Especial Reference to Sex Determination*, vol. 36(2) (Washington, DC: Carnegie Institution, 1906), 56.

36. Ibid.

37. Until the late 1970s, Stevens was largely forgotten, the only tributes being a 1912 memoriam by Morgan and a minor write-up in *Notable American Women*. Ogilvie and Choquette, "Nettie Maria Stevens"; see also Brush, "Nettie M. Stevens."

38. Garland E. Allen, "Thomas Hunt Morgan and the Problem of Sex Determination, 1903–1910," *Proceedings of the American Philosophical Society* 110, no. 1 (1966): 50.

39. Peter J. Bowler, *The Mendelian Revolution: The Emergence of Hereditarian Concepts in Modern Science and Society* (Baltimore: Johns Hopkins University Press, 1989), 132, my emphasis.

40. Elof Axel Carlson, *Mendel's Legacy: The Origin of Classical Genetics* (Cold Spring Harbor, NY: Cold Spring Harbor Laboratory Press, 2004), 91, 95, 96.

41. Ibid., 91, 94.

42. Farley, *Gametes and Spores*, 287n265.

43. Cytogeneticists were not able to reliably distinguish the X and Y from similarly sized chromosomes in human cells until the 1960s.

44. T. H. Morgan, "The Scientific Work of Miss N. M. Stevens," *Science* 36, no. 928 (1912): 469.

45. Edmund B. Wilson, "A Chromatoid Body Simulating an Accessory Chromosome in Pentatoma," *Biological Bulletin* 24, no. 6 (1913): 402–3.

46. See Wilson, "Studies on Chromosomes II," 521.

47. Ibid., 522.

48. See N. M. Stevens, "A Study of the Germ Cells of Aphis Rosae and Aphis Oenotherae," *Journal of Experimental Zoology* 2, no. 3 (1905); N. M. Stevens, "Color Inheritance and Sex Inheritance in Certain Aphids," *Science* 26, no. 659 (1907); NM Stevens, "An unpaired heterochromosome in the aphids," *Journal of Experimental Zoology* 6, no. 1 (1909); N. M. Stevens, "A Note on Reduction in the Maturation of Male Eggs in Aphis," *Biological Bulletin* 18, no. 2 (1910).

49. Edmund B. Wilson, "Croonian Lecture: The Bearing of Cytological Research on Heredity," *Proceedings of the Royal Society of London. Series B, Containing Papers of a Biological Character* 88, no. 603 (1914): 341.

50. T. H. Morgan, "A Biological and Cytological Study of Sex Determination in Phylloxerans and Aphids," *Journal of Experimental Zoology* 7, no. 2 (1909): 347.

51. McClung, "Accessory Chromosome," 74–75.

52. Ibid., 73, 75, 77, 80.

53. Edmund B. Wilson, "The Chromosomes in Relation to the Determination of Sex in Insects," *Science* 22, no. 564 (1905): 501, 502.

54. Edmund B. Wilson, "Selective Fertilization and the Relation of the Chromosomes to Sex-Production," *Science* 32, no. 816 (1910): 242–43.

55. Wilson, *Cell in Development*, 109.

56. Wilson, "Studies on Chromosomes III," 36, 35.

57. Ibid., 33, 38.

58. Wilson, *Cell in Development*, 145.

59. Wilson, "Studies on Chromosomes III," 36.

60. Ibid., 37n31, quoting Geddes and Thomson 1889 [1901 ed.].

61. Ibid., 36–37.

62. Geddes and Thomson, *Evolution of Sex*, 50, 33. Nathan Ha has recently shown how metabolic theories were also foundational for American biologist Oscar Riddle's genetic theory of sex determination: Nathan Q. Ha, "The Riddle of Sex: Biological Theories of Sexual Difference in the Early Twentieth-Century," *Journal of the History of Biology* 44, no. 3 (2011).

63. Patrick Geddes and J. Arthur Thomson, *Sex* (London: Williams & Norgate, 1914), 110. See also T. H. Morgan, "Chromosomes and Heredity," *American Naturalist* 44, no. 524 (1910); Thomas Hunt Morgan, *The Mechanism of Mendelian Heredity* (New York: Holt, 1915); Thomas Hunt Morgan and Calvin B. Bridges, *Sex-Linked Inheritance in Drosophila* (Washington, DC: Carnegie Institution, 1916); McClung, "Accessory Chromosome"; Edmund B. Wilson, "Studies on Chromosomes I. The Behavior of the Idiochromosomes in Hemiptera," *Journal of Experimental Zoology* 2, no. 3 (1905); Wilson, "Studies on Chromosomes II"; Wilson, "Studies on Chromosomes III"; A. D. Darbishire, "Recent Advances in the Study of Heredity. Lecture VII. Cytological and Other Evidence Relating to the Inheritance of Sex," *New Phytologist* 9, no. 1/2 (1910); J. T. Cunningham, *Sexual Dimorphism in the Animal Kingdom: A Theory of the Evolution of Secondary Sexual Characters* (London: Black, 1900); J. T. Cunningham, *Hormones and Heredity* (London: Constable, 1921); F. A. E. Crew, *The Genetics of Sexuality in Animals* (Cambridge: Cambridge University Press, 1927).

64. Michael F. Guyer, "Recent Progress in Some Lines of Cytology," *Transactions of the American Microscopical Society* 30, no. 2 (1911): 184, original emphasis.

Chapter 3

1. See "A Proposed Standard System on Nomenclature of Human Mitotic Chromosomes," *American Journal of Human Genetics* 12, no. 3 (1960). For more on this conference, see M. Susan Lindee, *Moments of Truth in Genetic Medicine* (Baltimore: Johns Hopkins University Press, 2005).

2. Thomas H. Montgomery, "Are Particular Chromosomes Sex Determinants?," *Biological Bulletin* 19, no. 1 (1910): 1. Montgomery listed the following names commonly used for the X and Y in 1910: "accessory, special, lagging, heterotropic, sex chromosomes, idiochromosomes, microchromosomes, diplosomes, gonochromosomes, chromatin nucleoli" (ibid.).

3. C. E. McClung, "A Peculiar Nuclear Element in the Male Reproductive Cells of

Insects," *Zoological Bulletin* 2, no. 4 (1899); Thomas H. Montgomery, "The Morphological Superiority of the Female Sex," *Proceedings of the American Philosophical Society* 43, no. 178 (1904); Edmund B. Wilson, "Studies on Chromosomes I. The Behavior of the Idiochromosomes in Hemiptera," *Journal of Experimental Zoology* 2, no. 3 (1905); Edmund B. Wilson, "Studies on Chromosomes II. The Paired Microchromosomes, Idiochromosomes and Heterotropic Chromosomes in Hemiptera," *Journal of Experimental Zoology* 2, no. 4 (1905); Thomas H. Montgomery, "The Terminology of Aberrant Chromosomes and Their Behavior in Certain Hemiptera," *Science* 23, no. 575 (1906); Edmund B. Wilson, "Studies on Chromosomes III. The Sexual Differences of the Chromosome-Groups in Hemiptera, with some Considerations on the Determination and Inheritance of Sex," *Journal of Experimental Zoology* 3, no. 1 (1906); Fernandus Payne, "On the Sexual Differences of the Chromosome Groups in Galgulus Oculatus," *Biological Bulletin* 14, no. 5 (1908).

4. See F. C. Paulmier, "The Spermatogenesis of *Anasa tristis*," *Journal of Morphology* 15 (1899); Michael F. Guyer, "Recent Progress in Some Lines of Cytology," *Transactions of the American Microscopical Society* 30, no. 2 (1911); Wilson, "Studies on Chromosomes III"; Fernandus Payne, "Some New Types of Chromosome Distribution and Their Relation to Sex—Continued," *Biological Bulletin* 16, no. 4 (1909).

5. L. J. Bachhuber, "The Behavior of the Accessory Chromosomes and of the Chromatoid Body in the Spermatogenesis of the Rabbit," *Biological Bulletin* 30, no. 4 (1916).

6. C. E. McClung, "The Accessory Chromosome: Sex Determinant?," *Biological Bulletin* 3, no. 1/2 (1902).

7. Montgomery, "Terminology of Aberrant Chromosomes," 38.

8. N. M. Stevens, *Studies in Spermatogenesis with Especial Reference to the Accessory Chromosome*, vol. 36(1) (Washington, DC: Carnegie Institution, 1905), 13, my emphasis.

9. See N. M. Stevens, "The Chromosomes in Diabrotica Vittata, Diabrotica Soror and Diabrotica 12-Punctata: A Contribution to the Literature on Heterochromosomes and Sex Determination," *Journal of Experimental Zoology* 5, no. 4 (1908): 453, 457–58.

10. Wilson, "Studies on Chromosomes II," 508n501, original emphasis.

11. Wilson, "Studies on Chromosomes I," 385.

12. Ibid., 375, my emphasis.

13. Ibid., 383, original emphasis.

14. See Montgomery, "Terminology of Aberrant Chromosomes," 36.

15. C. E. McClung, "Cytological Nomenclature," *Science* 37, no. 949 (1913).

16. Wilson, "Studies on Chromosomes III," 28. As late as the 1930s, researchers frequently referred to the "so-called" or "supposed sex chromosomes." See Edmund B. Wilson, "Sex Determination in Relation to Fertilization and Parthenogenesis," *Science* 25, no. 636 (1907): 378; W. W. Swingle, "The Accessory Chromosome in a Frog Possessing Marked Hermaphroditic Tendencies," *Biological Bulletin* 33, no. 2 (1917): 70; Edward C. Jeffrey and Edwin J. Haertl, "The Nature of Certain Supposed Sex Chromosomes," *American Naturalist* 72, no. 742 (1938).

17. Gregor von Mendel, *Versuche über Pflanzen-Hybriden, Vorgelegt in den Sitzungen vom 8. Februar und 8. März 1865*, translated to English in 1866 as *Experiments on Plant*

Hybridization. Hugo de Vries, Carl Correns, and Erich von Tschermak are typically credited with this rediscovery. See Robert C. Olby, *Origins of Mendelism*, 2nd ed. (Chicago: University of Chicago Press, 1985).

18. T. Boveri, "On Multiple Mitoses as a Means for the Analysis of the Cell Nucleus," in *The Chromosome Theory of Inheritance*, ed. Bruce R. Voeller (1902; New York: Appleton-Century-Crofts, 1968).

19. William Bateson, "Hybridisation and Cross-Breeding as a Method of Scientific Investigation, a Report of a Lecture Given at the RHS Hybrid Conference in 1899," *Journal of the Royal Horticultural Society* 24 (1900); William Bateson, "Problems of Heredity as a Subject for Horticultural Investigation," *Journal of the Royal Horticultural Society* 25 (1900).

20. Elof Axel Carlson, *Mendel's Legacy: The Origin of Classical Genetics* (Cold Spring Harbor, NY: Cold Spring Harbor Laboratory Press, 2004), 156.

21. W. E. Castle, "The Heredity of Sex," *Bulletin of the Museum of Comparative Zoology* 40, no. 4 (1903).

22. Montgomery, "Are Particular Chromosomes Sex Determinants?," 13.

23. Ibid., 14.

24. Ibid.

25. Ibid., 12.

26. Ibid., 15, 14.

27. Ibid., 13.

28. Ibid., 10.

29. T. H. Morgan, "Chromosomes and Heredity," *American Naturalist* 44, no. 524 (1910): 451–52.

30. Garland E. Allen, "Thomas Hunt Morgan and the Problem of Sex Determination, 1903–1910," *Proceedings of the American Philosophical Society* 110, no. 1 (1966): 51.

31. Carlson, *Mendel's Legacy*, 167; T. H. Morgan, "Recent Theories in Regard to the Determination of Sex," *Popular Science Monthly* 64 (1903): 116.

32. T. H. Morgan, "A Biological and Cytological Study of Sex Determination in Phylloxerans and Aphids," *Journal of Experimental Zoology* 7, no. 2 (1909): 339, original emphasis.

33. Thomas Hunt Morgan, *The Mechanism of Mendelian Heredity* (New York: Holt, 1915), 94.

34. Morgan, "Chromosomes and Heredity," 495, original emphasis.

35. Morgan, *Mechanism of Mendelian Heredity*, 133, original emphasis.

36. Ibid., 107, my emphasis.

37. T. H. Morgan, "Sex-Limited and Sex-Linked Inheritance," *American Naturalist* 48, no. 574 (1914): 583.

38. Ibid.

39. Walter S. Sutton, "The Chromosomes in Heredity," *Biological Bulletin* 4, no. 5 (1903); Boveri, "On Multiple Mitoses."

40. McClung's 1899 paper on the possible sex-determining role of the "accessory chromosome" was also the first to raise, as historian Elof Carlson has noted, "the possibility that a specific chromosome might be associated with a specific trait." Carlson, *Mendel's Legacy*, 82.

41. Walter S. Sutton, "The Spermatogonial Divisions in Brachystola Magna,"

Bulletin of the University of Kansas, Kansas University Quarterly 9, no. 2 (1900), original emphasis.

42. Walter S. Sutton, "On the Morphology of the Chromosome Group in Brachystola Magna," *Biological Bulletin* 4, no. 1 (1902): 37–38.

43. Ibid.

44. Wilson, "Studies on Chromosomes II," 510–41.

45. Edmund B. Wilson, "Croonian Lecture: The Bearing of Cytological Research on Heredity," *Proceedings of the Royal Society of London. Series B, Containing Papers of a Biological Character* 88, no. 603 (1914): 351.

46. Ibid., 337.

47. Ibid., 339, my emphasis.

48. Ibid., 342.

49. T. H. Morgan, "Sex Limited Inheritance in Drosophila," *Science* 32, no. 812 (1910).

50. Allen, "Thomas Hunt Morgan," 53, 54.

51. Sharon E. Kingsland, "Maintaining Continuity through a Scientific Revolution: A Rereading of E. B. Wilson and T. H. Morgan on Sex Determination and Mendelism," *Isis* 98, no. 3 (2007): 481.

52. Morgan, *Mechanism of Mendelian Keredity,* viii–ix.

53. Thomas Hunt Morgan and Calvin B. Bridges, *Sex-Linked Inheritance in Drosophila* (Washington, DC: Carnegie Institution, 1916), 8.

54. Edmund B. Wilson, *The Cell in Development and Heredity,* 3rd ed. (1925; New York: Macmillan, 1928), 745.

55. F. A. E. Crew, *The Genetics of Sexuality in Animals* (Cambridge: Cambridge University Press, 1927), 7.

56. Morgan and Bridges, *Sex-Linked Inheritance in Drosophila,* 7.

Chapter 4

1. Thank you to Michael Dietrich for alerting me to the existence of this item. It can be found in the front matter of the scanned Archive.org version of the University of California's copy of *Mechanism,* at http://archive.org/details/mechanismofmende00morgiala.

2. Thomas Walter Laqueur, *Making Sex: Body and Gender from the Greeks to Freud* (Cambridge, MA: Harvard University Press, 1990); Londa L. Schiebinger, *Nature's Body: Gender in the Making of Modern Science* (Boston: Beacon Press, 1993); Cynthia Eagle Russett, *Sexual Science: The Victorian Construction of Womanhood* (Cambridge, MA: Harvard University Press, 1989).

3. See, e.g., Adele Clarke, *Disciplining Reproduction: Modernity, American Life Sciences, and the Problems of Sex* (Berkeley: University of California Press, 1998); Vern L. Bullough, *Science in the Bedroom: A History of Sex Research* (New York: Basic Books, 1994); Nelly Oudshoorn, *Beyond the Natural Body: An Archaeology of Sex Hormones* (New York: Routledge, 1994); Chandak Sengoopta, *The Most Secret Quintessence of Life: Sex, Glands, and Hormones, 1850–1950* (Chicago: University of Chicago Press, 2006).

4. See, e.g., Eugen Steinach and Josef Löbel, *Sex and Life: Forty Years of Biological and Medical Experiments* (New York: Viking Press, 1940). The Steinach experiments

depicted in figure 4.1 are detailed in Eugen Steinach, "Willkürliche Umwandlung von Säugetiermännchen in Tiere mit ausgeprägt weiblichen Geschlechtscharacteren und weiblicher Psyche [Arbitrary Transformation of Male Mammals into Animals with Pronounced Female Sex Characters and Feminine Psyche]," *Pflügers Archiv* 144, no. 71 (1912).

5. Frank Rattray Lillie, "The Theory of the Free-martin," *Science* 43, no. 28 (1916); Frank Rattray Lillie, "Free-martin: A Study of the Action of Sex Hormones in the Foetal Life of Cattle," *Journal of Experimental Biology* 23, no. 5 (1917); Frank Rattray Lillie, "Sex-Determination and Sex-Differentiation in Mammals," *Proceedings of the National Academy of Sciences* 3 (1917).

6. Christer Nordlund, "Endocrinology and Expectations in 1930s America: Louis Berman's Ideas on New Creations in Human Beings," *British Journal of History of Science* 40, no. 1 (2007): 89.

7. Ibid., 90.

8. Sengoopta, *Most Secret Quintessence of Life*, 2.

9. Julia E. Rechter, "'The Glands of Destiny': A History of Popular, Medical and Scientific Views of the Sex Hormones in 1920s America" (PhD diss., University of California, 1997), xvi.

10. Serge Voronoff and George Gibier Rambaud, *The Conquest of Life* (New York: Brentano's, 1928); Eden Paul and Norman Haire, *Rejuvenation: Steinach's Researches on the Sex-Glands*, vol. 11, British Society for the Study of Sex Psychology (London: J. E. Francis, Athenaeum Press, 1923); Norman Haire, Eugen Steinach, and Serge Voronoff, *Rejuvenation, the Work of Steinach, Voronoff, and Others* (London: G. Allen & Unwin, 1924).

11. Sengoopta, *Most Secret Quintessence of Life*, 87.

12. Anne Fausto-Sterling, *Sexing the Body: Gender Politics and the Construction of Sexuality* (New York: Basic Books, 2000), 171.

13. Bullough, *Science in the Bedroom*, 122.

14. Clarke, *Disciplining Reproduction*, 63.

15. Frank Rattray Lillie, "Suggestions for Organization and Conduct of Research on Problems of Sex," in *First Annual Report of the Committee for Research on Problems of Sex* (Washington, DC: CRPS, 1922), 1.

16. Lillie, "Theory of the Free-martin," 611.

17. Lillie, "Sex-Determination and Sex-Differentiation," 466.

18. Ibid., 465.

19. Lillie, "Theory of the Free-martin," 613.

20. Rechter, "'The Glands of Destiny,'" 109–10.

21. Lillie, "Theory of the Free-martin."

22. Lillie, "Sex-Determination and Sex-Differentiation," 465.

23. T. H. Morgan, "Sex-Limited and Sex-Linked Inheritance," *American Naturalist* 48, no. 574 (1914): 582.

24. Thomas Hunt Morgan, *The Mechanism of Mendelian Heredity* (New York: Holt, 1915).

25. T. H. Morgan, *The Genetic and the Operative Evidence Relating to Secondary Sexual Characters* (Washington, DC: Carnegie Institution, 1919), 62.

26. On Goldschmidt, see Garland E. Allen, "The Historical Development of the

'Time Law of Intersexuality' and Its Philosophical Implications," in *Richard Goldschmidt: Controversial Geneticist and Creative Biologist: A Critical Review of his Contributions*, ed. Leonie K. Piternick (Boston: Birkhauser, 1980); Helga Satzinger, "Racial Purity, Stable Genes and Sex Difference: Gender in the Making of Genetic Concepts by Richard Goldschmidt and Fritz Lenz, 1916–1936," in *The Kaiser Wilhelm Society under National Socialism*, ed. Susanne Heim, Carola Sachse, and Mark Walker (New York: Cambridge University Press, 2009); Michael R. Dietrich, "Richard Goldschmidt: Hopeful Monsters and Other 'Heresies,'" *Nature Reviews: Genetics* 4, no. 1 (2003).

27. In both Morgan's and Goldschmidt's writings on the interactions between hormonal and genetic factors in sexual development, we find incipient a theory of how genes act in development, which Evelyn Fox Keller has called the "discourse of gene action." See Evelyn Fox Keller, *The Century of the Gene* (Cambridge, MA: Harvard University Press, 2000); Evelyn Fox Keller, "Genes, Gene Action, and Genetic Programs," chap. 4 in *Making Sense of Life: Explaining Biological Development with Models, Metaphors, and Machines* (Cambridge, MA: Harvard University Press, 2002). See also Lenny Moss, *What Genes Can't Do* (Cambridge, MA: MIT Press, 2002).

28. M. Susan Lindee, *Moments of Truth in Genetic Medicine* (Baltimore: Johns Hopkins University Press, 2005), 1.

29. Daniel J. Kevles, *In the Name of Eugenics: Genetics and the Uses of Human Heredity* (Cambridge, MA: Harvard University Press, 1995).

30. See Theophilus S. Painter, "The Sex Chromosomes of Man," *American Naturalist* 58, no. 659 (1924).

31. Thomas Hunt Morgan and Calvin B. Bridges, *Sex-Linked Inheritance in Drosophila* (Washington, DC: Carnegie Institution, 1916), 10; Edmund B. Wilson, "Notes on the Chromosome-Groups of Metapodius and Banasa," *Biological Bulletin* 12, no. 5 (1907): 304; Michael F. Guyer, "Recent Progress in Some Lines of Cytology," *Transactions of the American Microscopical Society* 30, no. 2 (1911): 182; N. M. Stevens, "Further Observations on Supernumerary Chromosomes, and Sex Ratios in Diabrotica Soror," *Biological Bulletin* 22, no. 4 (1912): 234.

32. Calvin B. Bridges, "Direct Proof through Non-disjunction That the Sex-Linked Genes of Drosophila Are Borne by the X-Chromosome," *Science* 40, no. 1020 (1914); Calvin B. Bridges, "Sex in Relation to Chromosomes and Genes," *American Naturalist* 59, no. 661 (1925).

33. F. A. E. Crew, *The Genetics of Sexuality in Animals* (Cambridge: Cambridge University Press, 1927), 25.

34. W. E. Castle, "A Mendelian View of Sex-Heredity," *Science* 29, no. 740 (1909): 398, 399.

35. Edmund B. Wilson, "Secondary Chromosome-Couplings and the Sexual Relations in Abraxas," *Science* 29, no. 748 (1909): 706.

36. T. H. Morgan, "Chromosomes and Heredity," *American Naturalist* 44, no. 524 (1910): 495.

37. T. H. Morgan, "Recent Results Relating to Chromosomes and Genetics," *Quarterly Review of Biology* 1, no. 2 (1926): 205.

38. Louis Berman, *The Glands Regulating Personality: A Study of the Glands of Internal Secretion in Relation to the Types of Human Nature* (New York: Macmillan, 1921),

135–36, my emphasis. See also Bridges, "Sex in Relation to Chromosomes and Genes"; Richard Goldschmidt, "The Quantitative Theory of Sex," *Science* 64, no. 1656 (1926).

39. "Stork to Take Orders for Boy or Girl Soon," *Chicago Daily Tribune*, 24 January 1922.

40. Charles Darwin, *The Descent of Man and Selection in Relation to Sex*, 2nd ed. (1871; New York: D. Appleton, 1897), 108.

41. See the chapter titled "The Variational Tendency of Men" in Havelock Ellis, *Man and Woman* (New York: Scribner, 1894). See also G. Stanley Hall, "The Contents of Children's Minds on Entering School," *Pedagogical Seminary* 1 (1891).

42. C. E. McClung, "A Peculiar Nuclear Element in the Male Reproductive Cells of Insects," *Zoological Bulletin* 2, no. 4 (1899); C. E. McClung, "The Accessory Chromosome: Sex Determinant?," *Biological Bulletin* 3, no. 1/2 (1902).

43. C. E. McClung, "Possible Action of the Sex-Determining Mechanism," *Proceedings of the National Academy of Sciences* 4, no. 6 (1918): 162.

44. R. E. Stevenson et al., "X-Linked Mental Retardation: The Early Era from 1943 to 1969," *American Journal of Medicine and Genetics* 51, no. 4 (1994): 838.

45. Stephanie A. Shields, "The Variability Hypothesis: The History of a Biological Model of Sex Differences in Intelligence," *Signs* 7, no. 4 (1982); Robert Gordon Lehrke, *Sex Linkage of Intelligence: The X-Factor* (Westport, CT: Praeger, 1997).

46. Helen Thompson Woolley, "The Psychology of Sex," *Psychological Bulletin* 11, no. 10 (1914): 354.

47. Berman, *Glands Regulating Personality*, 136.

48. Ashley Montagu, *The Natural Superiority of Women* (New York: Macmillan, 1953), 74.

49. Ibid., 76, 81.

50. Lindee, *Moments of Truth*, 10–11.

51. Ibid., 2.

52. Soraya de Chadarevian, "Mice and the Reactor: The 'Genetics Experiment' in 1950s Britain," *Journal of the History of Biology* 39 (2006).

53. Lindee, *Moments of Truth*, 1.

54. Researchers faithfully replicated the finding that humans have forty-eight chromosomes for a half-century before the number was revised to forty-six. Yet, it is well documented that much of the time, cytogeneticists were clearly looking at cells with forty-six chromosomes. Unlike much of today's biology, which relies on robotic instruments and computational tools, cytogenetics is an observational discipline, highly dependent on the trained eyes and skilled hands of the researcher. In the intertwined clusters of fuzzy stained chromosome bodies seen in the nucleus of a cell under a microscope, it is notoriously easy to see what you want to see. Some have suggested that misinterpretations of the X and Y chromosomes contributed to these persistent miscounts. See Peter S. Harper, *First Years of Human Chromosomes: The Beginnings of Human Cytogenetics* (Bloxham: Scion, 2006); Aryn Martin, "Can't Any Body Count? Counting as an Epistemic Theme in the History of Human Chromosomes," *Social Studies of Science* 34, no. 6 (2004); T. C. Hsu, *Human and Mammalian Cytogenetics: An Historical Perspective* (New York: Springer-Verlag, 1979); J. H. Tjio and A. Levan, "The Chromosome Number of Man," *Hereditas* 42 (1956).

55. M. L. Barr and E. G. Bertram, "A Morphological Distinction between Neurones of the Male and Female, and the Behaviour of the Nucleolar Satellite during Accelerated Nucleoprotein Synthesis," *Nature* 163, no. 4148 (1949).

56. Murray L. Barr, "Sex Chromatin and Phenotype in Man," *Science* 130, no. 3377 (1959): 681–82.

57. C. E. Ford, "A Sex Chromosome Anomaly in a Case of Gonadal Dysgenesis (Turner's Syndrome)," *Lancet* 273, no. 7075 (1959); P. A. Jacobs and J. A. Strong, "A Case of Human Intersexuality Having a Possible XXY Sex-Determining Mechanism," *Nature* 183, no. 4657 (1959).

58. See de Chadarevian, "Mice and the Reactor"; Harper, *First Years of Human Chromosomes*.

Chapter 5

1. Bryan Sykes, *Adam's Curse: A Future without Men* (New York: Bantam Press, 2003), 19, 29–30.

2. For a sensitive study of the cultural and symbolic narratives of Y chromosome ancestry testing, see Catherine Nash, "Genetic Kinship," *Cultural Studies* 18, no. 1 (2004).

3. Sykes, *Adam's Curse*, 187.

4. Ibid., 237.

5. Steve Jones, *Y: The Descent of Men* (London: Little, Brown, 2002), x, 4.

6. Ibid., 194.

7. J. Craig Venter, *A Life Decoded: My Genome, My Life* (New York: Viking, 2007). See also Thomas Hayden, "He Figured Out Y, but Not 'So What?,'" *Washington Post*, 25 October 2007; Oliver Morton, "A Life Decoded by J. Craig Venter," *Sunday Times*, 21 October 2007.

8. Some males do not have a Y, of course. This includes XX individuals with an *SRY* gene transplanted to one of their X chromosomes; individuals with disorders of sexual development, such as congenital adrenal hyperplasia, who are XX but may develop male gonads and secondary sexual characters; and transgender individuals who have transitioned from female-bodied to male-identified.

9. C. E. Ford, "A Sex Chromosome Anomaly in a Case of Gonadal Dysgenesis (Turner's Syndrome)," *Lancet* 273, no. 7075 (1959).

10. P. A. Jacobs and J. A. Strong, "A Case of Human Intersexuality Having a Possible XXY Sex-Determining Mechanism," *Nature* 183, no. 4657 (1959); N. Maclean et al., "Sex-Chromosome Abnormalities in Newborn Babies," *Lancet* 283, no. 7328 (1964); N. Maclean et al., "Survey of Sex-Chromosome Abnormalities among 4514 Mental Defectives," *Lancet* 279, no. 7224 (1962); W. M. Court Brown, *Abnormalities of the Sex Chromosome Complement in Man* (London: H. M. Stationery Office, 1964).

11. P. A. Jacobs et al., "Aggressive Behavior, Mental Sub-normality and the XYY Male," *Nature* 208, no. 5017 (1965).

12. Ibid., 1351.

13. Peter S. Harper, *First Years of Human Chromosomes: The Beginnings of Human Cytogenetics* (Bloxham: Scion, 2006), 89.

14. National Institutes of Health and Saleem Alam Shah, *Report on the XYY Chromosomal Abnormality* (Chevy Chase, MD: US GPO, 1970).

15. To obtain these citation statistics, I analyzed 1,656 PubMed citations between 1960 and 1985 matching the search terms "XYY" and "Y chromosome," restricted for human research and pruned to remove duplicates and nonresearch articles. The articles were spot-checked in full-text form to ensure accurate representation of the research contribution as indicated by title and abstract. Because of the use of the PubMed database, which focuses on medical and bioscience literature, these statistics may fully not represent the literature on XYY in the social science and behavioral science fields. For information on the comprehensiveness of the PubMed archive, see National Institutes of Health, "PMC FAQs," last updated 31 October 2012, http://www.ncbi.nlm.nih.gov/pmc/about/faq/.

16. Jeremy Green, "Media Sensationalization and Science: The Case of the Criminal Chromosome," in *Expository Science: Forms and Functions of Popularization*, ed. Terry Shinn and Richard Whitley, *Sociology of the Sciences* (Boston: D. Reidel, 1985), 144. In more recent science fiction, such as Gwyneth Jones's *Life*, the film *Alien 3* with its colony of "double Y-chromos," and the comic book series *Y: The Last Man*, the Y chromosome is a trope for examining masculinity, exploring different ideas about the relationship between biology and sex roles, and imagining other kinds of sex-gender systems. See Gwyneth A. Jones, *Life: A Novel* (Seattle: Aqueduct Press, 2004); *Alien 3*, directed by David Fincher (Los Angeles: Twentieth Century Fox, 1992); DC Comics Inc., *Y: The Last Man* (New York: DC Comics, 2002–).

17. Edgar Berman, *The Compleat Chauvinist: A Survival Guide for the Bedeviled Male* (New York: Macmillan, 1982).

18. H. A. Witkin, D. R. Goodenough, and K. Hirschhorn, "XYY Men: Are They Criminally Aggressive?," *Sciences (New York)* 17, no. 6 (1977); H. A. Witkin et al., "Criminality in XYY and XXY Men," *Science* 193, no. 4253 (1976).

19. As the 1970 NIMH report noted, "if one studies the frequency of XYY anomalies among tall males detained in institutions because of mental illness, criminal or violent behavior, then quite obviously all that would be determined would be the prevalence of that anomaly among tall mentally ill, criminal or violent institutionalized males." NIMH and Shah, *Report*, 20.

20. David T. Wasserman and Robert Samuel Wachbroit, *Genetics and Criminal Behavior*, Cambridge Studies in Philosophy and Public Policy (New York: Cambridge University Press, 2001), 9, my emphasis.

21. "The XYY Controversy: Researching Violence and Genetics," *Hastings Center Report* 10, no. 4 (1980): 6.

22. Ibid., 19.

23. Marcia Baron, "Crime, Genes, and Responsibility," in *Genetics and Criminal Behavior*, ed. David T. Wasserman and Robert Samuel Wachbroit (New York: Cambridge University Press, 2001), 218.

24. Wasserman and Wachbroit, *Genetics and Criminal Behavior*, 9.

25. James D. Watson, *Recombinant DNA*, 2nd ed. (New York: Scientific American Books, 1992), 559.

26. Green, "Media Sensationalization and Science," 155.

27. Ibid., 147.

28. Ibid., 153.

29. John L. Fuller and William Robert Thompson, *Behavior Genetics* (New York: Wiley, 1960).

30. Arthur R. Jensen, "How Much Can We Boost IQ and Scholastic Achievement?," *Harvard Educational Review* 39 (1969). For connections between the race and IQ debates and XYY research, see Nathaniel Weyl, "Genetics, Brain Damage and Crime," *Mankind Quarterly* 10 (1969); Robert E. Kuttner, "Chromosomes and Intelligence," *Mankind Quarterly* 12 (1971). For a discussion of the fallout of the Jensen controversy for the field of behavioral genetics, see Mark Snyderman and Stanley Rothman, *The IQ Controversy, the Media and Public Policy* (New Brunswick, NJ: Transaction Books, 1988).

31. The rise of brain and genetic theories of mental illness in the 1960s and 1970s has been widely documented. For a recent example, see Jonathan Metzl, *The Protest Psychosis: How Schizophrenia Became a Black Disease* (Boston: Beacon Press, 2009).

32. The behavioral geneticist Robert Plomin argues that it was not until the 1980s that behavioral genetics gained mainstream acceptance beyond a core group of practitioners. See R. Plomin and R. Rende, "Human Behavioral Genetics," *Annual Review of Psychology* 42 (1991).

33. Angela K. Turner, "Genetic and Hormonal Influences on Male Violence," in *Male Violence*, ed. John Archer (New York: Routledge, 1994).

34. For example, see S. Kessler and R. H. Moos, "XYY Chromosome: Premature Conclusions," *Science* 165, no. 3892 (1969).

35. For a representative statement of the ambition, explanatory scope, and momentum of cytogenetics research at this time, see Court Brown, *Abnormalities of the Sex Chromosome Complement*, vii. For a recent oral history of midcentury human cytogenetics, see Harper, *First Years of Human Chromosomes*.

36. Richard C. Lewontin, Steven P. R. Rose, and Leon J. Kamin, *Not in Our Genes: Biology, Ideology, and Human Nature* (New York: Pantheon Books, 1984), 157, 237.

37. Ibid., 135.

38. Desmond Morris, *The Naked Ape* (New York: McGraw-Hill, 1967); Lionel Tiger and Robin Fox, *The Imperial Animal* (New York: Holt, 1971); Edward O. Wilson, *On Human Nature* (Cambridge, MA: Harvard University Press, 1978); Edward O. Wilson, *Sociobiology: The New Synthesis* (Cambridge, MA: Belknap Press of Harvard University Press, 1975); David P. Barash, *The Whisperings Within* (New York: Harper & Row, 1979).

39. This point is eloquently developed in Donna Jeanne Haraway, "In the Beginning Was the Word: The Genesis of Biological Theory," in *Simians, Cyborgs, and Women: The Reinvention of Nature* (1981; New York: Routledge, 1991), 73.

40. Quoted in Richard Lyons, "Genetic Abnormality Is Linked to Crime," *New York Times*, 21 April 1968.

41. Quoted in Robert Stock, "The XYY and the Criminal," *New York Times*, 20 October 1968.

42. L. F. Jarvik, V. Klodin, and S. S. Matsuyama, "Human Aggression and the Extra Y Chromosome: Fact or Fantasy?," *American Psychology* 28, no. 8 (1973): 680.

43. Harper, *First Years of Human Chromosomes*, 90.

44. Friedrich Vogel and Arno G. Motulsky, *Human Genetics: Problems and Approaches* (New York: Springer-Verlag, 1979), 502.

45. Victor Cohn, "A Criminal by Heredity?," *Washington Post*, 7 August 1968; Lyons, "Genetic Abnormality Is Linked to Crime."

46. H. Eldon Sutton, *An Introduction to Human Genetics* (New York: Holt, 1965), 41.

47. Jarvik, Klodin, and Matsuyama, "Human Aggression and the Extra Y Chromosome," 678.

48. P. A. Jacobs, "XYY Genotype," *Science* 189, no. 4208 (1975): 1044.

49. Witkin et al., "Criminality in XYY and XXY Men," 554.

50. I use the term "gendered schema" in this book solely to refer to the way in which scientists thought of the X as the chromosome for feminine traits and the Y as the chromosome for masculine traits. No reference to "gender schema theory" is intended.

51. W. M. Court Brown, *Human Population Cytogenetics* (Amsterdam: North-Holland, 1967), 42.

52. Ibid., 68, 71.

53. Edward Novitski, *Human Genetics* (New York: Macmillan, 1977), 287.

54. Ibid., 286.

55. H. Eldon Sutton, *An Introduction to Human Genetics*, 3rd ed. (Philadelphia: Saunders College, 1980), 397, 513.

56. P. A. Jacobs, "Human Population Cytogenetics: The First Twenty-five Years," *American Journal of Human Genetics* 34, no. 5 (1982): 694, 689.

57. Ibid., 695. Later, Jacobs returned to the United Kingdom, and today she is a celebrated professor of human genetics at the University of Southampton.

58. Witkin, Goodenough, and Hirschhorn, "XYY Men"; Witkin et al., "Criminality in XYY and XXY Men."

59. Witkin et al., "Criminality in XYY and XXY Men," 549.

60. Ibid., 554.

61. "The XYY Controversy," 8.

62. Quoted in ibid., 18.

63. For discussions of the work of Science for the People in the broader context of science activism and debates over biological determinism in the mid-to-late twentieth century, see Scott Frickel and Kelly Moore, *The New Political Sociology of Science: Institutions, Networks, and Power* (Madison: University of Wisconsin Press, 2006); Kelly Moore, *Disrupting Science: Social Movements, American Scientists, and the Politics of the Military, 1945–1975* (Princeton, NJ: Princeton University Press, 2008); W. R. Albury, "Politics and Rhetoric in the Sociobiology Debate," *Social Studies of Science* 10 (1980); Neil Jumonville, "The Cultural Politics of the Sociobiology Debate," *Journal of the History of Biology* 35 (2002); Michael Yedell and Rob Desalle, "Sociobiology: Twenty-five Years Later," *Journal of the History of Biology* 33, no. 3 (2000).

64. Kelly Moore and Nicole Hala, "Organizing Identity: The Creation of Science for the People," *Research in the Sociology of Organizations* 19 (2002).

65. Jacobs, "Human Population Cytogenetics," 693.

66. Harper, *First Years of Human Chromosomes*, 90.

67. S. Walzer, P. S. Gerald, and S. A. Shah, "The XYY Genotype," *Annual Review of Medicine* 29 (1978): 568.

68. Ibid.

69. "The XYY Controversy," 10.

70. Quoted in ibid., 20.

71. H. Skaletsky et al., "The Male-Specific Region of the Human Y Chromosome Is a Mosaic of Discrete Sequence Classes," *Nature* 423, no. 6942 (2003).

72. Ibid., 825, original emphasis.

73. D. A. Hay, "Y Chromosome and Aggression in Mice," *Nature* 255, no. 5510 (1975); M. K. Selmanoff et al., "Evidence for a Y Chromosomal Contribution to an Aggressive Phenotype in Inbred Mice," *Nature* 253, no. 5492 (1975).

74. L. M. Kunkel, K. D. Smith, and S. H. Boyer, "Human Y-Chromosome-Specific Reiterated DNA," *Science* 191, no. 4232 (1976).

75. Ibid., 1190.

76. "In Pursuit of the Y Chromosome," *Nature* 226, no. 5249 (1970).

77. Quoted in "The XYY Controversy," 26.

Chapter 6

1. Natalie Angier, "For Motherly X Chromosome, Gender Is Only the Beginning," *New York Times*, 1 May 2007.

2. Natalie Angier, *Woman: An Intimate Geography* (Boston: Houghton Mifflin, 1999), 25.

3. David Bainbridge, *The X in Sex: How the X Chromosome Controls Our Lives* (Cambridge, MA: Harvard University Press, 2003), 127–29.

4. Ibid., 130, my emphasis.

5. Ibid., 151.

6. See Bruce R. Voeller, ed., *The Chromosome Theory of Inheritance: Classic Papers in Development and Heredity* (New York: Appleton-Century-Crofts, 1968), 78–80.

7. Thomas Hunt Morgan, *The Mechanism of Mendelian Heredity* (New York: Holt, 1915), 7; Edmund B. Wilson, *The Cell in Development and Heredity*, 3rd ed. (1925; New York: Macmillan, 1928).

8. Morgan, *Mechanism of Mendelian Heredity*, 78–79.

9. Theophilus S. Painter, "The Sex Chromosomes of Man," *American Naturalist* 58, no. 659 (1924): 509, 522.

10. Fiona Alice Miller, "Dermatoglyphics and the Persistence of 'Mongolism': Networks of Technology, Disease and Discipline," *Social Studies of Science* 33, no. 1 (2003): 76.

11. H. H. Turner, "A Syndrome of Infantilism, Congenital Webbed Neck, and Cubitus Valgus," *Endocrinology* 23 (1938).

12. H. F. Klinefelter, E. C. Reifenstein, and F. Albright, "Syndrome Characterized by Gynecomastia, Aspermatogenesis without Aleydigism, and Increased Excretion of Follicle Stimulating Hormone," *Journal of Clinical Endocrinology and Metabolism* 2 (1942).

13. Fiona Alice Miller, "'Your True and Proper Gender': The Barr Body as a Good Enough Science of Sex," *Studies in History and Philosophy of Biological and Biomedical Sciences* 37, no. 3 (2006).

14. Peter S. Harper, *First Years of Human Chromosomes: The Beginnings of Human Cytogenetics* (Bloxham: Scion, 2006), 79.

15. H. Eldon Sutton, *An Introduction to Human Genetics* (New York: Holt, 1965), 44.

16. Friedrich Vogel and Arno G. Motulsky, *Human Genetics: Problems and Approaches* (New York: Springer-Verlag, 1979), 500.

17. P. A. Jacobs and J. A. Strong, "A Case of Human Intersexuality Having a Possible XXY Sex-Determining Mechanism," *Nature* 183, no. 4657 (1959): 302.

18. Ibid.

19. Jane Brody, "If Her Chromosomes Add Up, A Woman Is Sure to Be a Woman," *New York Times*, 16 September 1967, 28.

20. Alice Domurat Dreger, *Hermaphrodites and the Medical Invention of Sex* (Cambridge, MA: Harvard University Press, 1998); Alice Domurat Dreger, *Intersex in the Age of Ethics* (Hagerstown, MD: University, 1999).

21. Robert Bock, "Understanding Klinefelter Syndrome: A Guide for XXY Males and Their Families" (Adolescence Section), NIH Pub. No. 93-3202, Office of Research Reporting (Washington, DC: NICHD, 1993).

22. D. Zenaty et al., "Le Syndrome De Turner: Quoi De Neuf Dans La Prise En Charge?," *Archives de Pediatrie* 18, no. 12 (2011); C. H. Gravholt, "Epidemiological, Endocrine and Metabolic Features in Turner Syndrome," *European Journal of Endocrinology* 151, no. 6 (2004).

23. Mary F. Lyon, "Some Milestones in the History of X-Chromosome Inactivation," *Annual Review of Genetics* 26 (1992); Mary F. Lyon, "Gene Action in the X-Chromosome of the Mouse," *Nature* 190 (1961).

24. In biology, a genetic mosaic is distinct from a genetic chimera. Mosaics carry two different types of cells, whereas chimeras are made up of fused cells of two individuals or species. In the literature on female X mosaicism, however, mosaic and chimera are used interchangeably, blending these two connotations.

25. Joshua Lederberg, "Poets Knew It All Along: Science Finally Finds Out That Girls Are Chimerical; You Know, Xn/Xa," *Washington Post*, 18 December 1966. Lederberg also notes, however, that the case of XXY males "complicates the myth that chimerism is femininity."

26. "Research Makes It Official: Women Are Genetic Mosaics," *Time*, 4 January 1963.

27. Wade quoted in Maureen Dowd, "X-Celling over Men," *New York Times*, 20 March 2005.

28. "Men and Women: The Differences Are in the Genes," *ScienceDaily* (2005), http://www.sciencedaily.com/releases/2005/03/050323124659.htm.

29. Dowd, "X-Celling over Men."

30. Julianna Kettlewell, "Female Chromosome Has X Factor," *BBCNEWS.com*, 16 March 2005, http://news.bbc.co.uk/2/hi/science/nature/4355355.stm.

31. E. A. Grosz, *Volatile Bodies: Toward a Corporeal Feminism* (Bloomington: Indiana University Press, 1994).

32. *Cat People*, directed by Jacques Tourneur (Los Angeles: RKO, 1942). This is not to say that males are never depicted as changeable—after all, there is also the werewolf, the vampire, and Dr. Jekyll and Mr. Hyde.

33. W. E. Castle, *Genetics and Eugenics* (Cambridge, MA: Harvard University Press, 1916), 176, original emphasis.

34. Iris Marion Young, "Pregnant Embodiment: Subjectivity and Alienation," in *Throwing Like a Girl and Other Essays in Feminist Philosophy and Social Theory* (Bloomington: Indiana University Press, 1990), 160–61.

35. See Elizabeth Lunbeck, *The Psychiatric Persuasion: Knowledge, Gender, and Power in Modern America* (Princeton, NJ: Princeton University Press, 1994).

36. Louis Berman, *The Glands Regulating Personality: A Study of the Glands of Internal Secretion in Relation to the Types of Human Nature* (New York: Macmillan, 1921), 142.

37. Anne Fausto-Sterling, *Myths of Gender: Biological Theories about Women and Men* (New York: Basic Books, 1985), 91.

38. For epidemiological statistics on male and female incidence and prevalence of autoimmune diseases, see M. D. Lockshin, "Sex Differences in Autoimmune Disease," *Lupus* 15, no. 11 (2006); P. A. McCombe, J. M. Greer, and I. R. Mackay, "Sexual Dimorphism in Autoimmune Disease," *Current Molecular Medicine* 9, no. 9 (2009); G. S. Cooper, M. L. Bynum, and E. C. Somers, "Recent Insights in the Epidemiology of Autoimmune Diseases: Improved Prevalence Estimates and Understanding of Clustering of Diseases," *Journal of Autoimmunity* 33, no. 3–4 (2009); D. L. Jacobson et al., "Epidemiology and Estimated Population Burden of Selected Autoimmune Diseases in the United States," *Clinical Immunology and Immunopathology* 84, no. 3 (1997); W. W. Eaton et al., "Epidemiology of Autoimmune Diseases in Denmark," *Journal of Autoimmunity* 29, no. 1 (2007).

39. S. J. Walsh and L. M. Rau, "Autoimmune Diseases: A Leading Cause of Death among Young and Middle-Aged Women in the United States," *American Journal of Public Health* 90, no. 9 (2000): 1464.

40. R. E. Kast, "Predominance of Autoimmune and Rheumatic Diseases in Females," *Journal of Rheumatology* 4, no. 3 (1977); J. J. Stewart, "The Female X-Inactivation Mosaic in Systemic Lupus Erythematosus," *Immunology Today* 19, no. 8 (1998).

41. J. E. Oliver and A. J. Silman, "Why Are Women Predisposed to Autoimmune Rheumatic Diseases?," *Arthritis Research & Therapy* 11, no. 5 (2009); C. C. Whitacre, "Sex Differences in Autoimmune Disease," *Nature Immunology* 2, no. 9 (2001); McCombe, Greer, and Mackay, "Sexual Dimorphism in Autoimmune Disease."

42. Carlo Selmi et al., "The X Chromosome and the Sex Ratio of Autoimmunity," in "Gender, Sex Hormones, Pregnancy and Autoimmunity," ed. Yehuda Shoenfeld, Angela Tincani, and M. Eric Gershwin, special issue, *Autoimmunity Reviews* 11, no. 6–7 (2012); Claude Libert, Lien Dejager, and Iris Pinheiro, "The X Chromosome in Immune Functions: When a Chromosome Makes the Difference," *Nature Reviews Immunology* 10, no. 8 (2010).

43. T. F. Davies, "Editorial: X versus X—The Fight for Function within the Female Cell and the Development of Autoimmune Thyroid Disease," *Journal of Clinical Endocrinology and Metabolism* 90, no. 11 (2005): 6332.

44. "Sex, Genes and Women's Health," *Nature Genetics* 25, no. 1 (2000): 1, 2.

45. Krisha McCoy, "Women and Autoimmune Disorders," *Everydayhealth.com*, last updated 20 December 2010, http://www.everydayhealth.com/autoimmune-disorders/understanding/women-and-autoimmune-diseases.aspx.

46. Candace Tingen, "Science Mini-Lesson: X Chromosome Inactivation," *Women's Health Research Institute* (blog), Northwestern University, 21 October 2009, http://blog.womenshealth.northwestern.edu/2009/10/science-mini-lesson-x-chromosome-inactivation/.

47. Karl S. Kruszelnicki, "Hybrid Auto-Immune Women 3," *ABC Science In Depth*, 12 February 2004, http://www.abc.net.au/science/articles/2004/02/12/1002754.htm.

48. Donna Jeanne Haraway, "The Biopolitics of Postmodern Bodies: Constitutions of Self in Immune System Discourse," in *Simians, Cyborgs, and Women: The Reinvention of Nature* (New York: Routledge, 1991); Emily Martin, "The Egg and the Sperm: How Science Has Constructed a Romance Based on Stereotypical Male-Female Roles," *Signs* 16, no. 3 (1991); Lisa H. Weasel, "Dismantling the Self/Other Dichotomy in Science: Towards a Feminist Model of the Immune System," *Hypatia* 16, no. 1 (2001).

49. Haraway, "Biopolitics of Postmodern Bodies," 204, 223.

50. Emily Martin, "The Woman in the Flexible Body," in *Revisioning Women, Health and Healing: Feminist, Cultural, and Technoscience Perspectives*, ed. Adele Clarke and Virginia L. Olesen (New York: Routledge, 1999), 101.

51. Ibid., 101, 103, 102.

52. Weasel, "Dismantling the Self/Other Dichotomy," 30, 35.

53. Kast, "Predominance of Autoimmune and Rheumatic Diseases"; Stewart, "Female X-Inactivation Mosaic."

54. Z. Ozbalkan et al., "Skewed X Chromosome Inactivation in Blood Cells of Women with Scleroderma," *Arthritis & Rheumatism* 52, no. 5 (2005); T. Ozcelik et al., "Evidence from Autoimmune Thyroiditis of Skewed X-Chromosome Inactivation in Female Predisposition to Autoimmunity," *European Journal of Human Genetics* 14, no. 6 (2006).

55. P. Invernizzi et al., "X Monosomy in Female Systemic Lupus Erythematosus," *Annals of the New York Academy of Sciences* 1110 (2007); Accelerated Cure Project, "Analysis of Genetic Mutations or Alleles on the X or Y Chromosome as Possible Causes of Multiple Sclerosis," http://www.acceleratedcure.org/sites/default/files/curemap/phase2-genetics-xy-chromosomes.pdf (October 2006); G. P. Knudsen, "Gender Bias in Autoimmune Diseases: X Chromosome Inactivation in Women with Multiple Sclerosis," *Journal of the Neurological Sciences* 286, no. 1–2 (2009); G. P. Knudsen et al., "X Chromosome Inactivation in Females with Multiple Sclerosis," *European Journal of Neurology* 14, no. 12 (2007); S. Chitnis et al., "The Role of X-Chromosome Inactivation in Female Predisposition to Autoimmunity," *Arthritis Research* 2, no. 5 (2000); M. F. Seldin et al., "The Genetics Revolution and the Assault on Rheumatoid Arthritis," *Arthritis & Rheumatism* 42, no. 6 (1999); T. H. Brix et al., "No Link between X Chromosome Inactivation Pattern and Simple Goiter in Females: Evidence from a Twin Study," *Thyroid* 19, no. 2 (2009); E. Pasquier et al., "Strong Evidence that Skewed X-Chromosome Inactivation Is Not Associated with Recurrent Pregnancy Loss: An Incident Paired Case Control Study," *Human Reproduction* 22, no. 11 (2007).

56. P. Invernizzi, "The X Chromosome in Female-Predominant Autoimmune Diseases," *Annals of the New York Academy of Sciences* 1110 (2007); Y. Svyryd et al., "X Chromosome Monosomy in Primary and Overlapping Autoimmune Diseases," *Autoimmunity Reviews* 11, no. 5 (2012).

57. L. M. Russell et al., "X Chromosome Loss and Ageing," *Cytogenetics and*

Genome Research 116, no. 3 (2007). J. M. Amos-Landgraf et al., "X Chromosome-Inactivation Patterns of 1,005 Phenotypically Unaffected Females," *American Journal of Human Genetics* 79, no. 3 (2006): 497. See also Andrew Sharp, David Robinson, and Patricia Jacobs, "Age- and Tissue-Specific Variation of X Chromosome Inactivation Ratios in Normal Women," *Human Genetics* 107, no. 4 (2000).

58. See Lockshin, "Sex Differences in Autoimmune Disease"; M. D. Lockshin, "Nonhormonal Explanations for Sex Discrepancy in Human Illness," *Annals of the New York Academy of Sciences* 1193, no. 1 (2010); Oliver and Silman, "Why Are Women Predisposed."

59. For example, Quintero et al. note that "the inconsistencies found in some reports that fail to demonstrate a significant difference between the patterns of inactivation of patients compared to controls may be due to differences in etiology, pathophysiology and unknown mechanisms by means of which skewed X chromosome inactivation may influence female biased autoimmunity. Also, age differences, sample sizes, different definitions of cut-off points for skewed X inactivation might explain the discrepancies. Likewise, the possibility that tissues other than blood would be more appropriate for finding a significant difference in skewed X chromosome inactivation patterns has to be considered." Olga L. Quintero et al., "Autoimmune Disease and Gender: Plausible Mechanisms for the Female Predominance of Autoimmunity," in "Gender, Sex Hormones, Pregnancy and Autoimmunity," ed. Yehuda Shoenfeld, Angela Tincani, and M. Eric Gershwin, special issue, *Journal of Autoimmunity* 38, no. 2–3 (2012): J113.

60. C. Selmi, "The X in Sex: How Autoimmune Diseases Revolve around Sex Chromosomes," *Best Practice & Research: Clinical Rheumatology* 22, no. 5 (2008): 913; Z. Spolarics, "The X-Files of Inflammation: Cellular Mosaicism of X-Linked Polymorphic Genes and the Female Advantage in the Host Response to Injury and Infection," *Shock* 27, no. 6 (2007): 599–98.

61. Barbara R. Migeon, *Females Are Mosaics: X Inactivation and Sex Differences in Disease* (New York: Oxford University Press, 2007), 208.

62. Ibid., 211.

63. Ibid., 18.

64. Ibid., 44; Barbara R. Migeon, "X-Chromosome Inactivation: Molecular Mechanisms and Genetic Consequences," *Trends in Genetics* 10, no. 7 (1994): 230.

65. Barbara R. Migeon, "The Role of X Inactivation and Cellular Mosaicism in Women's Health and Sex-Specific Diseases," *JAMA* 295, no. 12 (2006): 1429.

66. Migeon, *Females Are Mosaics*, 208.

67. Barbara R. Migeon, "Non-random X Chromosome Inactivation in Mammalian Cells," *Cytogenetics and Cell Genetics* 80, no. 1–4 (1998): 147.

68. Migeon, *Females Are Mosaics*, 209; Migeon, "Role of X Inactivation and Cellular Mosaicism," 1432–33.

69. Migeon, "Role of X Inactivation and Cellular Mosaicism," 1432–33.

70. Migeon, *Females Are Mosaics*, 211.

71. Ibid., 17, 188.

72. N. Takagi, "The Role of X-Chromosome Inactivation in the Manifestation of Rett Syndrome," *Brain Development* 23 Suppl 1 (2001): S182.

73. A. Renieri et al., "Rett Syndrome: The Complex Nature of a Monogenic Dis-

ease," *Journal of Molecular Medicine* 81, no. 6 (2003). Migeon, however, maintains that skewed X inactivation can interact with the *MECP2* mutation to mediate Rett syndrome phenotype in females.

74. E. M. Maier et al., "Disease Manifestations and X Inactivation in Heterozygous Females with Fabry Disease," *Acta Paediatrica Supplement* 95, no. 451 (2006).

75. Migeon, "X-Chromosome Inactivation," 230.

76. Migeon, *Females Are Mosaics*, 211.

77. Ibid., 207; Migeon, "X-Chromosome Inactivation," 230.

78. Joel Zlotogora, "Germ Line Mosaicism," *Human Genetics* 102, no. 4 (1998).

79. A. Gimelbrant et al., "Widespread Monoallelic Expression on Human Autosomes," *Science* 318, no. 5853 (2007): 1139.

80. R. Ohlsson, "Genetics: Widespread Monoallelic Expression," *Science* 318, no. 5853 (2007): 1077.

81. Ibid.

82. Paul J. Bonthuis, Kimberly H. Cox, and Emilie F. Rissman, "X-Chromosome Dosage Affects Male Sexual Behavior," *Hormones and Behavior* 61, no. 4 (2012): 565, my emphasis.

83. Ibid., 571, my emphasis.

84. P. J. Wang et al., "An Abundance of X-Linked Genes Expressed in Spermatogonia," *Nature Genetics* 27, no. 4 (2001): 423, my emphasis; Seema Kumar, "Genes for Early Sperm Production Found to Reside on X Chromosome," press release, 4 April 2001, http://web.mit.edu/newsoffice/2001/sperm-0404.

85. Martin, "Egg and the Sperm."

86. Nellie Oudshoorn and Anne Fausto-Sterling have done similar work on the gendering of the sex steroids estrogen and testosterone. Nelly Oudshoorn, *Beyond the Natural Body: An Archaeology of Sex Hormones* (New York: Routledge, 1994); Anne Fausto-Sterling, *Sexing the Body: Gender Politics and the Construction of Sexuality* (New York: Basic Books, 2000).

87. Evelyn Fox Keller, "The Origin, History, and Politics of the Subject Called 'Gender and Science,'" in *Handbook of Science and Technology Studies*, ed. Sheila Jasanoff et al. (Thousand Oaks, CA: Sage, 1995), 87.

Chapter 7

1. Women's history is a large and methodologically diverse literature. For an introduction to its methods and contributions, see the work of Nancy Cott, in particular Nancy F. Cott, *No Small Courage: A History of Women in the United States* (New York: Oxford University Press, 2000). See also Joan Kelly, *Women, History & Theory: The Essays of Joan Kelly* (Chicago: University of Chicago Press, 1984); Joan W. Scott, *Gender and the Politics of History* (New York: Columbia University Press, 1988).

2. Steven Epstein, *Inclusion: The Politics of Difference in Medical Research* (Chicago: University of Chicago Press, 2007); Sandra Morgen, *Into Our Own Hands: The Women's Health Movement in the United States, 1969–1990* (New Brunswick, NJ: Rutgers University Press, 2002); Florence P. Haseltine and Beverly Greenberg Jacobson, *Women's Health Research: A Medical and Policy Primer* (Washington, DC: Health Press International, 1997).

3. See, for example, Patricia A. Gowaty, "Sexual Natures: How Feminism Changed Evolutionary Biology," *Signs* 28, no. 3 (2003); Angela N. H. Creager, Elizabeth Lunbeck, and Londa L. Schiebinger, *Feminism in Twentieth-Century Science, Technology, and Medicine* (Chicago: University of Chicago Press, 2001); Londa L. Schiebinger, *Gendered Innovations in Science and Engineering* (Stanford, CA: Stanford University Press, 2008); Joan M. Gero and Margaret Wright Conkey, *Engendering Archaeology: Women and Prehistory* (Cambridge, MA: Blackwell, 1991); Donna Jeanne Haraway, *Primate Visions: Gender, Race, and Nature in the World of Modern Science* (New York: Routledge, 1989); Nancy Tanner and Adrienne Zihlman, "Women in Evolution, Part I: Innovation and Selection in Human Origins," *Signs* 1, no. 3 (1976); Adrienne Zihlman, "Women in Evolution, Part II: Subsistence and Social Organization among Early Hominids " *Signs* 4, no. 1 (1978); Sarah Blaffer Hrdy, *The Woman that Never Evolved* (Cambridge, MA: Harvard University Press, 1981).

4. For a history of academic gender studies, see Marilyn J. Boxer, *When Women Ask the Questions: Creating Women's Studies in America* (Baltimore: Johns Hopkins University Press, 1998).

5. Helen E. Longino, "Can There Be a Feminist Science?," *Hypatia* 2, no. 3 (1987); Helen E. Longino, "Cognitive and Non-cognitive Values in Science: Rethinking the Dichotomy," in *Feminism and Philosophy of Science*, ed. Jack Nelson and Lynn Hankinson Nelson (Boston: Kluwer, 1996), 50.

6. For theoretical discussions in feminist science studies about gender analysis of science, see Sandra Harding, "Rethinking Standpoint Epistemology: What Is 'Strong Objectivity'?," in *Feminist Epistemologies*, ed. Linda Alcoff and Elizabeth Potter (New York: Routledge, 1993); Evelyn Fox Keller, "The Origin, History, and Politics of the Subject Called 'Gender and Science,'" in *Handbook of Science and Technology Studies*, rev. ed., ed. Sheila Jasanoff, Gerald E. Markle, James C. Petersen, and Trevor Pinch (Thousand Oaks, CA: Sage, 1995); Longino, "Can There Be a Feminist Science?"; Helen Longino, "Subjects, Power, Knowledge: Prescriptivism and Descriptivism in Feminist Philosophy of Science," in *Feminist Epistemologies*, ed. Linda Alcoff and Elizabeth Potter (New York: Routledge, 1992); Helen E. Longino and Evelynn Hammonds, "Conflicts and Tensions in the Feminist Study of Gender and Science," in *Conflicts in Feminism*, ed. Marianne Hirsch and Evelyn Fox Keller (New York: Routledge, 1990); Donna Jeanne Haraway, "Situated Knowledges: The Science Question in Feminism and the Privilege of Partial Perspective," in *Simians, Cyborgs, and Women: The Reinvention of Nature* (New York: Routledge, 1991).

7. Southern quoted in P. N. Goodfellow, *The Mammalian Y Chromosome: Molecular Search for the Sex-Determining Factor* (Cambridge: Company of Biologists, 1987).

8. Ibid., 1.

9. See Bryan Sykes, *Adam's Curse: A Future without Men* (New York: Bantam Press, 2003), 60–66.

10. Tom Wilkie, "At the Flick of a Genetic Switch," *London Independent*, 13 May 1991.

11. See Sykes, *Adam's Curse*, 60–66.

12. Ibid., 71.

13. D. C. Page et al., "The Sex-Determining Region of the Human Y Chromosome Encodes a Finger Protein," *Cell* 51, no. 6 (1987); P. Berta et al., "Genetic Evi-

dence Equating SRY and the Testis-Determining Factor," *Nature* 348, no. 6300 (1990); P. Koopman et al., "Male Development of Chromosomally Female Mice Transgenic for SRY," *Nature* 351, no. 6322 (1991); D. Vollrath et al., "The Human Y Chromosome: A 43-Interval Map Based on Naturally Occurring Deletions," *Science* 258, no. 5079 (1992).

14. Natalie Angier, "Scientists Say Gene on Y Chromosome Makes a Man a Man," *New York Times*, 19 July 1990; Nigel Williams, "So That's What Little Boys Are Made Of," *Guardian*, 20 July 1990.

15. Williams, "So That's What Little Boys Are Made Of." For a science communications analysis of the reportage of the *SRY* finding, see Molly J. Dingel and Joey Sprague, "Research and Reporting on the Development of Sex in Fetuses: Gendered from the Start," *Public Understanding of Science* 19, no. 2 (2010).

16. A. Jost, P. Gonse-Danysz, and R. Jacquot, "Studies on Physiology of Fetal Hypophysis in Rabbits and Its Relation to Testicular Function," *Journal of Physiology (Paris)* 45, no. 1 (1953).

17. C. E. Ford, "A Sex Chromosome Anomaly in a Case of Gonadal Dysgenesis (Turner's Syndrome)," *Lancet* 273, no. 7075 (1959).

18. See also Anne Fausto-Sterling, *Sexing the Body: Gender Politics and the Construction of Sexuality* (New York: Basic Books, 2000), 199–205.

19. Goodfellow, *Mammalian Y Chromosome*, 1.

20. Ibid.

21. Joan Fujimura makes a similar argument, using a more sociological approach, in Joan Fujimura, "'Sex Genes': A Critical Sociomaterial Approach to the Politics and Molecular Genetics of Sex Determination," *Signs* 32, no. 1 (2006).

22. "Mainstreaming" generally refers to explicit changes in social and institutional practices in which gender is systematically recognized as a category of analysis, as in changes to hiring and diversity practices, curricula, or the kind of scholarship recognized as prestigious in a field.

23. James C. Puffer, "Gender Verification of Female Olympic Athletes," *Medicine and Science in Sports and Exercise* 34, no. 10 (2002).

24. Alice Domurat Dreger, *Hermaphrodites and the Medical Invention of Sex* (Cambridge, MA: Harvard University Press, 1998); Alice Domurat Dreger, *Intersex in the Age of Ethics* (Hagerstown, MD: University, 1999).

25. See, for example, Joan Roughgarden, *Evolution's Rainbow: Diversity, Gender, and Sexuality in Nature and People* (Berkeley: University of California Press, 2004); Bruce Bagemihl, *Biological Exuberance: Animal Homosexuality and Natural Diversity* (New York: St. Martin's Press, 1999).

26. Anne Fausto-Sterling, "Life in the XY Corral," *Women's Studies International Forum* 12, no. 3 (1989): 326–27, 330.

27. Ibid., 329; E. M. Eicher and L. L. Washburn, "Genetic Control of Primary Sex Determination in Mice," *Annual Review of Genetics* 20 (1986).

28. Judith Butler, *Gender Trouble: Feminism and the Subversion of Identity* (New York: Routledge, 1990).

29. J. A. Graves and R. V. Short, "Y or X—Which Determines Sex?," *Reproduction, Fertility, and Development* 2, no. 6 (1990): 731.

30. Ken Reed and Jennifer A. Marshall Graves, *Sex Chromosomes and Sex-Determining Genes* (Langhorne, PA: Harwood, 1993), x, my emphasis.

31. Ibid., 375, original emphasis, identity of chair unknown.

32. Ibid., 384.

33. K. McElreavey et al., "A Regulatory Cascade Hypothesis for Mammalian Sex Determination: SRY Represses a Negative Regulator of Male Development," *Proceedings of the National Academy of Sciences USA* 90, no. 8 (1993).

34. Scott H. Podolsky and Alfred I. Tauber, *The Generation of Diversity: Clonal Selection Theory and the Rise of Molecular Immunology* (Cambridge, MA: Harvard University Press, 1997); Evelyn Fox Keller, *The Century of the Gene* (Cambridge, MA: Harvard University Press, 2000); Sohotra Sarkar, "From Genes as Determinants to DNA as Resource: Historical Notes on Development and Genetics," in *Genes in Development: Re-reading the Molecular Paradigm*, ed. Eva Neumann-Held and Christoph Rehmann-Sutter (Durham, NC: Duke University Press, 2006).

35. W. Just et al., "Absence of SRY in Species of the Vole Ellobius," *Nature Genetics* 11, no. 2 (1995).

36. Rosemary White, "Professor Jennifer A.M. Graves, FAA," *Women in Science Network Journal* 58 (2001).

37. A. H. Sinclair et al., "Sequences Homologous to ZFY, a Candidate Human Sex-Determining Gene, Are Autosomal in Marsupials," *Nature* 336, no. 6201 (1988); Steve Jones, *Y: The Descent of Men* (London: Little, Brown, 2002); Sykes, *Adam's Curse*; White, "Professor Jennifer A.M. Graves, FAA."

38. White, "Professor Jennifer A.M. Graves."

39. J. A. Graves, "Human Y Chromosome, Sex Determination, and Spermatogenesis—A Feminist View," *Biology of Reproduction* 63, no. 3 (2000): 667–68, 673.

40. Ibid., 674.

41. Ibid., 669.

42. Ibid.

43. Ibid., 667.

44. Derek Chadwick et al., *The Genetics and Biology of Sex Determination* (New York: John Wiley, 2002).

45. Ibid., 15.

46. Ibid., 247.

47. Ibid., 47, 49.

48. Ibid., 99.

49. Ibid., 40.

50. Ibid., 36, 12.

51. Ibid., 51.

52. Ibid., 253.

53. Ibid., 55.

54. E. Vilain, "Expert Interview Transcript," in *Rediscovering Biology: Molecular to Global Perspectives*, electronic media (Oregon Public Broadcasting and Annenberg-MediaLearner.org, 2004); D. C. Page, "Expert Interview Transcript," in *Rediscovering Biology*.

55. Vilain, "Expert Interview Transcript"; Holly Ingraham, "Expert Interview Transcript," in *Rediscovering Biology*; Page, "Expert Interview Transcript."

56. Vilain, "Expert Interview Transcript."

57. Page, "Expert Interview Transcript."

58. Vilain, "Expert Interview Transcript"; Ingraham, "Expert Interview Transcript"; Page, "Expert Interview Transcript."

59. Bob Beale, "The Sexes: New Insights into the X and Y Chromosomes," *Scientist*, 23 July 2001.

Chapter 8

1. D. C. Page, "Sexual Evolution: From X to Y," in *Sex Determination: Lecture Series* (Chevy Chase, MD: Howard Hughes Medical Institute, 2001).

2. Kate McDonald, "ICHG: Jenny Graves Is Talking About Sex—Again," *Australian Biotechnology News*, 21 July 2006.

3. See D. Haraway, "Situated Knowledges: The Science Question in Feminism and the Privilege of Partial Perspective," *Feminist Studies* 14, no. 3 (1988).

4. J. A. Graves and R. John Aitken, "The Future of Sex," *Nature* 415, no. 6875 (2002): 963.

5. J. A. Graves, "The Degenerate Y Chromosome—Can Conversion Save It?," *Reproduction, Fertility, and Development* 16, no. 5 (2004): 532.

6. J. A. Graves, "The Rise and Fall of SRY," *Trends in Genetics* 18, no. 5 (2002).

7. John Mangels, "Is the Gene Pool Shrinking Men Out of Existence?," *Cleveland Plain Dealer*, 30 May 2004; David Plotz, "The Male Malaise; Is the Y Chromosome Set to Self-Destruct?," *Washington Post*, 11 April 2004; Richard Pendlebury, "Men Are Doomed," *London Daily Mail*, 18 August 2003; e.g., "Study: 'Male' Chromosome to Stick Around," *CNN.com*, 31 August 2005.

8. Peter McAllister, *Manthropology: The Science of Why the Modern Male Is Not the Man He Used to Be* (New York: St. Martin's Press, 2010), 4.

9. Steve Jones, *Y: The Descent of Men* (London: Little, Brown, 2002), 257–58.

10. Bryan Sykes, *Adam's Curse: A Future without Men* (New York: Bantam Press, 2003), 277, 289, 292–93.

11. Ibid., 297, 302.

12. Gail Bederman, *Manliness & Civilization: A Cultural History of Gender and Race in the United States, 1880–1917* (Chicago: University of Chicago Press, 1995); David I. Macleod, *Building Character in the American Boy: The Boy Scouts, YMCA, and Their Forerunners, 1870–1920* (Madison: University of Wisconsin Press, 1983); Mark C. Carnes and Clyde Griffen, *Meanings for Manhood: Constructions of Masculinity in Victorian America* (Chicago: University of Chicago Press, 1990); Catherine Robson, *Men in Wonderland: The Lost Girlhood of the Victorian Gentleman* (Princeton, NJ: Princeton University Press, 2001).

13. Jane Mansbridge and Shauna L. Shames, "Toward a Theory of Backlash: Dynamic Resistance and the Central Role of Power," *Politics & Gender* 4, no. 4 (2008).

14. Susan Faludi, *Backlash: The Undeclared War against American Women* (New York: Crown, 1991); Ariel Levy, *Female Chauvinist Pigs: Women and the Rise of Raunch Culture* (New York: Free Press, 2005).

15. Yvonne Tasker and Diane Negra, *Interrogating Postfeminism: Gender and the Politics of Popular Culture* (Durham, NC: Duke University Press, 2007), 3.

16. Katie Roiphe, "The Naked and the Conflicted," *New York Times*, 31 December 2009.

17. Kerstin Aumann, Ellen Galinsky, and Kenneth Matos, "The New Male Mystique" (New York: Families and Work Institute, 2011).

18. The article that instigated interest in the declining quality of male sperm is E. Carlsen et al., "Evidence for Decreasing Quality of Semen During Past 50 Years," *BMJ* 305, no. 6854 (1992). These claims have been widely disputed, most recently in E. te Velde et al., "Is Human Fecundity Declining in Western Countries?," *Human Reproduction* 25, no. 6 (2010).

19. Ralph A. Catalano et al., "Temperature Oscillations May Shorten Male Lifespan via Natural Selection in Utero," *Climatic Change* (June 2011); Tim McDonnell, "Lost Boys: In a Warmer World, Will Males Die Sooner?," *Grist*, 30 June 2011, http://grist.org/climate-change/2011-06-30-lost-boys-warmer-world-males-die-sooner-global-warming/.

20. N. E. Skakkebaek, E. Rajpert-De Meyts, and K. M. Main, "Testicular Dysgenesis Syndrome: An Increasingly Common Developmental Disorder with Environmental Aspects," *Human Reproduction* 16, no. 5 (2001).

21. Hanna Rosin, "The End of Men," *Atlantic*, July/August 2010.

22. Ibid.

23. Martha McCaughey, *The Caveman Mystique: Pop-Darwinism and the Debates over Sex, Violence, and Science* (New York: Routledge, 2008), 18.

24. Michael S. Kimmel, *The Gendered Society* (New York: Oxford University Press, 2000), 146.

25. For these reasons, the mouse and chimpanzee sequencing projects excluded the Y.

26. H. Skaletsky et al., "The Male-Specific Region of the Human Y Chromosome Is a Mosaic of Discrete Sequence Classes," *Nature* 423, no. 6942 (2003).

27. Susumu Ohno, *Sex Chromosomes and Sex-Linked Genes* (New York: Springer-Verlag, 1967); Susumu Ohno, *Major Sex-Determining Genes* (1971; New York: Springer-Verlag, 1979).

28. Page quoted in Maureen Dowd, "X-Celling over Men," *New York Times*, 20 March 2005.

29. Gary Stix, "Geographer of the Male Genome," *Scientific American*, December 2004.

30. "David Page: The Evolution of Sex: Rethinking the Rotting Y Chromosome," in *MIT World* (Cambridge, MA: Whitehead Institute for Biomedical Research, 2003).

31. Kelli Whitlock, "The 'Y' Files," *Paradigm*, Fall 2003.

32. D. C. Page, "Save the Males!," *Nature Genetics* 17, no. 1 (1997); "David Page: The Evolution of Sex"; D. C. Page, "2003 Curt Stern Award Address: On Low Expectations Exceeded; or, The Genomic Salvation of the Y Chromosome," *American Journal of Human Genetics* 74, no. 3 (2004); Delia K. Cabe, "An Unfinished Story about the Genesis of Maleness," *HHMI Bulletin*, September 2000, 25.

33. Dowd, "X-Celling over Men."

34. Skaletsky et al., "Male-Specific Region of the Human Y Chromosome."

35. "David Page: The Evolution of Sex."

36. R. A. Fisher, "The Evolution of Dominance," *Biological Reviews* 6 (1931).

37. L. D. Hurst, "Embryonic Growth and the Evolution of the Mammalian Y Chromosome. I. The Y as an Attractor for Selfish Growth Factors," *Heredity* 73, pt. 3 (1994); L. D. Hurst, "Embryonic Growth and the Evolution of the Mammalian Y Chromosome. II. Suppression of Selfish Y-Linked Growth Factors May Explain Escape from X-Inactivation and Rapid Evolution of SRY," *Heredity* 73, pt. 3 (1994).

38. B. T. Lahn and D. C. Page, "Functional Coherence of the Human Y Chromosome," *Science* 278, no. 5338 (1997): 363; P. S. Burgoyne, "The Mammalian Y Chromosome: A New Perspective," *Bioessays* 20, no. 5 (1998).

39. Lahn and Page, "Functional Coherence of the Human Y Chromosome," 675, 679.

40. Ibid., 678.

41. R. Saxena et al., "The DAZ Gene Cluster on the Human Y Chromosome Arose from an Autosomal Gene that Was Transposed, Repeatedly Amplified and Pruned," *Nature Genetics* 14, no. 3 (1996): 297.

42. Ibid., 298.

43. Ibid., 292.

44. Ibid., 298.

45. B. T. Lahn and D. C. Page, "Retroposition of Autosomal mRNA Yielded Testis-Specific Gene Family on Human Y Chromosome," *Nature Genetics* 21, no. 4 (1999): 432.

46. Ibid., 429, 432.

47. Skaletsky et al., "Male-Specific Region of the Human Y Chromosome."

48. Ibid., 825; regarded as "anecdotal," Lahn and Page, "Functional Coherence of the Human Y Chromosome."

49. Skaletsky et al., "Male-Specific Region of the Human Y Chromosome," 835.

50. Ibid., 834.

51. R. Scott Hawley, "The Human Y Chromosome: Rumors of Its Death Have Been Greatly Exaggerated," *Cell* 113 (2003).

52. Page, "2003 Curt Stern Award Address."

53. Dowd, "X-Celling over Men"; Page, "Sexual Evolution: From X to Y"; "David Page: The Evolution of Sex."

54. Skaletsky et al., "Male-Specific Region of the Human Y Chromosome," 836.

55. Burgoyne, "Mammalian Y Chromosome," 365.

56. J. W. Foster et al., "Evolution of Sex Determination and the Y Chromosome: SRY-Related Sequences in Marsupials," *Nature* 359, no. 6395 (1992); J. W. Foster and J. A. Graves, "An SRY-Related Sequence on the Marsupial X Chromosome: Implications for the Evolution of the Mammalian Testis-Determining Gene," *Proceedings of the National Academy of Sciences USA* 91, no. 5 (1994).

57. J. A. Graves, "Sex Chromosomes and Sex Determination in Weird Mammals," *Cytogenetics and Genome Research* 96 (2002): 165.

58. Foster and Graves, "SRY-Related Sequence," 1927.

59. Graves, "Rise and Fall of SRY," 262.

60. W. Just et al., "Absence of SRY in Species of the Vole Ellobius," *Nature Genetics* 11, no. 2 (1995).

61. J. A. Graves, "The Evolution of Mammalian Sex Chromosomes and the Origin of Sex Determining Genes," *Philosophical Transactions of the Royal Society of London B: Biological Sciences* 350, no. 1333 (1995): 306.

62. Ibid.

63. Ibid., 306, 307.

64. "Junk DNA" is Graves's term. Graves may be using the term generically to refer to pseudogenes or other inactive genetic material. The view that so-called junk DNA has no role in gene action because it does not code for proteins is contradicted by evidence that modules located in the noncoding DNA play a central role in gene regulation, maintenance, and repair.

65. Graves, "Evolution of Mammalian Sex Chromosomes," 310.

66. M. L. Delbridge and J. A. Graves, "Mammalian Y Chromosome Evolution and the Male-Specific Functions of Y Chromosome-Borne Genes," *Reviews of Reproduction* 4, no. 2 (1999): 101.

67. Published as J. A. Graves, "Human Y Chromosome, Sex Determination, and Spermatogenesis—A Feminist View," *Biology of Reproduction* 63, no. 3 (2000).

68. Use of the term "selfish" to describe the sex-antagonistic model of the origin of Y chromosome genes originates with Hurst, who first developed the model. See Hurst, "Embryonic Growth and the Evolution of the Mammalian Y Chromosome. I. The Y as an Attractor for Selfish Growth Factors"; J. A. Graves and S. Shetty, "Sex from W to Z: Evolution of Vertebrate Sex Chromosomes and Sex Determining Genes," *Journal of Experimental Zoology* 290 (2001); Lahn and Page, "Functional Coherence of the Human Y Chromosome"; J. A. Graves, "Sex Chromosomes and the Future of Men," in *National Science Week* (Canberra: Australian National University College of Science, 2006).

69. Graves, "Human Y Chromosome, Sex Determination, and Spermatogenesis," 673.

70. Ibid., 674.

71. Ibid., 675.

72. J. A. Graves, "Sex, Genes and Chromosomes: A Feminist View," *Women in Science Network Journal* 59 (2002); Graves and Shetty, "Sex from W to Z," 454.

73. McDonald, "ICHG: Jenny Graves Is Talking About Sex—Again."

74. Graves, "Sex Chromosomes and Sex Determination in Weird Mammals."

75. Graves, "Evolution of Mammalian Sex Chromosomes," 312; J. A. Graves, "Sex Chromosomes and the Future of Men," in *Gender and Genomics: Sex, Science and Society* (Los Angeles: UCLA Center for Society and Genetics, 2005); J. A. Graves, "Sex Chromosome Specialization and Degeneration in Mammals," *Cell* 124 (2006): 906.

76 Graves, "Human Y Chromosome, Sex Determination, and Spermatogenesis," 668, 674.

77. D. C. Page et al., "Abundant Gene Conversion Between Arms of Palindromes in Human and Ape Y Chromosomes," *Nature* 423, no. 6942 (2003): 875.

78. Page, "2003 Curt Stern Award Address," 400.

79. E.g., David Bainbridge, "He and She: What's the Real Difference? A New Study of the Y Chromosome Suggests that the Genetic Variation Between Men and Women Is Greater than We Thought," *Boston Globe*, 6 July 2003.

80. Elizabeth Pennisi, "Mutterings from the Silenced X Chromosome," *Science* 307, no. 5716 (2005).

81. Hawley, "Human Y Chromosome," 825, 827.

82. D. C. Page et al., "Conservation of Y-Linked Genes during Human Evolution Revealed by Comparative Sequencing in Chimpanzee," *Nature* 437, no. 7055 (2005): 101, 103.

83. Ibid., 102.

84. "Study: 'Male' Chromosome to Stick Around."

85. Graves, "Degenerate Y Chromosome," 532.

86. Ibid.

87. Ibid., 531.

88. "David Page: The Evolution of Sex."

89. Graves, "Sex Chromosome Specialization," 910. See also D. T. Gerrard and D. A. Filatov, "Positive and Negative Selection on Mammalian Y Chromosomes," *Molecular Biology and Evolution* 22, no. 6 (2005); Page et al., "Conservation of Y-Linked Genes."

90. Graves, "Sex Chromosome Specialization," 911.

91. Stix, "Geographer of the Male Genome"; Dowd, "X-Celling over Men."

92. Note that what distinguishes this episode from cases such as XYY theories of male aggression and X mosaicism theories of female biology is not the presence or absence of feminist or antifeminist perspectives. Rather, it is the critical, open, and reflexive role of competing gender conceptions in the development of rival scientific models. This is an important point. As part of a critical and open debate, feminist perspectives may offer valuable contributions to our thinking about sex and gender and make evident the presence of biasing assumptions in science. However, the presence of a feminist perspective in science does not, in itself, render that science unbiased. "Bias" may be present when feminist views are openly invoked in science, and bad, biased science does not necessarily result when feminist perspectives are not welcome.

Chapter 9

1. L. Carrel and H. F. Willard, "X-Inactivation Profile Reveals Extensive Variability in X-Linked Gene Expression in Females," *Nature* 434, no. 7031 (2005). Although a rough cut of the human genome was released in 2001, contrary to common understanding this was not a "complete" sequence of the genome. Each individual chromosome sequencing consortium released "completed" sequences over the next several years. The X chromosome consortium published its results in March 2005. See also Mark T. Ross, "The DNA Sequence of the Human X Chromosome," *Nature* 434, no. 7031 (2005).

2. See "Variation in Women's X Chromosomes May Explain Difference among Individuals, between Sexes," press release, 16 March 2005, http://www.genome.duke.edu/press/news/03–16-2005/; Fred Guterl, "The Truth about Gender," *Newsweek*, 28 March 2005.

3. Robert Lee Hotz, "Women Are Very Much Not Alike, Gene Study Finds," *Los*

Angeles Times, 17 March 2005; Maureen Dowd, "X-Celling over Men," *New York Times*, 20 March 2005.

4. Elizabeth Pennisi, "Mutterings from the Silenced X Chromosome," *Science* 307, no. 5716 (2005): 1708.

5. H. Skaletsky et al., "The Male-Specific Region of the Human Y Chromosome Is a Mosaic of Discrete Sequence Classes," *Nature* 423, no. 6942 (2003): 836.

6. Ibid.

7. David Bainbridge, "He and She: What's the Real Difference? A New Study of the Y Chromosome Suggests that the Genetic Variation Between Men and Women Is Greater than We Thought," *Boston Globe*, 6 July 2003.

8. L. J. Shapiro et al., "Non-inactivation of an X-Chromosome Locus in Man," *Science* 204, no. 4398 (1979); Carolyn J. Brown, Laura Carrel, and Huntington F. Willard, "Expression of Genes from the Human Active and Inactive X Chromosomes," *American Journal of Human Genetics* 60, no. 6 (1997).

9. Carrel and Willard, "X-Inactivation Profile," 403.

10. Ibid.

11. Jonathan Marks, *What It Means to Be 98% Chimpanzee: Apes, People, and Their Genes* (Berkeley: University of California Press, 2002).

12. Ibid., 70.

13. Londa L. Schiebinger, *Nature's Body: Gender in the Making of Modern Science* (Boston: Beacon Press, 1993), 97–98.

14. Stephen Jay Gould, *The Mismeasure of Man*, rev. and expanded ed. (New York: Norton, 1996), 104; Schiebinger, *Nature's Body*, 158.

15. Carl Vogt, *Lectures on Man: His Place in Creation, and in the History of the Earth* (London: Longman, Green, Longman, and Roberts, 1864), 191–92.

16. Cynthia Eagle Russett, *Sexual Science: The Victorian Construction of Womanhood* (Cambridge, MA: Harvard University Press, 1989), 56–57.

17. M. C. King and A. C. Wilson, "Evolution at Two Levels in Humans and Chimpanzees," *Science* 188, no. 4184 (1975); Marks, *What It Means*.

18. J. L. Slightom et al., "Reexamination of the African Hominoid Trichotomy with Additional Sequences from the Primate Beta-Globin Gene Cluster," *Molecular Pharmocogenetics and Evolution* 1, no. 4 (1992); Marks, *What It Means*; "Initial Sequence of the Chimpanzee Genome and Comparison with the Human Genome," *Nature* 437, no. 7055 (2005): 69; T. Ryan Gregory, "Genome Size Evolution in Animals," in *The Evolution of the Genome*, ed. T. Ryan Gregory (New York: Elsevier, 2005).

19. Ajit Varki and Tasha K. Altheide, "Comparing the Human and Chimpanzee Genomes: Searching for Needles in a Haystack," in *Genomes: Perspectives from the 10th Anniversary Issue of Genome Research*, ed. Hillary E. Sussman and Maria A. Smit (Cold Spring Harbor, NY: Cold Spring Harbor Laboratory Press, 2006), 364; H. Kehrer-Sawatzki and D. N. Cooper, "Understanding the Recent Evolution of the Human Genome: Insights from Human-Chimpanzee Genome Comparisons," *Human Mutations* 28, no. 2 (2007): 116. See also "Initial Sequence of the Chimpanzee Genome"; Gregory, "Genome Size Evolution in Animals."

20. Kehrer-Sawatzki and Cooper, "Understanding the Recent Evolution"; Marks, *What It Means*.

21. "Initial Sequence of the Chimpanzee Genome"; Kehrer-Sawatzki and Cooper,

"Understanding the Recent Evolution." The expanding significance for human genomics of structural variations in the genome, such as copy number variation, inversions, and translocations, is discussed by Adam Bostanci, "Two Drafts, One Genome? Human Diversity and Human Genome Research," *Science as Culture* 15, no. 3 (2006): 195.

22. For a discussion of the merits and perils of gene expression microarray technologies, see Peter Keating and Alberto Cambrosio, "Too Many Numbers: Microarrays in Clinical Cancer Research," *Studies in History and Philosophy of Science Part C Studies in History and Philosophy of Biological and Biomedical Sciences* 43, no. 1 (2012).

23. Precise estimates of the number of genes on a chromosome should be viewed, of course, with skepticism. These estimates are constantly changing and depend on the operative definition of what counts as a gene. The traditional understanding of a gene as a discrete open reading frame that codes for a protein has been repeatedly undermined by new discoveries about gene regulation and transcription over the past several decades. See, e.g., Barry Barnes and John Dupré, *Genomes and What to Make of Them* (Chicago: University of Chicago Press, 2008); Evelyn Fox Keller, *The Century of the Gene* (Cambridge, MA: Harvard University Press, 2000).

24. A "functional homologue" may be a fully identical gene with equal dosage to its partner on the X or a structurally homologous gene that is expressed, but which has a different gene product.

25. Ross, "DNA Sequence," 330–31.

26. Z. Talebizadeh, S. D. Simon, and M. G. Butler, "X Chromosome Gene Expression in Human Tissues: Male and Female Comparisons," *Genomics* 88, no. 6 (2006): 680; D. K. Nguyen and C. M. Disteche, "Dosage Compensation of the Active X Chromosome in Mammals," *Nature Genetics* 38, no. 1 (2006): 48–49, 51.

27. I. W. Craig et al., "Application of Microarrays to the Analysis of the Inactivation Status of Human X-Linked Genes Expressed in Lymphocytes," *European Journal of Human Genetics* 12, no. 8 (2004); Talebizadeh, Simon, and Butler, "X Chromosome Gene Expression."

28. Mary F. Lyon, "No Longer 'All-or-None,'" *European Journal of Human Genetics* 13, no. 7 (2005).

29. Talebizadeh, Simon, and Butler, "X Chromosome Gene Expression," 680.

30. Nguyen and Disteche, "Dosage Compensation."

31. Talebizadeh, Simon, and Butler, "X Chromosome Gene Expression."

32. R. R. Delongchamp et al., "Genome-wide Estimation of Gender Differences in the Gene Expression of Human Livers: Statistical Design and Analysis," *BMC Bioinformatics* 6 Suppl 2 (2005): S13.

33. J. A. Graves and C. M. Disteche, "Does Gene Dosage Really Matter?," *Journal of Biology* 6, no. 1 (2007), http://jbiol.com/content/6/1/1 .

34. Ibid.

35. Rinn and Snyder, "Sexual Dimorphism," 300, 301.

36. This concept of the "human genome" is also surely a construct, even an idealization. But it is an accurate idealization, facilitating genetic analysis and reasoning without introducing distortions or leaving out essential features of genetic ontology. See Bostanci, "Two Drafts, One Genome?"

37. Ibid., 185.

38. Delongchamp et al., "Genome-wide Estimation"; Rinn and Snyder, "Sexual Dimorphism"; Talebizadeh, Simon, and Butler, "X Chromosome Gene Expression"; D. K. Nguyen and C. M. Disteche, "High Expression of the Mammalian X Chromosome in Brain," *Brain Research* 1126, no. 1 (2006).

39. Gregory, "Genome Size Evolution in Animals," 3.

40. John Dupré, *The Disorder of Things: Metaphysical Foundations of the Disunity of Science* (Cambridge, MA: Harvard University Press, 1993); Brent D. Mishler and Robert N. Brandon, "Individuality, Pluralism, and the Phylogenetic Species Concept," in *The Philosophy of Biology*, ed. David L. Hull and Michael Ruse (1987; New York: Oxford University Press, 1998); Kevin de Queiroz and Michael J. Donogue, "Phylogenetic Systematics and the Species Problem," in Hull and Ruse, *Philosophy of Biology*.

41. Dupré, *Disorder of Things*.

42. Ibid., 79, 80, 72, 73.

43. Evelyn Fox Keller, *Secrets of Life, Secrets of Death: Essays on Language, Gender, and Science* (New York: Routledge, 1992), 138.

44. Joan Roughgarden, *Evolution's Rainbow: Diversity, Gender, and Sexuality in Nature and People* (Berkeley: University of California Press, 2004); Joan Roughgarden, *The Genial Gene: Deconstructing Darwinian Selfishness* (Berkeley: University of California Press, 2009).

45. In developing a critique of the "discourse of reproductive autonomy," Keller's questions were different from mine. Keller was intervening in the so-called units of selection debate. Her concern was with the way in which human evolutionary biology and population genetics models sidestep questions of sexual difference by localizing reproduction in the desexed individual, thereby making sex and reproduction invisible and obscuring an important counterargument to the predominant genic selectionist view of evolution by natural selection.

46. Keller, *Secrets of Life*, 130, 142.

47. Ibid., 140. See Sarah S. Richardson, "Sexes, Species, and Genomes: Why Males and Females Are Not Like Humans and Chimpanzees," *Biology and Philosophy* 25, no. 5 (2010).

48. While the debate over whether species are best conceived as "classes" or "individuals" cannot be said to be settled in philosophy of biology, work by Hull and Ghiselin on the substantial ways in which species are like individuals is sufficient to sustain the distinction that I wish to draw here. See David L. Hull, "A Matter of Individuality," *Philosophy of Science* 45, no. 3 (1978); Michael J. Ghiselin, "A Radical Solution to the Species Problem," *Systematic Zoology* 23, no. 4 (1974).

49. Ghiselin, "Radical Solution"; Hull, "Matter of Individuality"; Mishler and Brandon, "Individuality, Pluralism, and the Phylogenetic Species Concept."

50. Nikolaos A. Patsopoulos, Athina Tatsioni, and John P. A. Ioannidis, "Claims of Sex Differences: An Empirical Assessment in Genetic Associations," *JAMA* 298, no. 8 (2007): 887–88.

51. Nancy Jay, "Gender and Dichotomy," *Feminist Studies* 7, no. 1 (1981); Ruth Bleier, *Science and Gender: A Critique of Biology and Its Theories on Women* (New York: Pergamon Press, 1984); Anne Fausto-Sterling, *Myths of Gender: Biological Theories about Women and Men* (New York: Basic Books, 1985); Donna Jeanne Haraway, "Science, Technology, and Socialist-Feminism in the Late Twentieth Century," in *Simians, Cy-*

borgs and Women: The Reinvention of Nature (1980; New York: Routledge, 1991); Anne Fausto-Sterling, *Sexing the Body: Gender Politics and the Construction of Sexuality* (New York: Basic Books, 2000).

52. "Dutch Scientists Sequence Female Genome," *Biotechniques Weekly*, 29 May 2008.

Chapter 10

1. N. Henriette Uhlenhaut et al., "Somatic Sex Reprogramming of Adult Ovaries to Testes by FOXL2 Ablation," *Cell* 139, no. 6 (2009); Clinton K. Matson et al., "DMRT1 Prevents Female Reprogramming in the Postnatal Mammalian Testis," *Nature* 476, no. 7358 (2011).

2. D. J. Fairbairn and D. A. Roff, "The Quantitative Genetics of Sexual Dimorphism: Assessing the Importance of Sex-Linkage," *Heredity* 97, no. 5 (2006): 320–21.

3. This conception of postgenomics was initially developed in Sarah S. Richardson, "Race and IQ in the Postgenomic Age: The Microcephaly Case," *BioSocieties* 6 (2011).

4. Jonathan Kahn, "Exploiting Race in Drug Development," *Social Studies of Science* 38, no. 5 (2008); Adele E. Clarke et al., "Biomedicalizing Genetic Health, Diseases and Identities," in *Handbook of Genetics and Society: Mapping the New Genomic Era*, ed. Paul Atkinson, Peter Glasner, and Margaret Lock (London: Routledge, 2009).

5. Jonathan Kahn, "Patenting Race in a Genomic Age," in *Revisiting Race in a Genomic Age*, ed. Barbara A. Koenig, Sandra Soo-Jin Lee, and Sarah S. Richardson (New Brunswick, NJ: Rutgers University Press, 2008); Catherine Bliss, "Genome Sampling and the Biopolitics of Race," in *A Foucault for the 21st Century: Governmentality, Biopolitics and Discipline in the New Millennium*, ed. Sam Binkley and Jorge Capetillo (Cambridge, MA: Cambridge Scholars, 2009); Duana Fullwiley, "The Molecularization of Race: U.S. Health Institutions, Pharmacogenetics Practice, and Public Science after the Genome," in *Revisiting Race in a Genomic Age*, ed. Barbara A. Koenig, Sandra Soo-Jin Lee, and Sarah S. Richardson (New Brunswick, NJ: Rutgers University Press, 2008); Joan H. Fujimura, Troy Duster, and Ramya Rajagopalan, eds., "Race, Genomics, and Biomedicine," special issue, *Social Studies of Science* 38, no. 5 (2008).

6. Part of SWHR's aim in initiating this shift was to expand the terrain of women's health activism to include all women, not just those who can become pregnant. Equally in play, however, was a strategic desire to find what SWHR founder Florence P. Haseltine called "safe" issues, such as menopause, that were not highly politicized in the way that reproductive rights have historically been. See Florence P. Haseltine and Beverly Greenberg Jacobson, *Women's Health Research: A Medical and Policy Primer* (Washington, DC: Health Press International, 1997).

7. Sheryl Burt Ruzek, *The Women's Health Movement: Feminist Alternatives to Medical Control* (New York: Praeger, 1978); Sandra Morgen, *Into Our Own Hands: The Women's Health Movement in the United States, 1969–1990* (New Brunswick, NJ: Rutgers University Press, 2002).

8. See Steven Epstein, *Inclusion: The Politics of Difference in Medical Research* (Chicago: University of Chicago Press, 2007). The SWHR was previously known as the Society for the Advancement of Women's Health Research.

9. Theresa M. Wizemann and Mary-Lou Pardue, eds., *Exploring the Biological Contributions to Human Health: Does Sex Matter?* (Washington, DC: National Academy Press, 2001).

10. A decade later in 2011, the IOM released a tenth anniversary report, updating the report's original conclusions and restating its aims.

11. "About *Biology of Sex Differences*: Aims & Scope," Biology of Sex Differences, http://www.bsd-journal.com/about#aimsscope (accessed 10 February 2013).

12. Kahn, "Patenting Race in a Genomic Age"; Kahn, "Exploiting Race in Drug Development."

13. "Corporate Advisory Council," Society for Women's Health Research, http://www.womenshealthresearch.org/site/PageServer?pagename=about_partners_cac (accessed 10 February 2013).

14. It is possible that these companies are also eager to attach their names to the moniker of "women's health" by funding the efforts of the SWHR. The phenomenon of corporate "pinkwashing" through support of women's health issues has been most recently discussed in relation to breast cancer activism in the United States. See Barbara Ehrenreich, *Bright-Sided: How the Relentless Promotion of Positive Thinking Has Undermined America* (New York: Metropolitan Books, 2009); Samantha King, *Pink Ribbons, Inc.: Breast Cancer and the Politics of Philanthropy* (Minneapolis: University of Minnesota Press, 2006); Gayle A. Sulik, *Pink Ribbon Blues: How Breast Cancer Culture Undermines Women's Health* (New York: Oxford University Press, 2011).

15. A. P. Arnold and A. J. Lusis, "Understanding the Sexome: Measuring and Reporting Sex Differences in Gene Systems," *Endocrinology* 153, no. 6 (2012).

16. On the rise of the hormonal science of sex, gender, and sexuality, see Rebecca M. Jordan-Young, *Brain Storm: The Flaws in the Science of Sex Differences* (Cambridge, MA: Harvard University Press, 2010).

17. Margaret M. McCarthy et al., "Sex Differences in the Brain: The Not So Inconvenient Truth," *Journal of Neuroscience* 32, no. 7 (2012): 2246.

18. Ibid., 2243.

19. R. M. McCarthy and A. P. Arnold, "Reframing Sexual Differentiation of the Brain," *Nature Neuroscience* 14 (2011): 677.

20. Ibid., 682, 679.

21. Ibid., 678, 679.

22. G. J. De Vries et al., "A Model System for Study of Sex Chromosome Effects on Sexually Dimorphic Neural and Behavioral Traits," *Journal of Neuroscience* 22, no. 20 (2002); featured image is from McCarthy et al., "Sex Differences in the Brain."

23. Epstein, *Inclusion*, 243.

24. Anne Fausto-Sterling, "Bare Bones of Sex: Part I, Sex & Gender," *Signs* 30, no. 2 (2005): 1498. For a critical discussion of the focus on sex differences in the new field of "gender-specific medicine," another new field closely allied with sex-based biology, see E. Annandale and A. Hammarstrom, "Constructing the 'Gender-Specific Body': A Critical Discourse Analysis of Publications in the Field of Gender-Specific Medicine," *Health (London)* 15, no. 6 (2011).

25. See, e.g., Paul Atkinson, Peter E. Glasner, and Margaret M. Lock, *Handbook of Genetics and Society: Mapping the New Genomic Era* (New York: Routledge, 2009).

26. The parallels between race and sex as variables in biomedical research are treated in detail by Epstein in his discussion of the work of the SWHR. See Epstein, *Inclusion*. For more on the analogy of race and sex, including its limits, see Nancy Leys Stepan, "Race and Gender: The Role of Analogy in Science," *Isis* 77, no. 2 (1986); Sally Haslanger, "Gender and Race: (What) Are They? (What) Do We Want Them to Be?," *Nous* 34, no. 1 (2000).

27. Epstein, *Inclusion*. See also Nancy Krieger, "Genders, Sexes, and Health: What Are the Connections—and Why Does It Matter?," *International Journal of Epidemiology* 32, no. 4 (2003).

28. Nikolaos A. Patsopoulos, Athina Tatsioni, and John P. A. Ioannidis, "Claims of Sex Differences: An Empirical Assessment in Genetic Associations," *JAMA* 298, no. 8 (2007): 881.

29. Ibid., 888.

30. Ibid., 890.

31. Ibid., 887, table 883.

32. Lack of replication is here defined as failure to corroborate the finding in further studies. However, it should also be noted that studies are often never repeated because funding agencies do not want to pay twice for the same study. In a competitive research funding environment, it is frequently difficult to receive support to carry out research currently funded for another research group or that has already been done.

33. Patsopoulos, Tatsioni, and Ioannidis, "Claims of Sex Differences," 889.

34. As represented in the Thomson Reuters ISI Web of Science database.

35. D. Karasik and S. L. Ferrari, "Contribution of Gender-Specific Genetic Factors to Osteoporosis Risk," *Annals of Human Genetics* 72 (2008): 696.

36. Ibid., 698.

37. Ibid., 696.

38. Fausto-Sterling, "Bare Bones of Sex," 1495.

39. Krieger, "Genders, Sexes, and Health," 653.

40. Ibid., 654.

41. For a discussion of standards for use of gene expression profiling to determine sex differences, see A. L. Tarca, R. Romero, and S. Draghici, "Analysis of Microarray Experiments of Gene Expression Profiling," *American Journal of Obstetrics and Gynecology* 195, no. 2 (2006).

42. Y. J. Zhang et al., "Transcriptional Profiling of Human Liver Identifies Sex-Biased Genes Associated with Polygenic Dyslipidemia and Coronary Artery Disease," *Plos One* 6, no. 8 (2011): 13.

43. Ibid., 11.

44. Janet Shibley Hyde, "The Gender Similarities Hypothesis," *American Psychologist* 60, no. 6 (2005).

45. This point is also emphasized by Margaret McCarthy in McCarthy et al., "Sex Differences in the Brain."

46. Ibid.

47. Joerg Isensee et al., "Sexually Dimorphic Gene Expression in the Heart of Mice and Men," *Journal of Molecular Medicine* 86, no. 1 (2008): 72–73.

48. Patsopoulos, Tatsioni, and Ioannidis, "Claims of Sex Differences," 880. They cite "A Randomized Trial of Aspirin and Sulfinpyrazone in Threatened Stroke. The Canadian Cooperative Study Group," *New England Journal of Medicine* 299, no. 2 (1978).

49. Patsopoulos, Tatsioni, and Ioannidis, "Claims of Sex Differences," 891.

50. On this point, see also Anne Hammarstrom and Ellen Annandale, "A Conceptual Muddle: An Empirical Analysis of the Use of 'Sex' and "Gender' in 'Gender-Specific Medicine' Journals," *Plos One* 7, no. 4 (2012).

51. Stephen Jay Gould, *The Mismeasure of Man*, rev. and expanded ed. (New York: Norton, 1996). See also Cynthia Eagle Russett, *Sexual Science: The Victorian Construction of Womanhood* (Cambridge, MA: Harvard University Press, 1989); Londa L. Schiebinger, *Nature's Body: Gender in the Making of Modern Science* (Boston: Beacon Press, 1993).

BIBLIOGRAPHY

"About *Biology of Sex Differences*: Aims & Scope." Biology of Sex Differences. Accessed 10 February 2013. http://www.bsd-journal.com/about#aimsscope.

Accelerated Cure Project. "Analysis of Genetic Mutations or Alleles on the X or Y Chromosome as Possible Causes of Multiple Sclerosis." October 2006. http://www.acceleratedcure.org/sites/default/files/curemap/phase2-genetics-xy-chromosomes.pdf.

Albury, W. R. "Politics and Rhetoric in the Sociobiology Debate." *Social Studies of Science* 10 (1980): 519–36.

Alien 3. Directed by David Fincher. Los Angeles: Twentieth Century Fox, 1992.

Allen, Garland E. "The Historical Development of the 'Time Law of Intersexuality' and Its Philosophical Implications." In *Richard Goldschmidt: Controversial Geneticist and Creative Biologist: A Critical Review of His Contributions*, edited by Leonie K. Piternick, 41–48. Boston: Birkhauser, 1980.

———. "Thomas Hunt Morgan and the Problem of Sex Determination, 1903–1910." *Proceedings of the American Philosophical Society* 110, no. 1 (1966): 48–57.

Amos-Landgraf, J. M., et al. "X Chromosome-Inactivation Patterns of 1,005 Phenotypically Unaffected Females." *American Journal of Human Genetics* 79, no. 3 (2006): 493–99.

Angier, Natalie. "For Motherly X Chromosome, Gender Is Only the Beginning." *New York Times*, 1 May 2007, 1.

———. "Scientists Say Gene on Y Chromosome Makes a Man a Man." *New York Times*, 19 July 1990, 1.

———. *Woman: An Intimate Geography*. Boston: Houghton Mifflin, 1999.

Annandale, E., and A. Hammarstrom. "Constructing the 'Gender-Specific Body': A Critical Discourse Analysis of Publications in the Field of Gender-Specific Medicine." *Health (London)* 15, no. 6 (2011): 571–87.

Antony, Louise M. "Quine as Feminist: The Radical Import of Naturalized Epistemology." In *A Mind of One's Own: Feminist Essays on Reason and Objectivity*, edited by Louise M. Antony and Charlotte Witt, 185–225. Boulder, CO: Westview Press, 1993.

Arnold, A. P., and A. J. Lusis. "Understanding the Sexome: Measuring and Reporting Sex Differences in Gene Systems." *Endocrinology* 153, no. 6 (2012): 2551–55.

Atkinson, Paul, Peter E. Glasner, and Margaret M. Lock. *Handbook of Genetics and Society: Mapping the New Genomic Era*. New York: Routledge, 2009.

Aumann, Kerstin, Ellen Galinsky, and Kenneth Matos. "The New Male Mystique." New York: Families and Work Institute, 2011.

Bachhuber, L. J. "The Behavior of the Accessory Chromosomes and of the Chromatoid Body in the Spermatogenesis of the Rabbit." *Biological Bulletin* 30, no. 4 (1916): 294–311.

Bagemihl, Bruce. *Biological Exuberance: Animal Homosexuality and Natural Diversity*. New York: St. Martin's Press, 1999.

Bainbridge, David. "He and She: What's the Real Difference? A New Study of the Y Chromosome Suggests that the Genetic Variation between Men and Women Is Greater than We Thought." *Boston Globe*, 6 July 2003, H1.

———. *The X in Sex: How the X Chromosome Controls Our Lives*. Cambridge, MA: Harvard University Press, 2003.Barash, David P. *The Whisperings Within*. New York: Harper & Row, 1979.

Barnes, Barry, and John Dupré. *Genomes and What to Make of Them*. Chicago: University of Chicago Press, 2008.

Baron, Marcia. "Crime, Genes, and Responsibility." In *Genetics and Criminal Behavior*, edited by David T. Wasserman and Robert Samuel Wachbroit, 199–224. New York: Cambridge University Press, 2001.

Barr, Murray L. "Sex Chromatin and Phenotype in Man." *Science* 130, no. 3377 (1959): 679–85.

Barr, M. L., and E. G. Bertram. "A Morphological Distinction between Neurones of the Male and Female, and the Behaviour of the Nucleolar Satellite during Accelerated Nucleoprotein Synthesis." *Nature* 163, no. 4148 (1949): 676.

Bateson, William. "Hybridisation and Cross-Breeding as a Method of Scientific Investigation, a Report of a Lecture Given at the RHS Hybrid Conference in 1899." *Journal of the Royal Horticultural Society* 24 (1900): 59–66.

———. "Problems of Heredity as a Subject for Horticultural Investigation." *Journal of the Royal Horticultural Society* 25 (1900): 54–61.

Beale, Bob. "The Sexes: New Insights into the X and Y Chromosomes." *Scientist*, 23 July 2001, 18.

Bederman, Gail. *Manliness & Civilization: A Cultural History of Gender and Race in the United States, 1880–1917*. Chicago: University of Chicago Press, 1995.

Berman, Edgar. *The Compleat Chauvinist: A Survival Guide for the Bedeviled Male*. New York: Macmillan, 1982.

Berman, Louis. *The Glands Regulating Personality: A Study of the Glands of Internal Secretion in Relation to the Types of Human Nature*. New York: Macmillan, 1921.

Bernard, Pascal, and Vincent R. Harley. "WNT4 Action in Gonadal Development and

Sex Determination." *International Journal of Biochemistry & Cell Biology* 39, no. 1 (2007): 31–43.

Berta, P., et al. "Genetic Evidence Equating SRY and the Testis-Determining Factor." *Nature* 348, no. 6300 (1990): 448–50.

Blackwell, Antoinette Louisa Brown. *The Sexes Throughout Nature*. New York: G. P. Putnam, 1875.

Bleier, Ruth. *Science and Gender: A Critique of Biology and Its Theories on Women*. New York: Pergamon Press, 1984.

Bliss, Catherine. "Genome Sampling and the Biopolitics of Race." In *A Foucault for the 21st Century: Governmentality, Biopolitics and Discipline in the New Millennium*, edited by Sam Binkley and Jorge Capetillo, 322–39. Cambridge, MA: Cambridge Scholars, 2009.

Bock, Robert. "Understanding Klinefelter Syndrome: A Guide for XXY Males and Their Families" (Adolescence Section). NIH Pub. No. 93-3202. Office of Research Reporting. Washington, DC: National Institute of Child Health and Human Development, 1993. http://www.nichd.nih.gov/publications/pubs/klinefelter.cfm.

Bonthuis, Paul J., Kimberly H. Cox, and Emilie F. Rissman. "X-Chromosome Dosage Affects Male Sexual Behavior." *Hormones and Behavior* 61, no. 4 (2012): 565–72.

Bostanci, Adam. "Two Drafts, One Genome? Human Diversity and Human Genome Research." *Science as Culture* 15, no. 3 (2006): 183–98.

Boveri, T. "On Multiple Mitoses as a Means for the Analysis of the Cell Nucleus" (1902). In *The Chromosome Theory of Inheritance*, edited by Bruce R. Voeller, 87–94. New York: Appleton-Century-Crofts, 1968.

Bowler, Peter J. *The Mendelian Revolution: The Emergence of Hereditarian Concepts in Modern Science and Society*. Baltimore: Johns Hopkins University Press, 1989.

Boxer, Marilyn J. *When Women Ask the Questions: Creating Women's Studies in America*. Baltimore: Johns Hopkins University Press, 1998.

Bridges, Calvin B. "Direct Proof through Non-disjunction that the Sex-Linked Genes of Drosophila Are Borne by the X-Chromosome." *Science* 40, no. 1020 (1914): 107–9.

———. "Sex in Relation to Chromosomes and Genes." *American Naturalist* 59, no. 661 (1925): 127–37.

Brix, T. H., et al. "No Link between X Chromosome Inactivation Pattern and Simple Goiter in Females: Evidence from a Twin Study." *Thyroid* 19, no. 2 (2009): 165–69.

Brody, Jane. "If Her Chromosomes Add Up, a Woman Is Sure to Be a Woman." *New York Times*, 16 September 1967, 28.

Brown, Carolyn J., Laura Carrel, and Huntington F. Willard. "Expression of Genes from the Human Active and Inactive X Chromosomes." *American Journal of Human Genetics* 60, no. 6 (1997): 1333–43.

Brush, Stephen G. "Nettie M. Stevens and the Discovery of Sex Determination by Chromosomes." *Isis* 69, no. 2 (1978): 162–72.

Bullough, Vern L. *Science in the Bedroom: A History of Sex Research*. New York: Basic Books, 1994.

Burgoyne, P. S. "The Mammalian Y Chromosome: A New Perspective." *Bioessays* 20, no. 5 (1998): 363–66.

Butler, Judith. *Gender Trouble: Feminism and the Subversion of Identity.* New York: Routledge, 1990.
Cabe, Delia K. "An Unfinished Story about the Genesis of Maleness." *HHMI Bulletin,* September 2000, 20–25.
Capanna, Ernesto. "Chromosomes Yesterday: A Century of Chromosome Studies." *Chromosomes Today* 13 (2000).
Carlsen, E., et al. "Evidence for Decreasing Quality of Semen during Past 50 Years." *BMJ* 305, no. 6854 (1992): 609–13.
Carlson, Elof Axel. *Mendel's Legacy: The Origin of Classical Genetics.* Cold Spring Harbor, NY: Cold Spring Harbor Laboratory Press, 2004.
Carnes, Mark C., and Clyde Griffen. *Meanings for Manhood: Constructions of Masculinity in Victorian America.* Chicago: University of Chicago Press, 1990.
Carrel, L. "'X'-Rated Chromosomal Rendezvous." *Science* 311, no. 5764 (2006): 1107–9.
Carrel, L., and H. F. Willard. "X-Inactivation Profile Reveals Extensive Variability in X-Linked Gene Expression in Females." *Nature* 434, no. 7031 (2005): 400–404.
Castle, W. E. *Genetics and Eugenics.* Cambridge, MA: Harvard University Press, 1916.
———. "The Heredity of Sex." *Bulletin of the Museum of Comparative Zoology* 40, no. 4 (1903): 189–219.
———. "A Mendelian View of Sex-Heredity." *Science* 29, no. 740 (1909): 395–400.
Catalano, Ralph A., et al. "Temperature Oscillations May Shorten Male Lifespan via Natural Selection in Utero." *Climatic Change* (June 2011).
Cat People. Directed by Jacques Tourneur. Los Angeles: RKO, 1942.
Chadwick, Derek, et al. *The Genetics and Biology of Sex Determination.* New York: John Wiley, 2002.
Chitnis, S., et al. "The Role of X-Chromosome Inactivation in Female Predisposition to Autoimmunity." *Arthritis Research* 2, no. 5 (2000): 399–406.
Clarke, Adele. *Disciplining Reproduction: Modernity, American Life Sciences, and the Problems of Sex.* Berkeley: University of California Press, 1998.
Clarke, Adele E., et al. "Biomedicalizing Genetic Health, Diseases and Identities." In *Handbook of Genetics and Society: Mapping the New Genomic Era,* edited by Paul Atkinson, Peter Glasner, and Margaret Lock, 21–40. London: Routledge, 2009.
Cohn, Victor. "A Criminal by Heredity?" *Washington Post,* 7 August 1968, 1.
Colapinto, John. *As Nature Made Him: The Boy Who Was Raised as a Girl.* New York: HarperCollins, 2000.
Cooper, G. S., M. L. Bynum, and E. C. Somers. "Recent Insights in the Epidemiology of Autoimmune Diseases: Improved Prevalence Estimates and Understanding of Clustering of Diseases." *Journal of Autoimmunity* 33, no. 3–4 (2009): 197–207.
"Corporate Advisory Council." Society for Women's Health Research. Accessed 10 February 2013. http://www.womenshealthresearch.org/site/PageServer?pagename=about_partners_cac.
Cott, Nancy F. *No Small Courage: A History of Women in the United States.* New York: Oxford University Press, 2000.
Court Brown, William M. *Abnormalities of the Sex Chromosome Complement in Man.* London: H. M. Stationery Office, 1964.
———. *Human Population Cytogenetics.* Amsterdam: North-Holland, 1967.

Craig, I. W., et al. "Application of Microarrays to the Analysis of the Inactivation Status of Human X-Linked Genes Expressed in Lymphocytes." *European Journal of Human Genetics* 12, no. 8 (2004): 639–46.

Creager, Angela N. H., Elizabeth Lunbeck, and Londa L. Schiebinger. *Feminism in Twentieth-Century Science, Technology, and Medicine*. Chicago: University of Chicago Press, 2001.

Crew, F. A. E. *The Genetics of Sexuality in Animals*. Cambridge: Cambridge University Press, 1927.

Cunningham, J. T. *Hormones and Heredity*. London: Constable, 1921.

———. *Sexual Dimorphism in the Animal Kingdom: A Theory of the Evolution of Secondary Sexual Characters*. London: Black, 1900.

Darbishire, A. D. "Recent Advances in the Study of Heredity. Lecture VII. Cytological and Other Evidence Relating to the Inheritance of Sex." *New Phytologist* 9, no. 1/2 (1910): 1–10.

Darwin, Charles. *The Descent of Man and Selection in Relation to Sex*. 2nd ed. New York: D. Appleton, 1897. First published 1871 by D. Appleton.

———. *On the Origin of Species, 1859*. Washington Square, NY: New York University Press, 1988.

———. *Variation of Animals and Plants under Domestication*. The Works of Charles Darwin. New York: New York University Press, 1988. First published 1868 by J. Murray.

"David Page: The Evolution of Sex: Rethinking the Rotting Y Chromosome." In *MIT World*. Cambridge, MA: Whitehead Institute for Biomedical Research, 2003.

Davies, T. F. "Editorial: X versus X—The Fight for Function within the Female Cell and the Development of Autoimmune Thyroid Disease." *Journal of Clinical Endocrinology & Metabolism* 90, no. 11 (2005): 6332–33.

DC Comics Inc. *Y: The Last Man*. New York: DC Comics, 2002–.

de Chadarevian, Soraya. "Mice and the Reactor: The 'Genetics Experiment' in 1950s Britain." *Journal of the History of Biology* 39 (2006): 707–35.

Delbridge, M. L., and J. A. Graves. "Mammalian Y Chromosome Evolution and the Male-Specific Functions of Y Chromosome-Borne Genes." *Reviews of Reproduction* 4, no. 2 (1999): 101–9.

Delongchamp, R. R., et al. "Genome-wide Estimation of Gender Differences in the Gene Expression of Human Livers: Statistical Design and Analysis." *BMC Bioinformatics* 6 Suppl 2 (2005): S13.

de Queiroz, Kevin, and Michael J. Donogue. "Phylogenetic Systematics and the Species Problem" (1988). In *The Philosophy of Biology*, edited by David L. Hull and Michael Ruse, 319–47. New York: Oxford University Press, 1998.

De Vries, G. J., et al. "A Model System for Study of Sex Chromosome Effects on Sexually Dimorphic Neural and Behavioral Traits." *Journal of Neuroscience* 22, no. 20 (2002): 9005–14.

Dietrich, Michael R. "Richard Goldschmidt: Hopeful Monsters and Other 'Heresies.'" *Nature Reviews Genetics* 4, no. 1 (2003): 68–74.

Dingel, Molly J., and Joey Sprague. "Research and Reporting on the Development of Sex in Fetuses: Gendered from the Start." *Public Understanding of Science* 19, no. 2 (2010): 181–96.

"Doctorate Recipients from U.S. Universities: 2009." United States National Science Foundation. http://www.nsf.gov/statistics/doctorates.
Dowd, Maureen. "X-Celling over Men." *New York Times*, 20 March 2005.
Dreger, Alice Domurat. *Hermaphrodites and the Medical Invention of Sex*. Cambridge, MA: Harvard University Press, 1998.
———. *Intersex in the Age of Ethics*. Hagerstown, MD: University, 1999.
Dupré, John. *The Disorder of Things: Metaphysical Foundations of the Disunity of Science*. Cambridge, MA: Harvard University Press, 1993.
"Dutch Scientists Sequence Female Genome." *Biotechniques Weekly*, 29 May 2008.
Eaton, W. W., et al. "Epidemiology of Autoimmune Diseases in Denmark." *Journal of Autoimmunity* 29, no. 1 (2007): 1–9.
Ehrenreich, Barbara. *Bright-Sided: How the Relentless Promotion of Positive Thinking Has Undermined America*. New York: Metropolitan Books, 2009.
Ehrenreich, Barbara, and Deirdre English. *For Her Own Good: Two Centuries of the Experts' Advice to Women*. New York: Anchor Books, 2005.
Eicher, E. M., and L. L. Washburn. "Genetic Control of Primary Sex Determination in Mice." *Annual Review of Genetics* 20 (1986): 327–60.
Ellis, Havelock. *Man and Woman*. New York: Scribner, 1894.
Epstein, Steven. *Inclusion: The Politics of Difference in Medical Research*. Chicago: University of Chicago, 2007.
Fairbairn, D. J., and D. A. Roff. "The Quantitative Genetics of Sexual Dimorphism: Assessing the Importance of Sex-Linkage." *Heredity* 97, no. 5 (2006): 319–28.
Faludi, Susan. *Backlash: The Undeclared War against American Women*. New York: Crown, 1991.
Farley, John. *Gametes and Spores: Ideas about Sexual Reproduction, 1750–1914*. Baltimore: Johns Hopkins University Press, 1982.
Fausto-Sterling, Anne. "Bare Bones of Sex: Part I, Sex & Gender." *Signs* 30, no. 2 (2005): 1491–1528.
———. "Life in the XY Corral." *Women's Studies International Forum* 12, no. 3 (1989): 319–31.
———. *Myths of Gender: Biological Theories about Women and Men*. New York: Basic Books, 1985.
———. *Sexing the Body: Gender Politics and the Construction of Sexuality*. New York: Basic Books, 2000.
Findlay, Deborah. "Discovering Sex: Medical Science, Feminism and Intersexuality." *Canadian Review of Sociology and Anthropology* 32, no. 1 (1995): 25–52.
Fine, Cordelia. *Delusions of Gender: How Our Minds, Society, and Neurosexism Create Difference*. New York: W. W. Norton, 2010.
Fisher, R. A. "The Evolution of Dominance." *Biological Reviews* 6 (1931): 345–68.
Fleck, Ludwik. *Genesis and Development of a Scientific Fact*. Chicago: University of Chicago Press, 1979. First published 1935, *Entstehung und Entwicklung einer wissenschaftlichen Tatsache. Einführung in die Lehre vom Denkstil und Denkkollektiv*. Basel: Schwabe und Co., Verlagsbuchhandlung.
Flemming, Walther. *Zellsubstanz, Kern Und Zelltheilung*. Leipzig: Vogel, 1882.
Fooden, Myra, Susan Gordon, and Betty Hughley. *The Second X and Women's Health*. New York: Gordian Press, 1983.

Ford, C. E. "A Sex Chromosome Anomaly in a Case of Gonadal Dysgenesis (Turner's Syndrome)." *Lancet* 273, no. 7075 (1959): 711–13.

Foster, J. W., et al. "Evolution of Sex Determination and the Y Chromosome: SRY-Related Sequences in Marsupials." *Nature* 359, no. 6395 (1992): 531–33.

Foster, J. W., and J. A. Graves. "An SRY-Related Sequence on the Marsupial X Chromosome: Implications for the Evolution of the Mammalian Testis-Determining Gene." *Proceedings of the National Academy of Sciences USA* 91, no. 5 (1994): 1927–31.

Franklin, Sarah, and Celia Roberts. *Born and Made: An Ethnography of Preimplantation Genetic Diagnosis*. Princeton, NJ: Princeton University Press, 2006.

Frickel, Scott, and Kelly Moore. *The New Political Sociology of Science: Institutions, Networks, and Power.* Madison: University of Wisconsin Press, 2006.

Fujimura, Joan. "'Sex Genes': A Critical Sociomaterial Approach to the Politics and Molecular Genetics of Sex Determination." *Signs* 32, no. 1 (2006): 49–82.

Fujimura, Joan H., Troy Duster, and Ramya Rajagopalan, eds. "Race, Genomics, and Biomedicine." Special issue, *Social Studies of Science* 38, no. 5 (2008): 643. doi: 10.1177/0306312708091926.

Fujimura, Joan H., and Ramya Rajagopalan. "Different Differences: The Use of 'Genetic Ancestry' Versus Race in Biomedical Human Genetic Research." *Social Studies of Science* 41, no. 1 (2011): 5–30.

Fuller, John L., and William Robert Thompson. *Behavior Genetics*. New York: Wiley, 1960.

Fullwiley, Duana. "The Molecularization of Race: U.S. Health Institutions, Pharmacogenetics Practice, and Public Science after the Genome." In *Revisiting Race in a Genomic Age*, edited by Barbara A. Koenig, Sandra Soo-Jin Lee and Sarah S. Richardson, 149–71. New Brunswick, NJ: Rutgers University Press, 2008.

Gamble, Eliza Burt. *The Evolution of Woman: An Inquiry into the Dogma of Her Inferiority to Man*. New York: G. P. Putnam's Sons, 1894.

———. *The Sexes in Science and History: An Inquiry into the Dogma of Woman's Inferiority to Man*. Westport, CT: Hyperion Press, 1976. First published 1916 by G. P. Putnam's Sons.

Geddes, Patrick, and John Arthur Thomson. *The Evolution of Sex*. London: Walter Scott, 1889.

———. *Sex*. London: Williams & Norgate, 1914.

Gero, Joan M., and Margaret Wright Conkey. *Engendering Archaeology: Women and Prehistory.* Social Archaeology. Cambridge, MA: Blackwell, 1991.

Gerrard, D. T., and D. A. Filatov. "Positive and Negative Selection on Mammalian Y Chromosomes." *Molecular Biology and Evolution* 22, no. 6 (2005): 1423–32.

Ghiselin, Michael J. "A Radical Solution to the Species Problem." *Systematic Zoology* 23, no. 4 (1974): 536–44.

Gilbert, Scott. "The Embryological Origins of the Gene Theory." *Journal of the History of Biology* 11, no. 2 (1978): 307–51.

Gilman, Charlotte Perkins, and Mary Armfield Hill. *The Man-Made World*. Amherst, NY: Humanity Books, 2001. First published 1911 by Charlton.

Gimelbrant, A., et al. "Widespread Monoallelic Expression on Human Autosomes." *Science* 318, no. 5853 (2007): 1136–40.

Goldberg, Steven. *The Inevitability of Patriarchy.* New York: Morrow, 1973.

Goldschmidt, Richard. "The Quantitative Theory of Sex." *Science* 64, no. 1656 (1926): 299–300.

Goodfellow, P. N. *The Mammalian Y Chromosome: Molecular Search for the Sex-Determining Factor.* Cambridge: Company of Biologists, 1987.

Gould, Stephen Jay. *The Mismeasure of Man.* Rev. and expanded ed. New York: Norton, 1996.

Gowaty, Patricia A. "Sexual Natures: How Feminism Changed Evolutionary Biology." *Signs* 28, no. 3 (2003): 901–21.

Graves, J. A. "The Degenerate Y Chromosome—Can Conversion Save It?" *Reproduction, Fertility, and Development* 16, no. 5 (2004): 527–34.

———. "The Evolution of Mammalian Sex Chromosomes and the Origin of Sex Determining Genes." *Philosophical Transactions of the Royal Society of London B: Biological Sciences* 350, no. 1333 (1995): 305–11.

———. "Human Y Chromosome, Sex Determination, and Spermatogenesis—A Feminist View." *Biology of Reproduction* 63, no. 3 (2000): 667–76.

———. "Recycling the Y Chromosome." *Science* 307 (2005).

———. "The Rise and Fall of SRY." *Trends in Genetics* 18, no. 5 (2002): 259–64.

———. "Sex Chromosomes and Sex Determination in Weird Mammals." *Cytogenetics and Genome Research* 96 (2002): 161–68.

———. "Sex Chromosomes and the Future of Men." In *Gender and Genomics: Sex, Science and Society.* Los Angeles: UCLA Center for Society and Genetics, 2005.

———. "Sex Chromosomes and the Future of Men." In *National Science Week.* Canberra: Australian National University College of Science, 2006.

———. "Sex Chromosome Specialization and Degeneration in Mammals." *Cell* 124 (2006): 901–14.

———. "Sex, Genes and Chromosomes: A Feminist View." *Women in Science Network Journal* 59 (2002): n.p.

Graves, J. A., and R. John Aitken. "The Future of Sex." *Nature* 415, no. 6875 (2002): 963.

Graves, J. A., and C. M. Disteche. "Does Gene Dosage Really Matter?" *Journal of Biology* 6, no. 1 (2007). http://jbiol.com/content/6/1/1.

Graves, J. A., J. Gecz, and H. Hameister. "Evolution of the Human X—A Smart and Sexy Chromosome that Controls Speciation and Development." *Cytogenetics and Genome Research* 99, no. 1–4 (2002): 141–45.

Graves, J. A., and S. Shetty. "Sex from W to Z: Evolution of Vertebrate Sex Chromosomes and Sex Determining Genes." *Journal of Experimental Zoology* 290 (2001): 449–62.

Graves, J. A., and R. V. Short. "Y or X—Which Determines Sex?" *Reproduction, Fertility and Development* 2, no. 6 (1990): 729–35.

Gravholt, C. H. "Epidemiological, Endocrine and Metabolic Features in Turner Syndrome." *European Journal of Endocrinology* 151, no. 6 (2004): 657–87.

Gray, John. *Men Are from Mars, Women Are from Venus: A Practical Guide for Improving Communication and Getting What You Want in Your Relationships.* New York: HarperCollins, 1992.

Green, Jeremy. "Media Sensationalization and Science: The Case of the Criminal Chromosome." In *Expository Science: Forms and Functions of Popularization*, edited by Terry Shinn and Richard Whitley, 139–61. Boston: D. Reidel, 1985.

Gregory, T. Ryan. "Genome Size Evolution in Animals." In *The Evolution of the Genome*, edited by T. Ryan Gregory, 3–87. New York: Elsevier, 2005.

Grosz, E. A. *Volatile Bodies: Toward a Corporeal Feminism*. Bloomington: Indiana University Press, 1994.

Gunter, C. "She Moves in Mysterious Ways." *Nature* 434, no. 7031 (2005): 279–80.

Guterl, Fred. "The Truth about Gender." *Newsweek*, 28 March 2005, 42.

Guyer, Michael F. "Recent Progress in Some Lines of Cytology." *Transactions of the American Microscopical Society* 30, no. 2 (1911): 145–90.

Ha, Nathan Q. "The Riddle of Sex: Biological Theories of Sexual Difference in the Early Twentieth-Century." *Journal of the History of Biology* 44, no. 3 (2011): 505–46.

Haire, Norman, Eugen Steinach, and Serge Voronoff. *Rejuvenation, the Work of Steinach, Voronoff, and Others*. London: G. Allen & Unwin, 1924.

Hall, G. Stanley. "The Contents of Children's Minds on Entering School." *Pedagogical Seminary* 1 (1891): 139–73.

Hammarström, Anne, and Ellen Annandale. "A Conceptual Muddle: An Empirical Analysis of the Use of 'Sex' and "Gender' in 'Gender-Specific Medicine' Journals." *Plos One* 7, no. 4 (2012): e34193.

Haraway, Donna Jeanne. "The Biopolitics of Postmodern Bodies: Constitutions of Self in Immune System Discourse" (1981). In *Simians, Cyborgs, and Women: The Reinvention of Nature*, 203–30. New York: Routledge, 1991.

———. "In the Beginning Was the Word: The Genesis of Biological Theory" (1981). In *Simians, Cyborgs, and Women: The Reinvention of Nature*, 71–80. New York: Routledge, 1991.

———. *Primate Visions: Gender, Race, and Nature in the World of Modern Science*. New York: Routledge, 1989.

———. "Science, Technology, and Socialist-Feminism in the Late Twentieth Century" (1980). In *Simians, Cyborgs and Women: The Reinvention of Nature*, 149–81. New York: Routledge, 1991.

———. "Situated Knowledges: The Science Question in Feminism and the Privilege of Partial Perspective." In *Simians, Cyborgs and Women: The Reinvention of Nature*, 183–201. New York: Routledge, 1991 [1988].

Harding, Sandra. "Rethinking Standpoint Epistemology: What Is 'Strong Objectivity'?" In *Feminist Epistemologies*, edited by Linda Alcoff and Elizabeth Potter, 49–82. New York: Routledge, 1993.

Harper, Peter S. *First Years of Human Chromosomes: The Beginnings of Human Cytogenetics*. Bloxham: Scion, 2006.

Haseltine, Florence P., and Beverly Greenberg Jacobson. *Women's Health Research: A Medical and Policy Primer*. Washington DC: Health Press International, 1997.

Haslanger, Sally. "Gender and Race: (What) Are They? (What) Do We Want Them to Be?" *Nous* 34, no. 1 (2000): 31–55.

Hausman, Bernice L. "Demanding Subjectivity: Transsexualism, Medicine, and the Technologies of Gender." *Journal of the History of Sexuality* 3, no. 2 (1992): 270–302.

Hawley, R. Scott. "The Human Y Chromosome: Rumors of Its Death Have Been Greatly Exaggerated." *Cell* 113 (2003): 825–28.

Hay, D. A. "Y Chromosome and Aggression in Mice." *Nature* 255, no. 5510 (1975): 658.

Hayden, Thomas. "He Figured Out Y, but Not 'So What?'" *Washington Post*, 25 October 2007, C3.

Henking, H. "Uber Spermatogenese Und Deren Beziehung Zur Entwicklung Bei Pyrrhocoris Apterus L." *Zeitschriftffur wissenschaftliche Zoologie* 51 (1891): 685–736.

Herschberger, Ruth. *Adam's Rib*. New York: Pellegrini & Cudahy, 1948.

Holme, Ingrid. "Beyond XX and XY: Living Genomic Sex." In *Governing the Female Body: Gender, Health, and Networks of Power*, edited by Lori Stephens Reed and Paula Saukko, 271–94. Albany: State University of New York Press, 2010.

Hotz, Robert Lee. "Women Are Very Much Not Alike, Gene Study Finds." *Los Angeles Times*, 17 March 2005, 18.

Hrdy, Sarah Blaffer. *The Woman that Never Evolved*. Cambridge, MA: Harvard University Press, 1981.

Hsu, T. C. *Human and Mammalian Cytogenetics: An Historical Perspective*. New York: Springer-Verlag, 1979.

Hubbard, Ruth, and Marian Lowe. *Pitfalls in Research on Sex and Gender*. New York: Gordian Press, 1979.

Hull, David L. "A Matter of Individuality." *Philosophy of Science* 45, no. 3 (1978): 335–60.

Hunter, Anne E., Catherine M. Flamenbaum, and Suzanne R. Sunday. *On Peace, War, and Gender: A Challenge to Genetic Explanations*. New York: Feminist Press, 1991.

Hurst, L. D. "Embryonic Growth and the Evolution of the Mammalian Y Chromosome. I. The Y as an Attractor for Selfish Growth Factors." *Heredity* 73, pt. 3 (1994): 223–32.

———. "Embryonic Growth and the Evolution of the Mammalian Y Chromosome. II. Suppression of Selfish Y-Linked Growth Factors May Explain Escape from X-Inactivation and Rapid Evolution of SRY" *Heredity* 73, pt. 3 (1994): 233–43.

Hyde, Janet Shibley. "The Gender Similarities Hypothesis." *American Psychologist* 60, no. 6 (2005): 581–92.

Ingraham, Holly. "Expert Interview Transcript." In *Rediscovering Biology: Molecular to Global Perspectives*. Electronic media. Oregon Public Broadcasting and Annenberg-MediaLearner.org, 2004.

"Initial Sequence of the Chimpanzee Genome and Comparison with the Human Genome." *Nature* 437, no. 7055 (2005): 69–87.

"In Pursuit of the Y Chromosome." *Nature* 226, no. 5249 (1970): 897.

Invernizzi, P. "The X Chromosome in Female-Predominant Autoimmune Diseases." *Annals of the New York Academy of Sciences* 1110 (2007): 57–64.

Invernizzi, P., et al. "X Monosomy in Female Systemic Lupus Erythematosus." *Annals of the New York Academy of Sciences* 1110 (2007): 84–91.

Isensee, Joerg, et al. "Sexually Dimorphic Gene Expression in the Heart of Mice and Men." *Journal of Molecular Medicine* 86, no. 1 (2008): 61–74.

Jacobs, P. A. "Human Population Cytogenetics: The First Twenty-Five Years." *American Journal of Human Genetics* 34, no. 5 (1982): 689–98.
———. "XYY Genotype." *Science* 189, no. 4208 (1975): 1040.
Jacobs, P. A., et al. "Aggressive Behavior, Mental Sub-Normality and the XYY Male." *Nature* 208, no. 5017 (1965): 1351–52.
Jacobs, P. A., and J. A. Strong. "A Case of Human Intersexuality Having a Possible XXY Sex-Determining Mechanism." *Nature* 183, no. 4657 (1959): 302–3.
Jacobson, D. L., et al. "Epidemiology and Estimated Population Burden of Selected Autoimmune Diseases in the United States." *Clinical Immunology and Immunopathology* 84, no. 3 (1997): 223–43.
Jarvik, L. F., V. Klodin, and S. S. Matsuyama. "Human Aggression and the Extra Y Chromosome. Fact or Fantasy?" *American Psychology* 28, no. 8 (1973): 674–82.
Jay, Nancy. "Gender and Dichotomy." *Feminist Studies* 7, no. 1 (1981): 38–56.
Jeffrey, Edward C., and Edwin J. Haertl. "The Nature of Certain Supposed Sex Chromosomes." *American Naturalist* 72, no. 742 (1938): 473–76.
Jensen, Arthur R. "How Much Can We Boost IQ and Scholastic Achievement?" *Harvard Educational Review* 39 (1969): 1–123.
Jones, Gwyneth A. *Life: A Novel*. Seattle: Aqueduct Press, 2004.
Jones, Steve. *Y: The Descent of Men*. London: Little, Brown, 2002.
Jordan, H. E. "Recent Literature Touching the Question of Sex-Determination." *American Naturalist* 44, no. 520 (1910): 245–52.
Jordan-Young, Rebecca M. *Brain Storm: The Flaws in the Science of Sex Differences*. Cambridge, MA: Harvard University Press, 2010.
Jost, A., P. Gonse-Danysz, and R. Jacquot. "Studies on Physiology of Fetal Hypophysis in Rabbits and Its Relation to Testicular Function." *Journal of Physiology (Paris)* 45, no. 1 (1953): 134–36.
Jumonville, Neil. "The Cultural Politics of the Sociobiology Debate." *Journal of the History of Biology* 35 (2002): 569–93.
Just, W., et al. "Absence of SRY in Species of the Vole Ellobius." *Nature Genetics* 11, no. 2 (1995): 117–18.
Kahn, Jonathan. "Exploiting Race in Drug Development." *Social Studies of Science* 38, no. 5 (2008): 727–58.
———. "Patenting Race in a Genomic Age." In *Revisiting Race in a Genomic Age*, edited by Barbara A. Koenig, Sandra Soo-Jin Lee and Sarah S. Richardson, 129–48. New Brunswick, NJ: Rutgers University Press, 2008.
Karasik, D., and S. L. Ferrari. "Contribution of Gender-Specific Genetic Factors to Osteoporosis Risk." *Annals of Human Genetics* 72 (2008): 696–714.
Karkazis, Katrina Alicia. *Fixing Sex: Intersex, Medical Authority, and Lived Experience*. Durham, NC: Duke University Press, 2008.
Kast, R. E. "Predominance of Autoimmune and Rheumatic Diseases in Females." *Journal of Rheumatology* 4, no. 3 (1977): 288–92.
Keating, Peter, and Alberto Cambrosio. "Too Many Numbers: Microarrays in Clinical Cancer Research." *Studies in History and Philosophy of Science Part C Studies in History and Philosophy of Biological and Biomedical Sciences* 43, no. 1 (2012): 37–51.
Kehrer-Sawatzki, H., and D. N. Cooper. "Understanding the Recent Evolution of the

Human Genome: Insights from Human-Chimpanzee Genome Comparisons." *Human Mutations* 28, no. 2 (2007): 99–130.

Keller, Evelyn Fox. *The Century of the Gene*. Cambridge, MA: Harvard University Press, 2000.

———. "The Origin, History, and Politics of the Subject Called 'Gender and Science.'" Chap. 4 in *Handbook of Science and Technology Studies*, rev. ed., edited by Sheila Jasanoff, Gerald E. Markle, James C. Petersen, and Trevor Pinch, 80–94. Thousand Oaks, CA: Sage, 1995.

———. *Making Sense of Life: Explaining Biological Development with Models, Metaphors, and Machines*. Cambridge, MA: Harvard University Press, 2002.

———. *Reflections on Gender and Science*. New Haven: Yale University Press, 1985.

———. *Secrets of Life, Secrets of Death: Essays on Language, Gender, and Science*. New York: Routledge, 1992.

Kelly, Joan. *Women, History & Theory: The Essays of Joan Kelly*. Chicago: University of Chicago Press, 1984.

Kessler, S., and R. H. Moos. "XYY Chromosome: Premature Conclusions." *Science* 165, no. 3892 (1969): 442.

Kessler, Suzanne J. *Lessons from the Intersexed*. New Brunswick, NJ: Rutgers University Press, 1998.

Kettlewell, Julianna. "Female Chromosome Has X Factor." *BBCNEWS.com*, 16 March 2005. http://news.bbc.co.uk/2/hi/science/nature/4355355.stm.

Kevles, Daniel J. *In the Name of Eugenics: Genetics and the Uses of Human Heredity*. Cambridge, MA: Harvard University Press, 1995.

Kimmel, Michael S. *The Gendered Society*. New York: Oxford University Press, 2000.

King, M. C., and A. C. Wilson. "Evolution at Two Levels in Humans and Chimpanzees." *Science* 188, no. 4184 (1975): 107–16.

King, Samantha. *Pink Ribbons, Inc.: Breast Cancer and the Politics of Philanthropy*. Minneapolis: University of Minnesota Press, 2006.

Kingsland, Sharon E. "Maintaining Continuity through a Scientific Revolution: A Rereading of E. B. Wilson and T. H. Morgan on Sex Determination and Mendelism." *Isis* 98, no. 3 (2007): 468–88.

Klinefelter, H. F., E. C. Reifenstein, and F. Albright. "Syndrome Characterized by Gynecomastia, Aspermatogenesis without Aleydigism, and Increased Excretion of Follicle Stimulating Hormone." *Journal of Clinical Endocrinology and Metabolism* 2 (1942): 615–27.

Knudsen, G. P. "Gender Bias in Autoimmune Diseases: X Chromosome Inactivation in Women with Multiple Sclerosis." *Journal of the Neurological Sciences* 286, no. 1–2 (2009): 43–46.

Knudsen, G. P., et al. "X Chromosome Inactivation in Females with Multiple Sclerosis." *European Journal of Neurology* 14, no. 12 (2007): 1392–96.

Koehler, Christopher. "The Sex Problem: Thomas Hunt Morgan, Richard Goldschmidt, and the Question of Sex and Gender in the Twentieth Century." PhD diss., University of Florida, 1998.

Koenig, Barbara A., Sandra Soo-Jin Lee, and Sarah S. Richardson. *Revisiting Race in a Genomic Age*. New Brunswick, NJ: Rutgers University Press, 2008.

Kohler, Robert E. *Partners in Science: Foundations and Natural Scientists, 1900–1945*. Chicago: University of Chicago Press, 1991.

Koopman, P., et al. "Male Development of Chromosomally Female Mice Transgenic for SRY." *Nature* 351, no. 6322 (1991): 117–21.

Kraus, Cynthia. "Naked Sex in Exile: On the Paradox of the 'Sex Question' in Feminism and in Science." *NWSA Journal* 12, no. 3 (2000): 151–77.

Krieger, Nancy. "Genders, Sexes, and Health: What Are the Connections—and Why Does It Matter?" *International Journal of Epidemiology* 32, no. 4 (2003): 652.

Kruszelnicki, Karl S. "Hybrid Auto-Immune Women 3." *ABC Science In Depth*. 12 February 2004. http://www.abc.net.au/science/articles/2004/02/12/1002754.htm.

Kuhn, Thomas S. *The Structure of Scientific Revolutions*. Chicago: University of Chicago Press, 1962.

Kumar, Seema. "Genes for Early Sperm Production Found to Reside on X Chromosome." Press release, 4 April 2001. http://web.mit.edu/newsoffice/2001/sperm-0404.

Kunkel, L. M., K. D. Smith, and S. H. Boyer. "Human Y-Chromosome-Specific Reiterated DNA." *Science* 191, no. 4232 (1976): 1189–90.

Kuttner, Robert E. "Chromosomes and Intelligence." *Mankind Quarterly* 12 (1971): 6–11.

Lahn, B. T., and D. C. Page. "Functional Coherence of the Human Y Chromosome." *Science* 278, no. 5338 (1997): 675–80.

———. "Retroposition of Autosomal mRNA Yielded Testis-Specific Gene Family on Human Y Chromosome." *Nature Genetics* 21, no. 4 (1999): 429–33.

Laqueur, Thomas Walter. *Making Sex: Body and Gender from the Greeks to Freud*. Cambridge, MA: Harvard University Press, 1990.

Lederberg, Joshua. "Poets Knew It All Along: Science Finally Finds Out that Girls Are Chimerical; You Know, Xn/Xa." *Washington Post*, 18 December 1966, E7.

Lehrke, Robert Gordon. *Sex Linkage of Intelligence: The X-Factor*. Westport, CT: Praeger, 1997.

Levy, Ariel. *Female Chauvinist Pigs: Women and the Rise of Raunch Culture*. New York: Free Press, 2005.

Lewontin, Richard C., Steven P. R. Rose, and Leon J. Kamin. *Not in Our Genes: Biology, Ideology, and Human Nature*. New York: Pantheon Books, 1984.

Libert, Claude, Lien Dejager, and Iris Pinheiro. "The X Chromosome in Immune Functions: When a Chromosome Makes the Difference." *Nature Reviews Immunology* 10, no. 8 (2010): 594–604.

Lillie, Frank Rattray. "Free-martin: A Study of the Action of Sex Hormones in the Foetal Life of Cattle." *Journal of Experimental Biology* 23, no. 5 (1917): 371–452.

———. "Sex-Determination and Sex-Differentiation in Mammals." *Proceedings of the National Academy of Sciences* 3 (1917): 464–70.

———. "Suggestions for Organization and Conduct of Research on Problems of Sex." In *First Annual Report of the Committee for Research in Problems of Sex*. Washington, DC: National Academies Press, 1922.

———. "The Theory of the Free-martin." *Science* 43, no. 28 (1916): 611–13.

Lindee, M. Susan. *Moments of Truth in Genetic Medicine*. Baltimore: Johns Hopkins University Press, 2005.

Lloyd, Elisabeth Anne. *The Case of the Female Orgasm: Bias in the Science of Evolution*. Cambridge, MA: Harvard University Press, 2005.
Lockshin, M. D. "Nonhormonal Explanations for Sex Discrepancy in Human Illness." *Annals of the New York Academy of Sciences* 1193, no. 1 (2010): 22–24.
———. "Sex Differences in Autoimmune Disease." *Lupus* 15, no. 11 (2006): 753–56.
Longino, Helen E. "Can There Be a Feminist Science?" *Hypatia* 2, no. 3 (1987): 51–64.
———. "Cognitive and Non-cognitive Values in Science: Rethinking the Dichotomy." In *Feminism and Philosophy of Science*, edited by Jack Nelson and Lynn Hankinson Nelson, 39–58. Boston: Kluwer, 1996.
———. "Subjects, Power, Knowledge: Prescriptivism and Descriptivism in Feminist Philosophy of Science." In *Feminist Epistemologies*, edited by Linda Alcoff and Elizabeth Potter, 385–404. New York: Routledge, 1992.
Longino, Helen E., and Ruth Doell. "Body, Bias and Behavior: A Comparative Analysis of Reasoning in Two Areas of Biological Science." *Signs* 9, no. 2 (1983): 206–27.
Longino, Helen E., and Evelynn Hammonds. "Conflicts and Tensions in the Feminist Study of Gender and Science." In *Conflicts in Feminism*, edited by Marianne Hirsch and Evelyn Fox Keller. New York: Routledge, 1990.
Lunbeck, Elizabeth. *The Psychiatric Persuasion: Knowledge, Gender, and Power in Modern America*. Princeton, NJ: Princeton University Press, 1994.
Lyon, Mary F. "Gene Action in the X-Chromosome of the Mouse." *Nature* 190 (1961): 372–73.
———. "No Longer 'All-or-None.'" *European Journal of Human Genetics* 13, no. 7 (2005): 796–97.
———. "Some Milestones in the History of X-Chromosome Inactivation." *Annual Review of Genetics* 26 (1992): 17–28.
Lyons, Richard. "Genetic Abnormality Is Linked to Crime." *New York Times*, 21 April 1968, 1.
M'Charek, Amade. "The Mitochondrial Eve of Modern Genetics: Of Peoples and Genomes, or the Routinization of Race." *Science as Culture* 14, no. 2 (2005): 161–83.
Maclean, N., et al. "Sex-Chromosome Abnormalities in Newborn Babies." *Lancet* 283, no. 7328 (1964): 286.
———. "Survey of Sex-Chromosome Abnormalities among 4514 Mental Defectives." *Lancet* 279, no. 7224 (1962): 293–96.
Macleod, David I. *Building Character in the American Boy: The Boy Scouts, YMCA, and Their Forerunners, 1870–1920*. Madison: University of Wisconsin Press, 1983.
Mahowald, Mary Briody. *Genes, Women, Equality*. New York: Oxford University Press, 2000.
Maienschein, Jane. "What Determines Sex? A Study of Converging Approaches, 1880–1916." *Isis* 75, no. 3 (1984): 456–80.
Maier, E. M., et al. "Disease Manifestations and X Inactivation in Heterozygous Females with Fabry Disease." *Acta Paediatrica Supplement* 95, no. 451 (2006): 30–38.
Mangels, John. "Is Gene Pool Shrinking Men Out of Existence?" *Cleveland Plain Dealer*, 30 May 2004.
Mansbridge, Jane, and Shauna L. Shames. "Toward a Theory of Backlash: Dynamic Resistance and the Central Role of Power." *Politics & Gender* 4, no. 4 (2008): 623–34.
Marchetti, M., and T. Raudma, eds. *Stocktaking: 10 Years of "Women in Science" Policy*

by the European Commission, 1999–2009. Luxembourg: Publications Office of the European Union, 2010.
Marks, Jonathan. *What It Means to Be 98% Chimpanzee: Apes, People, and Their Genes*. Berkeley: University of California Press, 2002.
Martin, Aryn. "Can't Any Body Count? Counting as an Epistemic Theme in the History of Human Chromosomes." *Social Studies of Science* 34, no. 6 (2004): 923–48.
Martin, Emily. "The Egg and the Sperm: How Science Has Constructed a Romance Based on Stereotypical Male-Female Roles." *Signs* 16, no. 3 (1991): 485–501.
———. "The Woman in the Flexible Body." In *Revisioning Women, Health and Healing: Feminist, Cultural, and Technoscience Perspectives*, edited by Adele Clarke and Virginia L. Olesen, 97–115. New York: Routledge, 1999.
Matson, Clinton K., et al. "DMRT1 Prevents Female Reprogramming in the Postnatal Mammalian Testis." *Nature* 476, no. 7358 (2011): 101–4.
McAllister, Peter. *Manthropology: The Science of Why the Modern Male Is Not the Man He Used to Be*. New York: St. Martin's Press, 2010.
McCarthy, Margaret M., et al. "Sex Differences in the Brain: The Not So Inconvenient Truth." *Journal of Neuroscience* 32, no. 7 (2012): 2241–47.
McCarthy, R. M., and A. P. Arnold. "Reframing Sexual Differentiation of the Brain." *Nature Neuroscience* 14 (2011): 677–83.
McCaughey, Martha. *The Caveman Mystique: Pop-Darwinism and the Debates over Sex, Violence, and Science*. New York: Routledge, 2008.
McClung, C. E. "The Accessory Chromosome: Sex Determinant?" *Biological Bulletin* 3, no. 1/2 (1902): 43–84.
———. "Cytological Nomenclature." *Science* 37, no. 949 (1913): 369–70.
———. "A Peculiar Nuclear Element in the Male Reproductive Cells of Insects." *Zoological Bulletin* 2, no. 4 (1899): 187–97.
———. "Possible Action of the Sex-Determining Mechanism." *Proceedings of the National Academy of Sciences* 4, no. 6 (1918): 160–63.
McCombe, P. A., J. M. Greer, and I. R. Mackay. "Sexual Dimorphism in Autoimmune Disease." *Current Molecular Medicine* 9, no. 9 (2009): 1058–79.
McCoy, Krisha. "Women and Autoimmune Disorders." *Everydayhealth.com*. Last updated 20 December 2010. http://www.everydayhealth.com/autoimmune-disorders/understanding/women-and-autoimmune-diseases.aspx.
McDonald, Kate. "ICHG: Jenny Graves Is Talking About Sex—Again." *Australian Biotechnology News*, 21 July 2006.
McDonnell, Tim. "Lost Boys: In a Warmer World, Will Males Die Sooner?" *Grist*, 30 June 2011. http://grist.org/climate-change/2011-06-30-lost-boys-warmer-world-males-die-sooner-global-warming/.
McElreavey, K., et al. "A Regulatory Cascade Hypothesis for Mammalian Sex Determination: SRY Represses a Negative Regulator of Male Development." *Proceedings of the National Academy of Sciences USA* 90, no. 8 (1993): 3368–72.
"Men and Women: The Differences Are in the Genes." *ScienceDaily*, 23 March 2005. http://www.sciencedaily.com/releases/2005/03/050323124659.htm.
Mendel, Gregor. *Versuche über Pflanzen-Hybriden,Vorgelegt in den Sitzungen vom 8. Februar und 8. März 1865*. Translated to English in 1866 as *Experiments on Plant Hybridization*.

Metzl, Jonathan. *The Protest Psychosis: How Schizophrenia Became a Black Disease*. Boston: Beacon Press, 2009.

Meyerowitz, Joanne J. *How Sex Changed: A History of Transsexuality in the United States*. Cambridge, MA: Harvard University Press, 2002.

Migeon, Barbara R. *Females Are Mosaics: X Inactivation and Sex Differences in Disease*. New York: Oxford University Press, 2007.

———. "Non-random X Chromosome Inactivation in Mammalian Cells." *Cytogenetics and Cell Genetics* 80, no. 1–4 (1998): 142–48.

———. "The Role of X Inactivation and Cellular Mosaicism in Women's Health and Sex-Specific Diseases." *JAMA* 295, no. 12 (2006): 1428–33.

———. "X-Chromosome Inactivation: Molecular Mechanisms and Genetic Consequences." *Trends in Genetics* 10, no. 7 (1994): 230–35.

———. "X Inactivation, Female Mosaicism, and Sex Differences in Renal Diseases." *Journal of the American Society of Nephrology* 19, no. 11 (2008): 2052–59.

Miller, Fiona Alice. "Dermatoglyphics and the Persistence of 'Mongolism': Networks of Technology, Disease and Discipline." *Social Studies of Science* 33, no. 1 (2003): 75–94.

———. "'Your True and Proper Gender': The Barr Body as a Good Enough Science of Sex." *Studies in History and Philosophy of Biological and Biomedical Sciences* 37, no. 3 (2006): 459–83.

Mishler, Brent D., and Robert N. Brandon. "Individuality, Pluralism, and the Phylogenetic Species Concept (1987)." In *The Philosophy of Biology*, edited by David L. Hull and Michael Ruse, 300–318. New York: Oxford University Press, 1998.

Money, John, and Anke A. Ehrhardt. *Man & Woman, Boy & Girl: The Differentiation and Dimorphism of Gender Identity from Conception to Maturity*. Baltimore: Johns Hopkins University Press, 1972.

Montagu, Ashley. *The Natural Superiority of Women*. New York: Macmillan, 1953.

Montgomery, Thomas H. "Are Particular Chromosomes Sex Determinants?" *Biological Bulletin* 19, no. 1 (1910): 1–17.

———. "The Morphological Superiority of the Female Sex." *Proceedings of the American Philosophical Society* 43, no. 178 (1904): 365–80.

———. "The Spermatogenesis in Pentatoma up to the Formation of the Spermatid." *Zoologische Jahrbucher* 12 (1898): 1–88.

———. "The Terminology of Aberrant Chromosomes and Their Behavior in Certain Hemiptera." *Science* 23, no. 575 (1906): 36–38.

Moore, Kelly. *Disrupting Science: Social Movements, American Scientists, and the Politics of the Military, 1945–1975*. Princeton, NJ: Princeton University Press, 2008.

Moore, Kelly, and Nicole Hala. "Organizing Identity: The Creation of Science for the People." *Research in the Sociology of Organizations* 19 (2002): 309–39.

Morgan, T. H. "A Biological and Cytological Study of Sex Determination in Phylloxerans and Aphids." *Journal of Experimental Zoology* 7, no. 2 (1909): 239–353.

———. "Chromosomes and Heredity." *American Naturalist* 44, no. 524 (1910): 449–96.

———. *The Genetic and the Operative Evidence Relating to Secondary Sexual Characters*. Washington, DC: Carnegie Institution, 1919.

———. *The Mechanism of Mendelian Heredity*. New York: Holt, 1915.

———. "Recent Results Relating to Chromosomes and Genetics." *Quarterly Review of Biology* 1, no. 2 (1926): 186–211.

———. "Recent Theories in Regard to the Determination of Sex." *Popular Science Monthly* 64 (1903): 97–116.

———. "The Scientific Work of Miss N. M. Stevens." *Science* 36, no. 928 (1912): 468–70.

———. "Sex-Limited and Sex-Linked Inheritance." *American Naturalist* 48, no. 574 (1914): 577–83.

———. "Sex Limited Inheritance in Drosophila." *Science* 32, no. 812 (1910): 120–22.

Morgan, Thomas Hunt, and Calvin B. Bridges. *Sex-Linked Inheritance in Drosophila.* Washington, DC: Carnegie Institution, 1916.

Morgen, Sandra. *Into Our Own Hands: The Women's Health Movement in the United States, 1969–1990.* New Brunswick, NJ: Rutgers University Press, 2002.

Morris, Desmond. *The Naked Ape.* New York: McGraw-Hill, 1967.

Morton, Oliver. "A Life Decoded by J Craig Venter." *Sunday Times*, 21 October 2007.

Moss, Lenny. *What Genes Can't Do.* Cambridge, MA: MIT Press, 2002.

Nash, Catherine. "Genetic Kinship." *Cultural Studies* 18, no. 1 (2004): 1–34.

National Institutes of Mental Health and Saleem Alam Shah. *Report on the XYY Chromosomal Abnormality.* Chevy Chase, MD: U.S. Government Printing Office, 1970.

Nguyen, D. K., and C. M. Disteche. "Dosage Compensation of the Active X Chromosome in Mammals." *Nature Genetics* 38, no. 1 (2006): 47–53.

———. "High Expression of the Mammalian X Chromosome in Brain." *Brain Research* 1126, no. 1 (2006): 46–49.

Nielsen, J., and U. Friedrich. "Length of the Y Chromosome in Criminal Males." *Clinical Genetics* 3, no. 4 (1972): 281–85.

Nordlund, Christer. "Endocrinology and Expectations in 1930s America: Louis Berman's Ideas on New Creations in Human Beings." *British Journal for the History of Science* 40, no. 1 (2007): 83–104.

Novitski, Edward. *Human Genetics.* New York: Macmillan, 1977.

Ogilvie, Marilyn Bailey, and Clifford J. Choquette. "Nettie Maria Stevens (1861–1912): Her Life and Contributions to Cytogenetics." *Proceedings of the American Philosophical Society* 125, no. 4 (1981): 292–311.

Ohlsson, R. "Genetics: Widespread Monoallelic Expression." *Science* 318, no. 5853 (2007): 1077–78.

Ohno, Susumu. *Major Sex-Determining Genes.* 1971; New York: Springer-Verlag, 1979.

———. *Sex Chromosomes and Sex-Linked Genes.* New York: Springer-Verlag, 1967.

Olby, Robert C. *Origins of Mendelism.* 2nd ed. Chicago: University of Chicago Press, 1985.

Oliver, J. E., and A. J. Silman. "Why Are Women Predisposed to Autoimmune Rheumatic Diseases?" *Arthritis Research & Therapy* 11, no. 5 (2009): 252.

Oudshoorn, Nelly. *Beyond the Natural Body: An Archaeology of Sex Hormones.* New York: Routledge, 1994.

Ozbalkan, Z., et al. "Skewed X Chromosome Inactivation in Blood Cells of Women with Scleroderma." *Arthritis & Rheumatism* 52, no. 5 (2005): 1564–70.

Ozcelik, T., et al. "Evidence from Autoimmune Thyroiditis of Skewed X-Chromosome

Inactivation in Female Predisposition to Autoimmunity." *European Journal of Human Genetics* 14, no. 6 (2006): 791–97.

Page, D. C. "2003 Curt Stern Award Address: On Low Expectations Exceeded; or, the Genomic Salvation of the Y Chromosome." *American Journal of Human Genetics* 74, no. 3 (2004): 399–402.

———. "Expert Interview Transcript." In *Rediscovering Biology: Molecular to Global Perspectives*. Electronic media. Oregon Public Broadcasting and AnnenbergMedia Learner.org, 2004.

———. "Sexual Evolution: From X to Y." In *Sex Determination: Lecture Series*: Chevy Chase, MD: Howard Hughes Medical Institute, 2001.

———. "Save the Males!" *Nature Genetics* 17, no. 1 (1997): 3.

Page, D. C., et al. "Abundant Gene Conversion between Arms of Palindromes in Human and Ape Y Chromosomes." *Nature* 423, no. 6942 (2003): 873–76.

———. "Conservation of Y-Linked Genes during Human Evolution Revealed by Comparative Sequencing in Chimpanzee." *Nature* 437, no. 7055 (2005): 101–4.

———. "The Sex-Determining Region of the Human Y Chromosome Encodes a Finger Protein." *Cell* 51, no. 6 (1987): 1091–1104.

Pain, Elisabeth. "A Genetic Battle of the Sexes." *ScienceNOW Daily News*, 22 March 2007. http://news.sciencemag.org/sciencenow/2007/03/22-04.html.

Painter, Theophilus S. "The Sex Chromosomes of Man." *American Naturalist* 58, no. 659 (1924): 506–24.

Pasquier, E., et al. "Strong Evidence that Skewed X-Chromosome Inactivation Is Not Associated with Recurrent Pregnancy Loss: An Incident Paired Case Control Study." *Human Reproduction* 22, no. 11 (2007): 2829–33.

Patsopoulos, Nikolaos A., Athina Tatsioni, and John P. A. Ioannidis. "Claims of Sex Differences: An Empirical Assessment in Genetic Associations." *JAMA* 298, no. 8 (2007): 880–93.

Paul, Eden, and Norman Haire. *Rejuvenation: Steinach's Researches on the Sex-Glands*. Vol. 11, British Society for the Study of Sex Psychology. London: J. E. Francis, Athenaeum Press, 1923.

Paulmier, F. C. "The Spermatogenesis of *Anasa tristis*." *Journal of Morphology* 15 (1899): 224–71.

Payne, Fernandus. "On the Sexual Differences of the Chromosome Groups in Galgulus Oculatus." *Biological Bulletin* 14, no. 5 (1908): 297–303.

———. "Some New Types of Chromosome Distribution and Their Relation to Sex—Continued." *Biological Bulletin* 16, no. 4 (1909): 153–66.

Pendlebury, Richard. "Men Are Doomed." *London Daily Mail*, 18 August 2003.

Pennisi, Elizabeth. "Mutterings from the Silenced X Chromosome." *Science* 307, no. 5716 (2005): 1708.

Plomin, R., and R. Rende. "Human Behavioral Genetics." *Annual Review of Psychology* 42 (1991): 161–90.

Plotz, David. "The Male Malaise; Is the Y Chromosome Set to Self-Destruct?" *Washington Post*, 11 April 2004.

Podolsky, Scott H., and Alfred I. Tauber. *The Generation of Diversity: Clonal Selection Theory and the Rise of Molecular Immunology*. Cambridge, MA: Harvard University Press, 1997.

Polani, P. E. "Abnormal Sex Chromosomes and Mental Disorders." *Nature* 223, no. 5207 (1969): 680–86.

Porter, Roy, and Lesley A. Hall. *The Facts of Life: The Creation of Sexual Knowledge in Britain, 1650–1950.* New Haven, CT: Yale University Press, 1995.

"A Proposed Standard System on Nomenclature of Human Mitotic Chromosomes." *American Journal of Human Genetics* 12, no. 3 (1960): 384–88.

Puffer, James C. "Gender Verification of Female Olympic Athletes." *Medicine and Science in Sports and Exercise* 34, no. 10 (2002): 1543.

Quintero, Olga L., et al. "Autoimmune Disease and Gender: Plausible Mechanisms for the Female Predominance of Autoimmunity." Special issue, *Gender, Sex Hormones, Pregnancy and Autoimmunity* 38, no. 2–3 (2012): J109–19.

"A Randomized Trial of Aspirin and Sulfinpyrazone in Threatened Stroke. The Canadian Cooperative Study Group." *New England Journal of Medicine* 299, no. 2 (1978): 53–59.

Rapp, Rayna. *Testing Women, Testing the Fetus: The Social Impact of Amniocentesis in America.* New York: Routledge, 1999.

Rechter, Julia E. "'The Glands of Destiny': A History of Popular, Medical and Scientific Views of the Sex Hormones in 1920s America." PhD diss., University of California, 1997.

"Recruit and Advance: Women Students and Faculty in Science and Engineering." In *National Academies Press.* Washington, DC: National Academy of Sciences, 2006.

Reed, Ken, and Jennifer A. Marshall Graves. *Sex Chromosomes and Sex-Determining Genes.* Langhorne, PA: Harwood Academic, 1993.

Renieri, A., et al. "Rett Syndrome: The Complex Nature of a Monogenic Disease." *Journal of Molecular Medicine* 81, no. 6 (2003): 346–54.

"Research Makes It Official: Women Are Genetic Mosaics." *Time,* 4 January 1963.

Richardson, Sarah S. "Feminist Philosophy of Science: History, Contributions, and Challenges." *Synthese* 177, no. 3 (2010): 337–62.

———. "Gendering the Genome: Sex Chromosomes in Twentieth Century Genetics." PhD diss., Stanford University, 2009.

———. "Race and IQ in the Postgenomic Age: The Microcephaly Case." *BioSocieties* 6 (2011): 420–46.

———. "Sexes, Species, and Genomes: Why Males and Females Are Not Like Humans and Chimpanzees." *Biology and Philosophy* 25, no. 5 (2010): 823–41.

———. "Sexing the X: How the X Became the 'Female Chromosome.'" *Signs* 37, no. 4 (2012): 909–33.

———. "When Gender Criticism Becomes Standard Scientific Practice: The Case of Sex Determination Genetics." In *Gendered Innovations in Science and Engineering,* edited by Londa Schiebinger, 22–42. Palo Alto, CA: Stanford University Press, 2008.

Ridley, Matt. *Genome: The Autobiography of a Species in 23 Chapters.* New York: HarperCollins, 1999.

Rinn, J. L., and M. Snyder. "Sexual Dimorphism in Mammalian Gene Expression." *Trends in Genetics* 21, no. 5 (2005): 298–305.

Roberts, Dorothy E. *Killing the Black Body: Race, Reproduction, and the Meaning of Liberty.* New York: Pantheon, 1997.

Robson, Catherine. *Men in Wonderland: The Lost Girlhood of the Victorian Gentleman*. Princeton, NJ: Princeton University Press, 2001.

Roiphe, Katie. "The Naked and the Conflicted." *New York Times*, 31 December 2009.

Rose, Nikolas. *The Politics of Life Itself: Biomedicine, Power, and Subjectivity in the Twenty-first Century*. Princeton, NJ: Princeton University Press, 2006.

Rosin, Hanna. "The End of Men." *Atlantic*, July/August 2010.

Ross, Mark T. "The DNA Sequence of the Human X Chromosome." *Nature* 434, no. 7031 (2005): 325–37.

Roughgarden, Joan. *Evolution's Rainbow: Diversity, Gender, and Sexuality in Nature and People*. Berkeley: University of California Press, 2004.

———. *The Genial Gene: Deconstructing Darwinian Selfishness*. Berkeley: University of California Press, 2009.

Rubin, Gayle. "The Traffic in Women: Notes on the 'Political Economy' of Sex." In *Toward an Anthropology of Women*, edited by Rayna Reiter, 157–210. New York: Monthly Review Press, 1975.

Russell, L. M., et al. "X Chromosome Loss and Ageing." *Cytogenetics and Genome Research* 116, no. 3 (2007): 181–85.

Russett, Cynthia Eagle. *Sexual Science: The Victorian Construction of Womanhood*. Cambridge, MA: Harvard University Press, 1989.

Ruzek, Sheryl Burt. *The Women's Health Movement: Feminist Alternatives to Medical Control*. New York: Praeger, 1978.

Sarkar, Sohotra. "From Genes as Determinants to DNA as Resource: Historical Notes on Development and Genetics." In *Genes in Development: Re-reading the Molecular Paradigm*, edited by Eva Neumann-Held and Christoph Rehmann-Sutter, 77–97. Durham, NC: Duke University Press, 2006.

Satzinger, Helga. *Differenz Und Vererbung: Geschlechterordnungen in Der Genetik Und Hormonforschung 1890–1950 [Heredity and Difference: Gender Orders in Genetics and Hormone Research, 1890–1950]*. Köln: Böhlau Verlag, 2009.

———. "Racial Purity, Stable Genes and Sex Difference: Gender in the Making of Genetic Concepts by Richard Goldschmidt and Fritz Lenz, 1916–1936." In *The Kaiser Wilhelm Society under National Socialism*, edited by Susanne Heim, Carola Sachse and Mark Walker, 145–70. New York: Cambridge University Press, 2009.

Saxena, R., et al. "The DAZ Gene Cluster on the Human Y Chromosome Arose from an Autosomal Gene that Was Transposed, Repeatedly Amplified and Pruned." *Nature Genetics* 14, no. 3 (1996): 292–99.

Schiebinger, Londa L. *Gendered Innovations in Science and Engineering*. Stanford, CA: Stanford University Press, 2008.

———. *Has Feminism Changed Science?* Cambridge, MA: Harvard University Press, 1999.

———. *Nature's Body: Gender in the Making of Modern Science*. Boston: Beacon Press, 1993.

Scott, Joan W. *Gender and the Politics of History*. New York: Columbia University Press, 1988.

Seldin, M. F., et al. "The Genetics Revolution and the Assault on Rheumatoid Arthritis." *Arthritis & Rheumatism* 42, no. 6 (1999): 1071–79.

Selmanoff, M. K., et al. "Evidence for a Y Chromosomal Contribution to an Aggressive Phenotype in Inbred Mice." *Nature* 253, no. 5492 (1975): 529–30.

Selmi, C. "The X in Sex: How Autoimmune Diseases Revolve around Sex Chromosomes." *Best Practice & Research: Clinical Rheumatology* 22, no. 5 (2008): 913–22.

Selmi, Carlo, et al. "The X Chromosome and the Sex Ratio of Autoimmunity." In "Gender, Sex Hormones, Pregnancy and Autoimmunity," edited by Yehuda Shoenfeld, Angela Tincani, and M. Eric Gershwin. Special issue, *Autoimmunity Reviews* 11, no. 67 (2012): A531–37.

Sengoopta, Chandak. *The Most Secret Quintessence of Life: Sex, Glands, and Hormones, 1850–1950*. Chicago: University of Chicago Press, 2006.

Serano, Julia. *Whipping Girl: A Transsexual Woman on Sexism and the Scapegoating of Femininity*. Emeryville, CA: Seal Press, 2007.

"Sex, Genes and Women's Health." *Nature Genetics* 25, no. 1 (2000): 1–2.

Shapiro, L. J., et al. "Non-inactivation of an X-Chromosome Locus in Man." *Science* 204, no. 4398 (1979): 1224–26.

Sharp, Andrew, David Robinson, and Patricia Jacobs. "Age- and Tissue-Specific Variation of X Chromosome Inactivation Ratios in Normal Women." *Human Genetics* 107, no. 4 (2000): 343–49.

Shields, Stephanie A. "The Variability Hypothesis: The History of a Biological Model of Sex Differences in Intelligence." *Signs* 7, no. 4 (1982): 769–97.

Sinclair, A. H., et al. "Sequences Homologous to ZFY, a Candidate Human Sex-Determining Gene, Are Autosomal in Marsupials." *Nature* 336, no. 6201 (1988): 780–83.

Skakkebaek, N. E., E. Rajpert-De Meyts, and K. M. Main. "Testicular Dysgenesis Syndrome: An Increasingly Common Developmental Disorder with Environmental Aspects." *Human Reproduction* 16, no. 5 (2001): 972–78.

Skaletsky, H., et al. "The Male-Specific Region of the Human Y Chromosome Is a Mosaic of Discrete Sequence Classes." *Nature* 423, no. 6942 (2003): 825–37.

Slightom, J. L., et al. "Reexamination of the African Hominoid Trichotomy with Additional Sequences from the Primate Beta-Globin Gene Cluster." *Molecular Pharmocogenetics and Evolution* 1, no. 4 (1992): 97–135.

Snyderman, Mark, and Stanley Rothman. *The IQ Controversy, the Media and Public Policy*. New Brunswick, NJ: Transaction Books, 1988.

Spanier, Bonnie. *Im/Partial Science: Gender Ideology in Molecular Biology*. Race, Gender, and Science. Bloomington: Indiana University Press, 1995.

Spolarics, Z. "The X-Files of Inflammation: Cellular Mosaicism of X-Linked Polymorphic Genes and the Female Advantage in the Host Response to Injury and Infection." *Shock* 27, no. 6 (2007): 597–604.

Steinach, Eugen. "Willkürliche Umwandlung Von Säugetiermännchen in Tiere Mit Ausgeprägt Weiblichen Geschlechtscharacteren Und Weiblicher Psyche (Arbitrary Transformation of Male Mammals into Animals with Pronounced Female Sex Characters and Feminine Psyche)." *Pflügers Archiv* 144, no. 71 (1912).

Steinach, Eugen, and Josef Löbel. *Sex and Life: Forty Years of Biological and Medical Experiments*. New York: Viking Press, 1940.

Stepan, Nancy Leys. "Race and Gender: The Role of Analogy in Science." *Isis* 77, no. 2 (1986): 261–77.

Stevens, N. M. "The Chromosomes in Diabrotica Vittata, Diabrotica Soror and Diabrotica 12-Punctata: A Contribution to the Literature on Heterochromosomes and Sex Determination." *Journal of Experimental Zoology* 5, no. 4 (1908): 453–70.

———. "Color Inheritance and Sex Inheritance in Certain Aphids." *Science* 26, no. 659 (1907): 216–18.

———. "Further Observations on Supernumerary Chromosomes, and Sex Ratios in Diabrotica Soror." *Biological Bulletin* 22, no. 4 (1912): 231–38.

———. "A Note on Reduction in the Maturation of Male Eggs in Aphis." *Biological Bulletin* 18, no. 2 (1910): 72–75.

———. *Studies in Spermatogenesis: A Comparative Study of the Heterochromosomes in Certain Species of Coleoptera, Hemiptera and Lepidoptera, with Especial Reference to Sex Determination.* Vol. 36(2). Washington, DC: Carnegie Institution, 1906.

———. *Studies in Spermatogenesis with Especial Reference to the Accessory Chromosome.* Vol. 36(1). Washington, DC: Carnegie Institution, 1905.

———. "A Study of the Germ Cells of *Aphis Rosae* and *Aphis Oenotherae*." *Journal of Experimental Zoology* 2, no. 3 (1905): 313–33.

———. "A Study of the Germ Cells of Certain Diptera, with Reference to the Heterochromosomes and the Phenomena of Synapsis." *Journal of Experimental Zoology* 5, no. 3 (1908): 359–74.

———. "An Unpaired Heterochromosome in the Aphids." *Journal of Experimental Zoology* 6, no. 1 (1909): 115–23.

Stevenson, R. E., et al. "X-Linked Mental Retardation: The Early Era from 1943 to 1969." *American Journal of Medical Genetics* 51, no. 4 (1994): 538–41.

Stewart, J. J. "The Female X-Inactivation Mosaic in Systemic Lupus Erythematosus." *Immunology Today* 19, no. 8 (1998): 352–57.

Stix, Gary. "Geographer of the Male Genome." *Scientific American*, December 2004, 40–42.

Stock, Robert. "The XYY and the Criminal." *New York Times*, 20 October 1968, SM30.

"Stork to Take Orders for Boy or Girl Soon." *Chicago Daily Tribune*, 24 January 1922.

"Study: 'Male' Chromosome to Stick Around." *CNN.com*, 31 August 2005.

Sulik, Gayle A. *Pink Ribbon Blues: How Breast Cancer Culture Undermines Women's Health.* New York: Oxford University Press, 2011.

Sunday, Suzanne R., and Ethel Tobach. *Violence against Women: A Critique of the Sociobiology of Rape.* New York: Gordian Press, 1985.

Sutton, H. Eldon. *An Introduction to Human Genetics.* New York: Holt, 1965.

———. *An Introduction to Human Genetics.* 3rd ed. Philadelphia: Saunders College, 1980.

Sutton, Walter S. "The Chromosomes in Heredity." *Biological Bulletin* 4, no. 5 (1903): 231–51.

———. "On the Morphology of the Chromosome Group in Brachystola Magna." *Biological Bulletin* 4, no. 1 (1902): 24–39.

———. "The Spermatogonial Divisions in Brachystola Magna." *Bulletin of the University of Kansas, Kansas University Quarterly* 9, no. 2 (1900): 152–54.

Svyryd, Y., et al. "X Chromosome Monosomy in Primary and Overlapping Autoimmune Diseases." *Autoimmunity Reviews* 11, no. 5 (2012): 301–4.

Swingle, W. W. "The Accessory Chromosome in a Frog Possessing Marked Hermaphroditic Tendencies." *Biological Bulletin* 33, no. 2 (1917): 70–86.

Sykes, Bryan. *Adam's Curse: A Future without Men.* New York: Bantam Press, 2003.

Takagi, N. "The Role of X-Chromosome Inactivation in the Manifestation of Rett Syndrome." *Brain Development* 23 Suppl 1 (2001): S182–85.

Talebizadeh, Z., S. D. Simon, and M. G. Butler. "X Chromosome Gene Expression in Human Tissues: Male and Female Comparisons." *Genomics* 88, no. 6 (2006): 675–81.

Tanner, Nancy, and Adrienne Zihlman. "Women in Evolution. Part I: Innovation and Selection in Human Origins." *Signs* 1, no. 3 (1976): 585–608.

Tarca, A. L., R. Romero, and S. Draghici. "Analysis of Microarray Experiments of Gene Expression Profiling." *American Journal of Obstetrics and Gynecology* 195, no. 2 (2006): 373–88.

Tasker, Yvonne, and Diane Negra. *Interrogating Postfeminism: Gender and the Politics of Popular Culture.* Durham, NC: Duke University Press, 2007.

te Velde, E., et al. "Is Human Fecundity Declining in Western Countries?" *Human Reproduction* 25, no. 6 (2010): 1348–53.

Thacker, Eugene. *The Global Genome: Biotechnology, Politics, and Culture.* Cambridge, MA: MIT Press, 2005.

Thompson, Charis. *Making Parents: The Ontological Choreography of Reproductive Technologies.* Cambridge, MA: MIT Press, 2005.

Thompson Woolley, Helen. "The Psychology of Sex." *Psychological Bulletin* 11, no. 10 (1914): 353–79.

———. "A Review of the Recent Literature on the Psychology of Sex." *Psychological Bulletin* 7 (1910): 335–42.

Tiger, Lionel, and Robin Fox. *The Imperial Animal.* New York: Holt, 1971.

Tingen, Candace. "Science Mini-Lesson: X Chromosome Inactivation." *Institute for Women's Health Research* (blog). Northwestern University. 21 October 2009. http://blog.womenshealth.northwestern.edu/2009/10/science-mini-lesson-x-chromosome-inactivation/.

Tjio, J. H., and A. Levan. "The Chromosome Number of Man." *Hereditas* 42 (1956): 1–6.

Tobach, Ethel, and Betty Rosoff. *Challenging Racism and Sexism: Alternatives to Genetic Explanations.* New York: Feminist Press at the City University of New York, 1994.

———. *Genetic Determinism and Children.* New York: Gordian Press, 1980.

Turner, Angela K. "Genetic and Hormonal Influences on Male Violence." In *Male Violence,* edited by John Archer, 233–54. New York: Routledge, 1994.

Turner, H. H. "A Syndrome of Infantilism, Congenital Webbed Neck, and Cubitus Valgus." *Endocrinology* 23 (1938): 566–74.

Uhlenhaut, N. Henriette, et al. "Somatic Sex Reprogramming of Adult Ovaries to Testes by FOX12 Ablation." *Cell* 139, no. 6 (2009): 1130–42.

Vallender, E. J., N. M. Pearson, and B. T. Lahn. "The X Chromosome: Not Just Her Brother's Keeper." *Nature Genetics* 37, no. 4 (2005): 343–45.

"Variation in Women's X Chromosomes May Explain Difference among Individuals, between Sexes." Press release, 16 March 2005. http://www.genome.duke.edu/press/news/03-16-2005/.

Varki, Ajit, and Tasha K. Altheide. "Comparing the Human and Chimpanzee Genomes: Searching for Needles in a Haystack." In *Genomes: Perspectives from the 10th Anniversary Issue of Genome Research*, edited by Hillary E. Sussman and Maria A. Smit, 357–93. Cold Spring Harbor, NY: Cold Spring Harbor Laboratory Press, 2006.

Venter, J. Craig. *A Life Decoded: My Genome, My Life*. New York: Viking, 2007.

Vilain, E. "Expert Interview Transcript." In *Rediscovering Biology: Molecular to Global Perspectives*. Electronic media. Oregon Public Broadcasting and AnnenbergMedia Learner.org, 2004.

Virchow, Rudolf. *Cellular Pathology: As Based upon Physiological and Pathological Histology. Twenty Lectures Delivered in the Pathological Institute of Berlin During the Months of February, March and April, 1858*. London, 1860.

Voeller, Bruce R. *The Chromosome Theory of Inheritance: Classic Papers in Development and Heredity*. New York: Appleton-Century-Crofts, 1968.

Vogel, Friedrich, and Arno G. Motulsky. *Human Genetics: Problems and Approaches*. New York: Springer-Verlag, 1979.

Vogt, Carl. *Lectures on Man: His Place in Creation, and in the History of the Earth*. London: Longman, Green, Longman, and Roberts, 1864.

Vollrath, D., et al. "The Human Y Chromosome: A 43-Interval Map Based on Naturally Occurring Deletions." *Science* 258, no. 5079 (1992): 52–59.

Voronoff, Serge, and George Gibier Rambaud. *The Conquest of Life*. New York: Brentano's, 1928.

Vroman, Georgine M., Dorothy Burnham, and Susan Gordon. *Women at Work: Socialization toward Inequality*. New York: Gordian Press, 1988.

Waldeyer, Heinrich. "Über Karyokinese Und Ihre Beziehungen Zu Den Befruchtungsvorgängen." *Archiv für mikroskopische Anatomie und Entwicklungsmechanik* 32 (1888): 1–122.

Walsh, S. J., and L. M. Rau. "Autoimmune Diseases: A Leading Cause of Death among Young and Middle-Aged Women in the United States." *American Journal of Public Health* 90, no. 9 (2000): 1463–66.

Walzer, S., P. S. Gerald, and S. A. Shah. "The XYY Genotype." *Annual Review of Medicine* 29 (1978): 568–570.

Wang, P. J., et al. "An Abundance of X-Linked Genes Expressed in Spermatogonia." *Nature Genetics* 27, no. 4 (2001): 422–26.

Wasserman, David T., and Robert Samuel Wachbroit. "Genetics and Criminal Behavior." In *Cambridge Studies in Philosophy and Public Policy*. New York: Cambridge University Press, 2001.

Watson, James D. *Recombinant DNA*. 2nd ed. New York: Scientific American Books, 1992.

Weasel, Lisa H. "Dismantling the Self/Other Dichotomy in Science: Towards a Feminist Model of the Immune System." *Hypatia* 16, no. 1 (2001): 27–44.

Weismann, August. *Das Keimplasma; Eine Theorie Der Vererbung*. Jena: Fischer, 1892.

Weyl, Nathaniel. "Genetics, Brain Damage and Crime." *Mankind Quarterly* 10 (1969): 100–109.

Whitacre, C. C. "Sex Differences in Autoimmune Disease." *Nature Immunology* 2, no. 9 (2001): 777–80.

White, Rosemary. "Professor Jennifer A.M. Graves, FAA." *Women in Science Network Journal* 58 (2001).

Whitlock, Kelli. "The 'Y' Files." *Paradigm*, Fall 2003, 24–29. http://wi.mit.edu/news/paradigm/archives.

Whitmarsh, Ian, and David S. Jones. *What's the Use of Race? Modern Governance and the Biology of Difference*. Cambridge, MA: MIT Press, 2010.

Wilkie, Tom. "At the Flick of a Genetic Switch." *London Independent*, 13 May 1991, 18.

Williams, Nigel. "So That's What Little Boys Are Made Of." *Guardian*, 20 July 1990.

Wilson, Elizabeth A. *Neural Geographies: Feminism and the Microstructure of Cognition*. New York: Routledge, 1998.

Wilson, Edmund B. *The Cell in Development and Heredity*. 3rd ed. New York: Macmillan, 1925.

———. *The Cell in Development and Inheritance*. 2nd ed. New York: Macmillan, 1906. First published 1896 by Macmillan.

———. "A Chromatoid Body Simulating an Accessory Chromosome in Pentatoma." *Biological Bulletin* 24, no. 6 (1913): 392–410.

———. "The Chromosomes in Relation to the Determination of Sex in Insects." *Science* 22, no. 564 (1905): 500–502.

———. "Croonian Lecture: The Bearing of Cytological Research on Heredity." *Proceedings of the Royal Society of London. Series B, Containing Papers of a Biological Character* 88, no. 603 (1914): 333–52.

———. "Notes on the Chromosome-Groups of Metapodius and Banasa." *Biological Bulletin* 12, no. 5 (1907): 303–13.

———. "Secondary Chromosome-Couplings and the Sexual Relations in Abraxas." *Science* 29, no. 748 (1909): 704–6.

———. "Selective Fertilization and the Relation of the Chromosomes to Sex-Production." *Science* 32, no. 816 (1910): 242–44.

———. "Sex Determination in Relation to Fertilization and Parthenogenesis." *Science* 25, no. 636 (1907): 376–79.

———. "Studies on Chromosomes I. The Behavior of the Idiochromosomes in Hemiptera." *Journal of Experimental Zoology* 2, no. 3 (1905): 371–405.

———. "Studies on Chromosomes II. The Paired Microchromosomes, Idiochromosomes and Heterotropic Chromosomes in Hemiptera." *Journal of Experimental Zoology* 2, no. 4 (1905): 507–45.

———. "Studies on Chromosomes III. The Sexual Differences of the Chromosome-Groups in Hemiptera, with Some Considerations on the Determination and Inheritance of Sex." *Journal of Experimental Zoology* 3, no. 1 (1906): 1–40.

Wilson, Edward O. *On Human Nature*. Cambridge: Harvard University Press, 1978.

———. *Sociobiology: The New Synthesis*. Cambridge, MA: Belknap Press of Harvard University Press, 1975.

Witkin, H. A., et al. "Criminality in XYY and XXY Men." *Science* 193, no. 4253 (1976): 547–55.

Witkin, H. A., D. R. Goodenough, and K. Hirschhorn. "XYY Men: Are They Criminally Aggressive?" *Sciences (New York)* 17, no. 6 (1977): 10–13.

Wizemann, Theresa M., and Mary-Lou Pardue, eds. *Exploring the Biological Contributions to Human Health: Does Sex Matter?* Washington, DC: National Academy Press, 2001.

The XX Factor: What Women Really Think. Slate (blog), 2007–9.

"The XYY Controversy: Researching Violence and Genetics." *Hastings Center Report* 10, no. 4 (1980): Suppl 1–32.

Yedell, Michael, and Rob Desalle. "Sociobiology: Twenty-five Years Later." *Journal of the History of Biology* 33, no. 3 (2000): 577–84.

Young, Iris Marion. "Pregnant Embodiment: Subjectivity and Alienation." In *Throwing Like a Girl and Other Essays in Feminist Philosophy and Social Theory*, 160–74. Bloomington: Indiana University Press, 1990.

Zenaty, D., et al. "Le Syndrome De Turner: Quoi De Neuf Dans La Prise En Charge?" *Archives de Pediatrie* 18, no. 12 (2011): 1338–42.

Zhang, Y. J., et al. "Transcriptional Profiling of Human Liver Identifies Sex-Biased Genes Associated with Polygenic Dyslipidemia and Coronary Artery Disease." *Plos One* 6, no. 8 (2011): e23506.

Zihlman, Adrienne. "Women in Evolution, Part II: Subsistence and Social Organization among Early Hominids." *Signs* 4, no. 1 (1978): 4–20.

Zlotogora, Joel. "Germ Line Mosaicism." *Human Genetics* 102, no. 4 (1998): 381–86.

INDEX

Page numbers followed by an *f* indicate a figure.